Soil Clays

Linking Geology, Biology, Agriculture, and the Environment

Soil Clays

Linking Geology, Biology, Agriculture, and the Environment

By
G. Jock Churchman and Bruce Velde

CRC Press
Taylor & Francis Group
Boca Raton London New York

CRC Press is an imprint of the
Taylor & Francis Group, an **informa** business

CRC Press
Taylor & Francis Group
6000 Broken Sound Parkway NW, Suite 300
Boca Raton, FL 33487-2742

First issued in paperback 2021

ISBN 13: 978-1-03-209195-2 (pbk)
ISBN 13: 978-1-4987-7005-7 (hbk)

Publisher's Note
The publisher has gone to great lengths to ensure the quality of this reprint but points out that some imperfections in the original copies may be apparent.

Visit the Taylor & Francis Web site at
http://www.taylorandfrancis.com

and the CRC Press Web site at
http://www.crcpress.com

GJC: To the memories of my father, Gordon Churchman, who tended his garden to support his family, like many millions worldwide; and my former colleague and friend, Kevin Tate, who introduced me to the importance and challenges of soil organic matter.

And to the futures of Ingrid and Cara, and all other children.

BV: I would like to acknowledge the great help and in fact inspiration that my young colleague Pierre Barré has given me during the preparation of this project. Pierre is beginning in a field that one can call biogeology where he is working on clays and organic matter in the soil zones. He is quite enthusiastic and this has helped me very much. He will do good things in the future.

Contents

Preface

The question we wish to pose and attempt to answer, above all, is: What can a soil scientist, environmental scientist, agronomist and/or land use planner gain from a knowledge of soil clays?

We humans require several services from soils that are basic to our continuing existence. In 1993, the US National Research Council formulated three functions for soils (Ewing and Singer, 2012), namely:

1. Soils make it possible for plants to grow.
2. Soils regulate and partition water flow through the environment.
3. Soils buffer environmental change.

As the number of humans grows from the current 7 billion towards an inevitable 9 or 10 billion, the demands on the limited supply of soils on Earth can only grow and intensify. With about 1 billion people chronically undernourished at present (FAO, 2012), the planet's soils must provide 50% more food than today to feed 2–3 billion more than now (Matson, 2011; Tilman et al., 2011; Hall, 2014). By far the most of human-consumed calories are produced in soils on land (Hall, 2014), so the vast bulk of the extra production of food must be achieved "on less land, with less fresh water, using less energy, fertilizer and pesticide" (Lenné, 2011, p. 14). And it will be needed by about 2030, a little over a decade from today (Lenné, 2011; Hall, 2014). To provide sufficient food for all by then, the productivity of soils will need to be raised, drastically, while enabling their conservation as a continuing fruitful resource to support plant growth for food production into the foreseeable future (Cribb, 2010; Hall, 2014). And all this has to happen while some soil is lost to agriculture under cities and roads and some by various forms of erosion (e.g. Kuhn, 2014), with the climate undergoing rapid change (e.g. Tate and Theng, 2014), and industrial development being accompanied by increased poisoning of lands through pollution (e.g. Hseu and Chen, 2014). Much will be demanded of "the genius of soils", as it was characterised by Publius Vergilius Maro in 29 BCE (Sposito 2014), if humanity is to be able to live at something like the current standard of living which some of us enjoy and to which most others aspire.

To put it another way, soils need to be made to perform as well as they can, far into the future. Some would characterise this as the need to raise, and maintain, the level of soil quality for useable soils. A common definition of the somewhat controversial concept of soil quality is "the capacity of a soil to function within ecosystem boundaries to sustain biological productivity, maintain environmental quality, and promote plant and animal health" (Doran and Parkin, 1994; Ewing and Singer, 2012). Some others would want to optimise "the flow of valuable ecosystem services from the natural capital stocks of the world's soils" (Clothier and Kirkham, 2014, p. 136). To further quote Clothier and Kirkham (2014, p. 137), "soil is a natural capital element, a critical component of ecological infrastructure, which supplies valuable ecosystem services". The main types of ecosystem services provided by soil to humanity are classified as "provisioning" and "regulating", while soils themselves are maintained by "supporting" processes and are devalued by "degradation" processes (Dominati et al., 2010; Clothier and Kirkham, 2014).

Following the framework established by Dominati et al. (2010), soils provide physical support for plants, and food, wood and fibre for humans. They regulate through flood mitigation; filtering of nutrients; biological control of pests and diseases; recycling of wastes and detoxification; carbon storage; and control of carbon dioxide (CO_2), nitrous oxide (N_2O) and methane (CH_4). They are supported through the processes of nutrient cycling, water cycling and biological activity. Their performance is optimised through a set of "manageable" properties, including soluble phosphorus (P), mineral nitrogen (N), soil organic matter (SOM), carbon content (C), temperature, pH, land cover, macroporosity, bulk density and strength and sizes of aggregates, where the latter two properties pertain particularly to topsoils. The extent to which these manageable properties can be manipulated and modified in any soil is dependent on the inherent properties of the soil. Along with such invariant characteristics as orientation, slope, depth and texture, these include types of clays.

Knowledge of the types of clays present arguably enables land users to most efficiently and economically manage and manipulate the manageable soil properties. This is because clays are the most reactive and interactive inorganic components in soils. They react with components of the soil solution, such as plant nutrients; with the soil biota, with its breakdown products and that of crops and animals, i.e. organic matter; and with other clays and also coarser minerals in soils. As minerals in soils, clays "are most usefully viewed as secondary inorganic compounds of clay (<2 μm) size within soils, regardless of their crystalline, nanocrystalline, or disordered character" (Churchman, 2010, p. 939). Among the manageable properties of soils listed earlier, they directly influence soluble P, mineral N, SOM and C content, and their manipulation, especially with SOM, can affect soil temperature and pH and also such physical properties as macroporosity, bulk density and the strength and sizes of aggregates in topsoils. To return briefly to a limited, but underlying concept of soil quality, that of soil physical quality (Dexter, 2004a,b,c), this concept, based on porosity, gives rise to an index which reflects the effect on a soil of tillage and is related to such properties of soils as density, root growth, friability, tilth, hard-setting and saturated hydraulic conductivity. Soil physical quality, as measured by this index, can also be altered by manipulation of interactions of clays in soils, principally with organic matter (e.g. Churchman, 2012).

To know soil clays is to enable their use towards achieving improvements in their management to enhance their performance of one or more of their three main functions of enabling plant growth, regulating water flow and buffering environmental change.

This book aims to help improve predictions of important properties of soils through a modern understanding of their highly reactive clay minerals as they are formed and occur in soils worldwide. Soils are essential for the sustenance of almost all plants and hence animals, including humans, but soils are virtually infinitely variable. On the other hand, the finest material, or clays, in soils offer the possibility of having a limited number of common components that dominate many of the most important soil properties.

The clays in soils contribute almost all of the surface area and a substantial portion – organic matter aside – of the electrical charge to soils. With the confounding factor of processed organic matter also to be considered, this book sets out to examine the proposition that major properties of soils can be reduced in large part to those of their clay minerals, helped by recent advances in the interpretation of their X-ray diffraction spectra. If so, soils with similar clay mineral compositions can be seen as equivalent for predicting the contribution of soils to plant growth, for instance.

Clay minerals in a soil reflect the process and environment of formation of their particular soils. Soils only occur through the breakdown of the rocks mediated by the actions of plants and their decomposition products. Compared with 'type' clays, those in soils are heterogeneous in their origins, and the environment for their formation may differ within metres within a single field. Therefore, in pursuit of our practical aims, we set out to first describe the conditions that lead to the formation of clay minerals in soils.

Our approach is necessarily both theoretical, describing the origin of soil clays, and also practical, describing their actual occurrences and the influence their mode of formation has on their natures. This approach follows that of the 16th century French philosopher, scientist and polymath Bernard Palissy, who featured a conversation between two characters, Théoretique and Practique (Palissy, 1580; Feller and Aeschlimann, 2014). We hope that we have achieved a true synthesis of their contrasting approaches. Only then can we examine whether important soil properties can be reduced to those of their constituent clay minerals, enabling predictions of how plants grow in different soils, and how soils can help solve important global challenges, including food security and excessive emissions of greenhouse gases.

We are grateful to many associates in different countries and to our past students for some of the material and many of the insights in this book. We particularly acknowledge the following colleagues for recent information: Professors David Lowe, in New Zealand, on soils on volcanic materials; Steve Hillier, in Scotland, on quantitative X-ray diffraction analyses; Rob Fitzpatrick, in Adelaide, on poorly crystalline iron compounds and also some applications of soil clays; and Alfred Hartemink, of Wisconsin, for the cover photograph. We also note their inspiration and encouragement over the years, together with those of Drs. Benny Theng in New Zealand; Jeff Wilson in Scotland; Hong Hanlie in Wuhan, China; Bob Gilkes in Perth; David Chittleborough in Adelaide; and Will Gates in Melbourne, among many others. We also thank our wives, Jan and Danielle, for their assistance, support and forbearance during the writing of the book.

REFERENCES

Churchman, G.J. 2010. Is the geological concept of clay minerals appropriate for soil science? *Journal of Physics and Chemistry of the Earth* 35: 927–940.

Churchman, G.J. 2012. Small heterogeneous associations and water retention link soil quality to carbon sequestration in soils: Philosophical and practical implications, p. 134–137. In L. Burkitt and L. Sparrow (eds.), *Proceedings of the 5th Joint Australian and New Zealand Soil Science Conference*, Hobart, Tasmania. Australian Society of Soil Science Inc., Warragul, Victoria, Australia.

Clothier, B., and M.B. Kirkham. 2014. Soil: Natural capital supplying valuable ecosystem services, p. 135–149. In G.J. Churchman and E.R. Landa (eds.), *The Soil Underfoot: Infinite Possibilities for a Finite Resource.* CRC Press/Taylor & Francis Group, Boca Raton, Florida.

Cribb J, 2010. *The Coming Famine: The Global Food Crisis and What We Can Do to Avoid It.* CSIRO Publishing, Melbourne.

Dexter, A.R. 2004a. Soil physical quality: Part I. Theory, effects of soil texture, density and organic matter, and effects on root growth. *Geoderma* 120: 201–214.

Dexter, A.R. 2004b. Soil physical quality: Part II. Friability, tillage, tilth and hard setting. *Geoderma* 120: 215–225.

Dexter, A.R. 2004c. Soil physical quality: Part III. Unsaturated hydraulic conductivity and general conclusions about S-theory. *Geoderma* 120: 227–239.

Dominati, E.M., M. Paterson, and A. Mackay. 2010. A draft framework for classifying and measuring soil natural capital and ecosystem services. *Ecological Economics* 69: 1858–1868.

Doran, J.W., and T.B. Parkin. 1994. Defining soil quality for a sustainable environment, p. 3–22. In J.W. Doran, D.C. Coleman, D.F. Bezdicek, and B.A. Stewart (eds.), *Special Publication 35.* Soil Science Society of America, Madison, Wisconsin.

Ewing, S.A., and M.J. Singer. 2012. Soil quality, p. 26-1–26-28. In P.M. Huang, Y. Li, and M.E. Sumner (eds.), *Handbook of Soil Sciences: Resource Management and Environmental Impacts.* CRC Press/Taylor & Francis Group, Boca Raton, Florida.

FAO. 2012. The state of food insecurity in the world. Economic growth is necessary but not sufficient to accelerate reduction of hunger and malnutrition. Food and Agriculture Organization of the United Nations, Rome.

Feller, C., and J.-P. Aeschlimann. 2014. Soils and salts in Bernard Palissy's (1510–1590) view: Was he the pioneer of the mineral theory of plant nutrition, p. 289–299. In G.J. Churchman and E.R. Landa (eds.), *The Soil Underfoot: Infinite Possibilities for a Finite Resource.* CRC Press/Taylor & Francis Group, Boca Raton, Florida.

Hall, S.J. 2014. Soils and the future of food: Challenges and opportunities for feeding nine billion people, p. 17–35. In G.J. Churchman and E.R. Landa (eds.), *The Soil Underfoot: Infinite Possibilities for a Finite Resource.* CRC Press/Taylor & Francis Group, Boca Raton, Florida.

Hseu, Z.-Y., and Z.-S. Chen. 2014. The finite soil resource for sustainable development: The case of Taiwan, p. 49–59. In G.J. Churchman and E.R. Landa (eds.), *The Soil Underfoot: Infinite Possibilities for a Finite Resource.* CRC Press/Taylor & Francis Group, Boca Raton, Florida.

Kuhn, N.J. 2014. Soil loss, p. 37–48. In G.J. Churchman and E.R. Landa (eds.), *The Soil Underfoot: Infinite Possibilities for a Finite Resource.* CRC Press/Taylor & Francis Group, Boca Raton, Florida.

Lenné, J.M. 2011. Food security and agrobiodiversity management, p. 12–25. In J.M. Lenné and D. Wood (eds.), *Agrobiodiversity Management for Food Security.* CABI Publishing, Wallingford, United Kingdom.

Matson, P.A. 2011. *Seeds of sustainability: Lessons from the Birthplace of the Green Revolution in Agriculture.* Island Press, Washington DC.

Palissy, B. 1580. Discours admirables de la nature des eaux et fontaines, tant naturelles qu'artificielles, des sels et salines, des pierres, des terres, du feu, des émaux, avec plusieurs autres excellents secrets des choses naturelles; plus un traité de la marne, fort utile et nécessaire à ceux qui se mellent d'agriculture; le tout dressé par dialogues, ès quels sont introduits la théorique et la pratique. Martin le jeune, Paris.

Sposito, G. 2014. Sustaining "the genius of soils", p. 395–408. In G.J. Churchman and E.R. Landa (eds.), *The Soil Underfoot: Infinite Possibilities for a Finite Resource.* CRC Press/Taylor & Francis Group, Boca Raton, Florida.

Stewart, C.E., A.F. Plante, K. Paustian, R.T. Conant, and J. Six. 2008. Soil carbon saturation: Linking concept and measurable carbon pools. *Soil Science Society of America Journal* 72: 379–392.

Tate, K.R., and B.K.G. Theng. 2014. Climate change: An underfoot perspective, p. 3–16. In G.J. Churchman and E.R. Landa (eds.), *The Soil Underfoot: Infinite Possibilities for a Finite Resource.* CRC Press/Taylor & Francis Group, Boca Raton, Florida.

Tilman, D., C. Balzer, J. Hill, and B.L. Befort. 2011. Global food demand and the sustainable intensification of agriculture. *Proceedings of the National Academy of Sciences of the United States of America* 108: 20260–20264.

Authors

G. Jock Churchman is adjunct senior lecturer at the University of Adelaide (Australia) and adjunct associate professor at the University of South Australia. Jock Churchman's clay interests began with a PhD in chemistry on halloysite at the University of Otago in his native New Zealand, followed by industrial ceramic research (1970–1971). He held a postdoctoral fellowship in soil science at the University of Wisconsin–Madison (1971–1973) and was employed at the New Zealand Soil Bureau (1973–89), then at CSIRO (1989–2003), the University of Adelaide (2003–2012) and the University of South Australia (2013–2014). He has also held visiting fellowships in soil science for one year at Reading University (UK) and for six months at the University of Western Australia. His research has encompassed halloysite; acid dissolution of montmorillonite; dust transport; clay mineral genesis; clay–organic complexes; the influence of clay mineralogy on soil physical properties; clays in sodic soils; the characteriation of bentonites and their industrial and environmental applications; and the philosophy of soil science.

He has published nearly 150 refereed papers and coedited four books, most recently *The Soil Underfoot: Infinite Possibilities for a Finite Resource* (CRC Press, 2014) and *Natural Mineral Nanotubes* (CRC Press, 2015). He is a former editor (now emeritus) of *Applied Clay Science.* He has received awards from the New Zealand Society of Soil Science, Soil Science Australia, the Association Internationale pour l'Étude des Argiles (AIPEA) and the Clay Minerals Society.

Bruce Velde is an emeritus researcher for the Centre Nationale de Recherche Scientifique at the Ecole Normale Supèrieure in Paris. He did his PhD at Montana State University (1962) under the direction of John Hower, then he did a postdoctoral study at the Carnegie Geophysical Laboratory in Washington DC (1962–1965) after which he joined the CNRS in Paris.

The initial research subjects treated were the evolution of clay minerals in sediments and sedimentary rocks, and their stability under different laboratory conditions of pressure and temperature. During the latter period, he published 237 refereed papers, authored and coauthored 8 books on clays and their chemical relations in natural situations and advised 22 PhD theses on these subjects. His books are *Clays and Clay Minerals in Natural and Synthetic Systems* (Springer, 1977); *Introduction to Clay Minerals: Chemistry, Uses and Environmental Significance* (Chapman & Hall, 1992); *Archaeological Ceramic Materials: Origin and Utilization* (Springer, 1999); *Clay Minerals: A Physico-Chemical Explanation of Their Occurrence* (Elsevier, 2000); *Illite: Origins, Evolution and Metamorphism* (Springer, 2004); *The Origin of Clay Minerals in Soils and Weathered Rocks* (Springer, 2008); *Soils, Plants and Clay Minerals: Mineral and Biologic Interactions* (Springer, 2009); *Origin and Mineralogy of Clays: Clays and the Environment* (edited) (Springer, 2013); and *Geochemistry at the Earth's Surface* (2016).

The evolution of his work was to understand the chemical and physical reasons for the variety and stability of clay mineral associations from depth towards the surface of the Earth. He also did work on the formation of clay-associated structures (aggregates) and surface cracking using image analysis.

1 Introduction and Definitions

> The soil is the great connector of lives, the source and destination of all. It is the healer and the resurrector, by which disease passes into health, age into youth, death into life. Without proper care for it we cannot have community, because without proper care for it we can have no life.
>
> **Wendell Berry, 1934**

1.1 SOIL

Soil is basically the material formed at the atmosphere, rock or sediment interface on the Earth, being largely affected by plant chemical action. There is no definition of "soil" that is accepted universally. Some conclude that this must be the case. R.E White, in his textbook *Principles and Practice of Soil Science* states: "There is little point in giving a rigorous definition of soil because of the complexity of its make-up, and of the physical, chemical and biological forces that act on it. Nor is it necessary to do so, for soil means different things to different users" (2006, p. 4). In the same manner, Hillel (2008, p. 2) states, "A precise definition (of soil) is elusive, for what is commonly called soil is anything but a homogeneous entity. ... Perhaps the best we can do at this stage is to define soil as the naturally occurring, fragmented, porous and relatively loose assemblage of mineral particles and organic matter that covers the surfaces of our planet's terrestrial domains", although this author then goes on to describe its formation and its fractions. Some have attempted definitions that simply describe its appearance. These include Birkeland (1974, p. 3), to whom soil is "a natural body consisting of layers or horizons of minerals and/or organic constituents of variable thickness, which differ from the parent material in their morphological, physical, chemical and mineralogical properties and their biological characteristics". This definition, highlighting horizons as the distinguishing compositional feature of soils, is endorsed by Weaver (1989). Furthermore, Chesworth (2008, p. 629) effectively defines soils as the products of the particular processes that have given rise to horizons. Thus:

> The gravitational movement of water within soil effects the downward transport of solid particles and dissolved species whereas the solar energy concentrates constituents at the surface, either indirectly, through plants, or directly, with the upward transport of water under evaporative conditions. This produces over a relatively short period (order of 10^2–10^3 years) a vertical differentiation that appears megascopically as a series of horizons more or less parallel to the land surface.

The World Reference Base for Soil Resources (WRB) has a definition of soil which is very similar to this horizon-based compositional definition (Canarache et al., 2006).

Nonetheless, most of the definitions that have been proposed include statements about both its composition and its functions with some putting more emphasis on its composition, or morphology, whereas others emphasise its functions, or properties. Some also include statements about the origins of soil. In 1975, the US Soil Survey Staff included the following as defining features of soils (Fanning and Fanning, 1989):

Natural bodies on the Earth's surface

- Contain living matter
- Support or are capable of supporting plants out of doors
- Have air or shallow water as an upper limit

- At their margins, grade to deep water or rock or ice
- Include horizons that differ widely as a result of interactions through time of climate, living components, parent materials and relief, etc.
- Normally have the lower limit of biological activity as their lower limit

To its inclusion of horizons and its ability to support rooted land plants, Schaetzl and Anderson (2005, p. 20–22) add that soils comprise solids, liquids and gases, are "essential to life through recycling of nutrients, carbon and oxygen" and are "nonrenewable in human timescales". In the 14th edition of their textbook, Brady and Weil (2008) defined soils by their main functions, i.e. as a medium for plant growth, as a regulator of water supplies, as a modifier of the atmosphere and as a habitat for soil organisms. Together with Canarache et al. (2006), these authors also rightly observed that soils are also defined as an engineering medium, but that particular application is beyond our concern here. Churchman (2010a) deduced from the literature that studies of soils are uniquely concerned with horizons, aggregates and distinctive colloidal material. Churchman and Lowe (2012) make note of the character of soils as "the most complex ecosystem on earth" and "a biological habitat and critically important repository for genes", as well as a provider of "ecosystem services" and as "natural capital".

Undoubtedly, however, an unequivocal, universally accepted definition of soils remains as a work in progress according to a recent contribution by Certini and Ugolini (2013). These particular authors widen the definition of soils to include the possibility that they occur on other planets. Their newly proposed definition requires no requirement for plants in their formation. In contrast, we believe that something of the essential nature of soils is derived from their processes of formation (see next Section). These almost always involve plants, except in extreme conditions on Earth such as the dry valleys of Antarctica and deserts. With these exceptions, it is consistent with the action of plants on weathering, including physical weathering, that we can say there is no soil without clay (or <2 μm particles).

1.2 THE ORIGIN OF SOILS AND CLAYS IN GEOLOGICAL TIME

In order to be able to appreciate differences between soil clays and those from other 'geological' origins, it helps to realise that soils are relative late-comers in the geological record, and also that weathering occurred and clay minerals were formed before there were any soils on Earth. Recently, the development of new minerals has come to be seen as an evolutionary process. Following an era of planetary accretion prior to 4.5 billion years (4.5 Ga) ago, volcanism occurred, with associated processes, followed by the formation of granites and pegmatites, until the inception of plate tectonics before 3 Ga led to subduction of a range of materials in a water-rich environment (Hazen and Ferry, 2010). New types of minerals appeared at each stage. Weathering occurred and modelling of possible pathways using irreversible thermodynamics shows that the conditions, which were reductive owing to low oxygen levels, could have led to some clay minerals, including kaolinite $Al_2Si_2O_5(OH)_4$, among just a few other minerals (Sverjensky and Lee, 2010).

However, as Hazen and Ferry (2010, p. 11) describe it, "the situation changed in a geological instant". They are describing the "Great Oxidation Event" involving the onset of an oxygen-rich atmosphere, which began about 2.4 Ga BP. This led to a profusion of new types of minerals formed as hydrated, oxidised products of previously existing minerals. Most known mineral species were formed following that event (Sverjensky and Lee, 2010). However, the onset of soils only occurred as a result of the advent of vascular land plants in the Silurian period about 440 million years BP. (Knoll and James, 1987). While lichens colonising rocks had been effecting some weathering before this time, it was the emergence of higher (vascular) plants with deep roots that led to the establishment of soils (Verboom and Pate, 2006).

Deep-rooted higher plants brought about an acceleration in the rate of production of soils through concomitant processes of mineral weathering and the deposition of plant products, including exudates and litter, to produce organic matter following processing by microorganisms (Lambers et al., 2009). Roots of these higher plants, and associated microbes, have co-evolved with soils (Verboom and Pate, 2006). There is much evidence available, especially in semi-arid environments in Australia (Verboom and Pate, 2006), to show that higher plants and microorganisms can play a proactive role in "bioengineering" pedogenic processes for their own benefit. Their effect can be both physical and chemical/biochemical. Among physical effects, roots create macropores and fragment primary minerals (e.g., Calvaruso et al., 2009). Among biochemical effects, metal-chelating root exudates and the activity of microorganisms are key bioengineering agents (Verboom and Pate, 2006). Thus, vertical channels and pores, ideal for transporting water to depths, may become lined with Fe or Si, and surface layers may be rendered hydrophobic, which, together with the creation of hardpans and texture contrast, or of pavements from compounds of Al, Si, Ca and Fe and also of carbonates from Ca, enable retention of water at depth for use by roots. The various concentrations of Al, Fe, Si and/or secondary carbonates also enable sequestration of concentrations of phosphorus for later supply of this essential nutrient to plants via various means, including mycorrhizal exchange of carbon for P and other forms of microbial 'mining'.

Vascular deep-rooted plants also played an important physical part in the sustained development of soils. Their deep roots into the soils they helped to create enabled these soils to withstand erosive forces, at least for considerable periods of time, and hence remain in place for a relatively long time. As well, fungi and root hairs confer stability to soils through their adhesion to soil particles, while various chemicals in root exudates, including phenolics, but especially polysaccharides, enhance processes of aggregation of the particles, as do cycles of wetting and drying (Hinsinger et al., 2009).

1.3 WEATHERING AS THE ORIGIN OF (MOST) SOILS

Weathering literally occurs when minerals ('primary minerals') in or from rocks are exposed to the weather at Earth–surface conditions. These minerals have developed in rocks by crystallisation out of magma as it cooled, or else have been deposited through recycling, hence sedimentation, with possible metamorphism from increases in temperature and/or pressure. Changes occur because minerals of igneous, metamorphic and even sedimentary origin are out of thermodynamic equilibrium with the environment in which they now reside. As expressed by Kittrick (1967, p. 315), "Fundamentally a mineral is a package for its elements. It will persist in nature only as long as it is the most stable package for those elements in its environment". For minerals of igneous origin, Goldich, in 1938, constructed a series of the relative stabilities to alteration by weathering of the most common primary minerals (Figure 1.1). Goldich's series is the exact inverse of a classic series devised by Bowen in 1922 to denote the relative order in which minerals crystallised out of magma on cooling (Bowen, 1922). For example, since mineral A which crystallised from magma at a higher temperature than mineral B was thereby more out of equilibrium with conditions at the Earth's surface than mineral B, mineral A would therefore be more vulnerable to breakdown by weathering than mineral B. Essentially, it is 'last in (to igneous rocks), first out (to breakdown on weathering)'.

Given the aforementioned considerations we will adopt the view that soils, and the clay materials found in them, have an intimate relationship with the biological activity at the surface. In our considerations, soil is the zone of contact between plants and mineral matter at the surface of the Earth. We follow the concept given by Jenny (1994) that there are several variables involved in the development of soils: climate, topography, parent material, time and organisms (the biological factor).

Weathering typically occurs within a depth profile. The basic ingredients for weathering are rocks and their constituent primary minerals, and water, usually containing oxygen and often also carbon dioxide. Weathering may occur with or without the involvement of biological agents. When biological agents, plants, microbes and fauna are involved, soils form. Typically, the surface parts of the profile in contact with the atmosphere are affected by their involvement and develop dark colours from organic matter that is incorporated following the breakdown and decomposition of

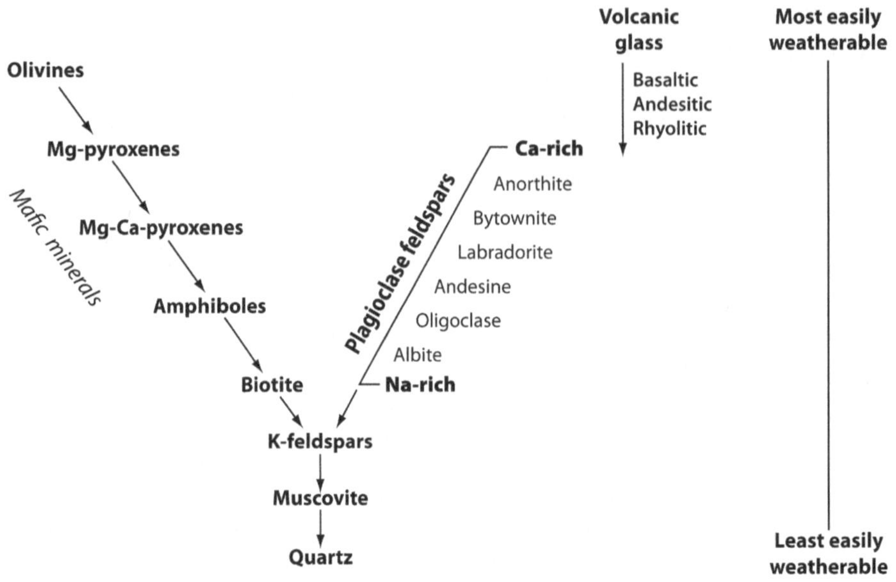

FIGURE 1.1 Stability series for the common primary minerals (after Goldich, 1938) and volcanic glass (not part of the Goldich series). Basaltic and other glasses, and olivines are normally the first phases altered by weathering (Wolff-Boenisch et al., 2004). (From Churchman, G.J., and D.J. Lowe, 2012, Alteration formation and occurrence of minerals in soils, p. 20.1–20.72, in P.M. Huang, Y. Li, and M.E. Sumner (eds.), *Handbook of Soil Sciences: Properties and Processes*, 2nd edn. CRC Press, Boca Raton.)

plants and other formerly living matter. However, the influence of plants and fauna tails off with depth, and, closer to the unaltered rock which remains, the colours are generally lighter, signifying much lower inputs of organic matter. In deep profiles, the material nearest the unaltered rocks may be essentially inorganic. In this respect, it resembles weathering profiles that develop without the involvement of biological agents. These profiles are regarded as regolith rather than soil, but within soil, there are deeper sections that are more properly labelled saprock, closest to the rock, or saprolite, still essentially within the inorganic zone but farther into an altered zone.

The parent materials giving rise to new minerals by weathering may themselves be secondary minerals resulting from former cycles of weathering that have led to sedimentary deposits or rocks. Also, the secondary minerals found in particular samples of soils may have originated by weathering in another location and transported into the situation of interest by wind or water or via biota. These particular secondary minerals are said to be 'inherited'.

1.4 FROM ROCK TO SOIL: THE BIOLOGICAL FACTOR IN THE INITIAL STAGES OF ROCK ALTERATION AND SOIL FORMATION

We would like to present a brief description of the processes of soil formation and rock alteration at the surface of the Earth at this point. It is most importantly concerned with the sequence of events that lead to the establishment of an alteration profile as described by soil scientists. The key to rock alteration is the prolonged contact with unsaturated water which can interact chemically with the solids of the rock. In order for this to occur there must be a retention of rainwater on the rock surface keeping it in contact with the high temperature minerals of the rock. Most rocks have small fractures present due either to past tectonic activity or the cracking of minerals due to thermal expansion and shrinkage of the highly asymmetric silicate minerals. Microcracks are usually present and this is where water–rock interaction can occur with the formation of clay minerals. This is an example of physical weathering, typical of desert climates. The result is the formation of small grains, sand or other sizes, which are not bound together and thus are easily moved by wind or water from their place of formation. Such reactions are slow due to the short periods of time of water–rock contact. Very few new mineral phases are present.

However, if the climate (rain and reasonable temperatures) permits, lichens and mosses fix on recently exposed rock surfaces. Here there is the beginning of clay formation at the plant–rock interface. The adhesion of pioneer plants to the rock surface and their penetration in less coherent areas of the rock cause physical disaggregation and fragmentation of the mineral surface. Chemical weathering of the rock minerals is initially due to the excretion of organic acids from the plants. Depending on the nature of the minerals, etching patterns and decomposition features are formed by biosolubilisation processes. The interaction permits the formation of a porous zone and the retention of rainwater for some time allowing it to interact with the rock materials and eventually to form new minerals stable under surface conditions. Increasing interaction allows the formation of a very thin but important layer of essentially soil materials: clays, oxides and organic material derived from the plants present. Lichens, the earliest interaction agents, can produce many of the clay minerals found in developed soils (Adamo and Violante, 2000). Other surface biological agents such as mosses or rooted plants interact with rock minerals to also produce soil clays (Carter and Arocena, 2000) especially through the interaction of mycorrhizae fixed on roots.

Mosses and eventually higher plant types produce clays, oxides and altered minerals typical of developed soils (Carter and Arocena, 2000) (Figures 1.2 and 1.3).

The formation of the initial soil layer on a rock permits the retention of water which interacts with the rock and continues to produce new minerals such as silicate clays. Below the organic zone the rocks continue to alter and produce other clays due to the presence of water infiltrated after being retained by the soil zone materials. The chemically altered rock constitutes the 'alterite' zone. The water gradually moves downward to the alterite–rock interface where it interacts to produce new clay minerals. This is the water–rock interaction zone which produces an alterite zone where clay minerals dominate.

As the alterite zone (water–rock interface) expands into the underlying rock by the production of new clays, a more porous zone is created which is favourable to the development of plant roots. This tends to expand the soil zone (from plant interaction) downward. As the soil zone (the site of organic interactions) descends into the alterite zone, the newly formed alterite minerals (from water–rock interaction) are frequently changed by transformation due to the chemistry engendered by the plants. Different types of plants produce different chemical regimes in the soils and produce

FIGURE 1.2 Lichens on 19th century gravestone, Brittany, France.

FIGURE 1.3 **(See colour insert.)** Mosses and grass on a 16th century granite church wall northern France.

slightly different minerals. These new minerals are those found in soils. Thus, the soil zone forms its specific mineralogy depending upon the type of plants present from the rock–water interaction zone, below which new minerals develop at increasingly greater depth with time. The end result is an alteration profile which includes the upper plant–soil zone, an alterite zone of water–rock interaction products and the (physically) altered rock below.

Thus, the interaction of atmospheric chemical forces with newly exposed rocks forms a two-part structure: one where rock is altered and transformed into new minerals through water–rock interaction, the alterite zone; and another where plants modify the materials produced by alteration in the soil zone. The critical point in this description is the importance of plants in stabilising the alteration material formed from rocks and maintaining a humid zone of mineral–water interaction. The soils, through plant action, keep the alterite material at the surface where it maintains humidity and is gradually organised into a layer that is more and more useful to plants with roots and more advanced types of structures (Figure 1.4).

The establishment of a stable layer of fine-grained material held in place by plants at the surface is the basis of water–rock alteration processes. Once the rainwater is held for some time it can react, slowly, with the rock substrate to form new minerals stable at the surface (usually hydrated or OH-containing). Gradually this zone deepens below the soil surface forming an alteration profile. This material is described as an altered rock zone and the alterite zone. The overall stability of these materials is guaranteed by the soil cover which is not easily eroded, or not at all, and which allows the slow movement of rainwater that is unsaturated with silicate material but slightly acidic due to the dissolution of CO_2.

In other words, first comes soil (from rock–plant interaction) and then comes chemical alteration (from water–rock interaction). Without vegetal cover the interaction of rainwater with rock is very limited and fractures of rock grains render the initial altered reaction material vulnerable to displacement by rain or wind forces. Very little fine-grained material is found on the altering rock.

In Figure 1.5 the soil zone is indicated as moving downward into the alterite zone and the alterite zone is seen as moving downward into the rock.

The fine-grained clays in the soil zone are formed by reworking alterite zone minerals and transforming them chemically to a certain extent. These transformations render the alterite zone minerals more amenable to plant growth, depending of course on the types of plants present in the soil zone and the chemical characteristics of the underlying bedrock.

FIGURE 1.4 **(See colour insert.)** Alteration profile on granite (Massif Central France) showing soil, alterite (altered clay-rich zone) and altered rock basement.

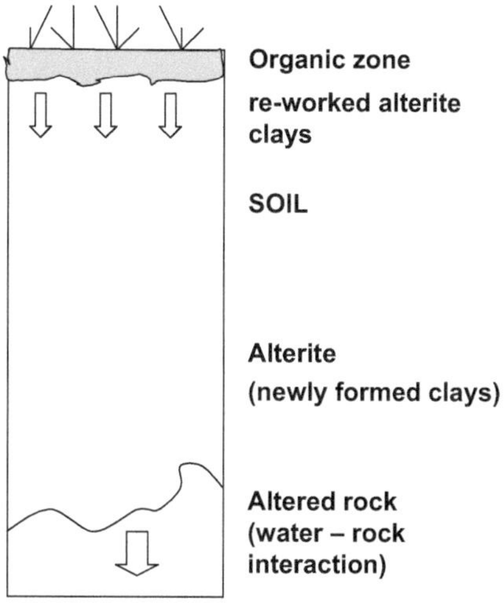

FIGURE 1.5 Schematic diagram of soil clay systems.

1.5 SOIL SCIENCE

Soils are important to humanity for many reasons. At base, they are the source of all food derived from the land, which amounts to the greater part of all human-consumed calories (FAO, 2009, cited by Hall, 2014). The most pressing current global challenges have soils front and centre. They are

- To provide enough food to enable adequate sustenance for a projected global population of 9 billion to 10 billion

- To ensure that the soil base is maintained, if not improved, to support humanity long into the future
- To provide a sink for continuing excessive emissions into the atmosphere of carbon dioxide and other greenhouse gases, so as to suppress their deleterious effects on climate

Of course, soils are central actors in other continuing global concerns. These include the disposal and remediation of contaminants, the provision of enough good quality water for burgeoning urban centres, and the provision of many of the materials for fibre and shelter. Through the biofuels industry, they are also major components in the provision of alternative energy sources to fossil fuels. No doubt other problems and challenges are yet to arise that will call upon soils to help provide answers.

Soil science – the scientific study of soils – has often been focused on questions of how soils can meet practical problems and challenges. By way of reaction, many soil scientists (see Churchman, 2010a, for a listing) have called for more effort to be expended on fundamental research on soils. In this regard, Sposito and Regnato (1992, p. 7) suggested that the simple basic question, What is soil?, needed to be addressed, and knowledge of the essential properties of soils should be advanced though studies of this kind. This knowledge may well enable improved solutions to unforeseen soil-related problems. Even so, the demands of tight financial accountability nowadays usually mean that soil scientists are employed to do research only when they focus first and foremost on practical problems. And economic considerations have also driven research into soils in the past. Even the widely acknowledged 'father' of soil science, Vasily V. Dokuchaev, in 1877, was sponsored by the Free Economic Society of St. Petersburg to study crop failure that occurred in a drought. However, the lesson for soil scientists then and now is that Dokuchaev's practical study enabled him to establish the fundamental principle that soil was an independent natural-historical body reflecting the influence of a number of environmental factors (Krupenikov, 1993). These fundamental principles that he drew from a practical study persist today as the founding principles of soil science (Churchman, 2010a).

Soil materials are very active chemically, where ionic exchange from aqueous solution is a major force. The exchange is the basis of nutrient supply for plants. Plants use chemical control of the solution chemistry to extract and store some chemical nutrients. The clay–organic associations forming microporosity allows a continuum of this reaction mechanism. There is overall chemical change in the soil clay zone, reflected in clay mineral compositions, that occurs over periods of decades, years or even one growing season. Soil systems are extremely dynamic in a chemical sense.

1.6 CLAYS

Simple mechanical disturbance of a soil by, for example, a garden fork, breaks it into units that might come in a variety of sizes – and shapes – but most are at least several millimetres, if not centimetres, long and wide. In some soils, generally regarded as poor soils, most units are seen to be of sand size, each particle just a small number of millimetres across. Clay, which is widely defined as being smaller than 2 micrometres (μm), hence individually invisible to the naked eye (or even optical microscope), does not actually occur as such in undisturbed soils. If it did, it would be blown away with the first decent wind, if not washed away in the next rainfall.

But clay from soils is widely studied after extraction from soil in the laboratory as the suspended material from sedimentation following strong disaggregation treatments. It is generally studied as an inorganic component, comprising so-called clay minerals. Compared with the other main size fractions of gravel-free soil, namely sand (0.002–2 mm) and silt (0.0002–0.002 mm or 2–20 μm), it contributes almost all (99%, according to (Lowe, 1995) of the surface area (Figure 1.6) and also a substantial portion – organic matter aside – of the electrical charge to soils. While there is no universally agreed definition of soil (see earlier), it is most widely thought of as a medium for growing plants. Together with air and mechanical support, plants require water and

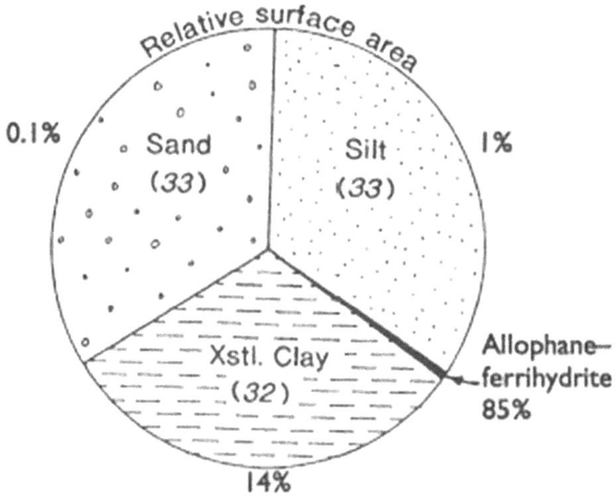

FIGURE 1.6 Surface areas as percentages of the whole soil calculated for sand, silt and (crystalline) clay sizes, and nanominerals (allophane, ferrihydrite). Particles assumed to be spherical. (From Lowe, D.J., 1995, Teaching clays: From ashes to allophane, p. 19–23, in G.J. Churchman, R.W. Fitzpatrick, and R.A. Eggleton (eds.), *Clays: Controlling the Environment.* Proceedings of the 10th International Clay Conference, Adelaide, Australia (1993), CSIRO Publishing, Melbourne With permission from CSIRO Publishing).

nutrients which usually come into plants as charged species (ions) dissolved in the water. It is the extensive surfaces of clay minerals (Figure 1.6) that can hold sufficient water against gravity and easy evaporation to enable plants to draw on it as required. Water retention can occur on clay surfaces but more important in capillary cavities in clay-organically associated materials which form microporous soil aggregates. This capillary water is a major resource for plants and obtaining and retaining a stable soil structure with additions of organic matter has been the object of farmers for centuries. By the same token, it is the charges on or within the clay minerals which hold nutrient ions, usually of the opposite charge to the clay surfaces. Certain sites, especially those within the clay structures, protect some of the cation nutrients against leaching but maintain them for easy availability to the plants.

1.7 CLAY MINERAL FORMATION

Clay minerals typically form as a result of the action of rainwater upon minerals from rocks (i.e. primary minerals) and are therefore known as secondary minerals. The extent of change is due to the intensity of rainfall and plant activity which determines the amount of chemical exchange between solids and solutions and also the duration of the alteration process. This action is known as weathering. Some also form hydrothermally. Most of the characterisation of clay minerals has been done using samples from large scale, often commercial, deposits. These have formed by inorganic reactions including weathering and/or precipitation of single types of clay minerals from solutions in environments in which both organic matter and other minerals such as oxides, hydroxides and oxyhydroxides of iron, aluminium and manganese are largely absent. Such deposits include those of kaolinites in Georgia (United States) and Cornwall (UK), and that of montmorillonite in Wyoming (United States). The clay minerals formed there show a degree of uniformity within each deposit, and they have served as examples of the type of mineral occurring in other environments, including soils. This approach is justified by the observation of a similar set of diffraction patterns for clays from both soil and non-soil sources, when X-ray diffraction was first applied to soil clays (by Hendricks and Fry, 1930) soon after its application to clay minerals from non-soil ('geological') sources (by A.R. Hadding in 1923 and F. Rinne in 1924, according to Bergaya et al., 2013).

This suggested that soil clays were crystalline and had a number of similar crystal structures to those of geological clays. This similarity provided the cue for the great bulk of studies in soil clay mineralogy ever since to infer that clay minerals in soils encompassed the same types found in other environments.

Accordingly, it has been largely assumed that clay minerals in soils are phyllosilicates, with layered structures. This means that there is a defined limit to the number of types of clay minerals that can occur, in soils as well as in other earth environments. Following Bergaya and Lagaly (2013, their tables 1.2 and 1.3), this would amount to clay minerals belonging to one of up to 17 groups, comprising altogether some 86 separately defined species. While these are large numbers, the distinctions between them are well demarcated. If some of them, at least, also appear in soils, then the clay minerals in soils will be well defined, so that there is the prospect of being able to understand, explain and predict the useful properties of soils, from these, their most reactive inorganic components. Of course, this leaves aside as a first approximation the challenge of accounting for the contributions to properties of soils from organic matter and also metal oxides, hydroxides and oxy-hydroxides, their other most reactive components. With soils being notoriously heterogeneous both between and even within profiles, let alone between soil types and different geographical areas, there is an allure in trying to compare across soils on the basis of a limited number of apparently common highly reactive components, especially when clay contents are high and those of organic matter are low. Central to this volume is an examination of the basis of the underlying assumption of a close identity between clay minerals in soils and exemplars of the same types from other, generally geological, environments.

An example of the differences between clay minerals of geological origin and those of soils can be seen in the development of bentonite, the type material for laboratory investigation of smectites. To highlight the difference in mineralogy and behaviour of classical clay minerals and those found in soils we give an example of a soil profile formed under prairie conditions on a bentonite layer. Bentonite is the name of a material formed from volcanic ash which contains almost exclusively smectite minerals. Such material can be found in near surface deposits or in sedimentary rock sequences. The behaviour of bentonite smectite in the laboratory is most often taken to represent that of similar materials found in soils. However, in one example given here, a bentonite material forming a soil in Argentina, the bentonite layer and the alterite (largely inorganic, see later) zone above it show the same mineralogy in X-ray diffraction (XRD) studies (Figure 1.7). Peak positions and widths, which indicate the average grain size, are virtually identical. However, in the soil (organic) zone the smectite peak is modified and can be represented as indicating groups of minerals, one similar to the initial material and another of greater peak width showing a smaller grain size and a slight tendency of have some non-expanding layers (indicated by a slightly lower peak position, 15.06 instead of 15.15 angstroms). The smectite material similar to that in the bentonite layer forms roughly 30% of the clays, while the smaller grain size and slightly altered material is a majority of the sample. This suggests that the type mineral bentonite smectite is in fact not that found in soils because it is modified when in contact with the chemistry of the surface material controlled by organic materials and incoming rainwater. Type minerals are often not those found in soils and alteration zones even though they can have very similar mineralogical characteristics as seen by traditional laboratory methods of identification.

We wish to propose an operational definition of soil phyllosilicate clay minerals. This is based upon recent observations of silicate mineral change observed in soil systems. Essentially there are two types of silica-containing phyllosilicate clay minerals in soils: (1) those with three layers of strongly covalent ions linked by oxygen anions, the 2:1 minerals, where silica is the predominant ion in the exterior layers; and (2) the 1:1 minerals, with almost exclusively silica and aluminium coordinated layer structures. The 1:1 minerals have virtually no other cations within the structure, whereas the 2:1 minerals have cations of different sorts ionically bonded to the clay layer surfaces. The composition of these ions has given rise to a mineral terminology with many terms mainly based upon minerals of higher temperature origin. In fact, the norm of soil phyllosilicate clays is

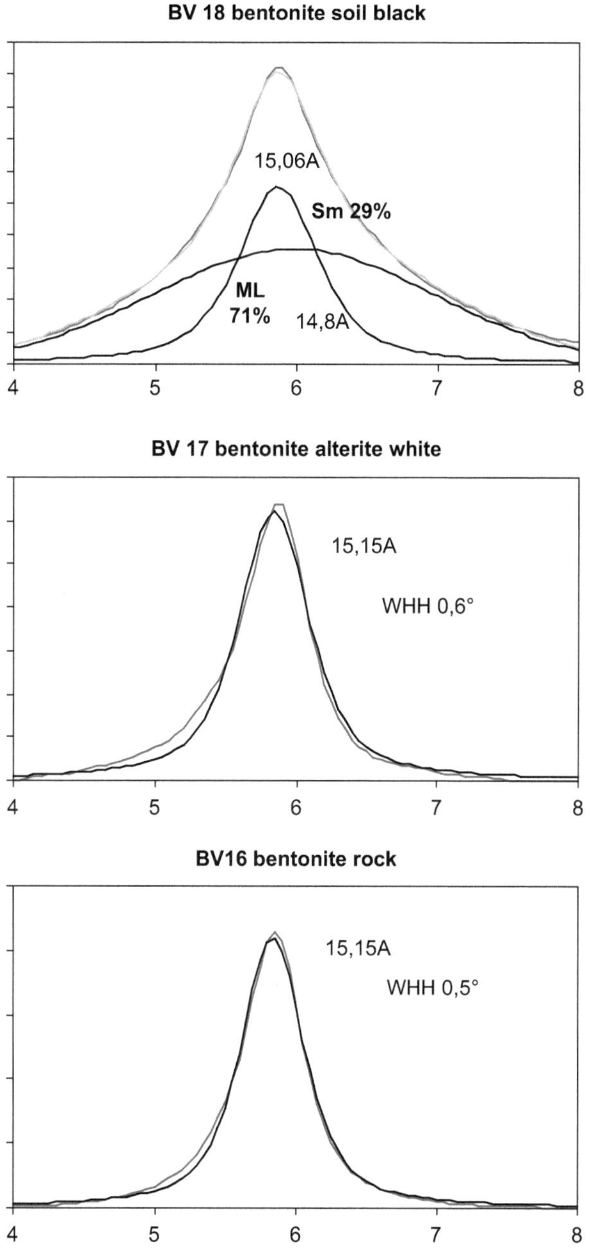

FIGURE 1.7 Evolution of a sedimentary bentonite layer as affected by soil action. X-ray diffraction spectra are decomposed into their basic components with width at half height (WHH) and peak position given in angstroms. Sm, smectite; ML, mixed-layer mineral.

that of species of variable compositions in a continuum of elemental variation, called a solid solution. The 2:1 minerals are compositionally variable, whereas the 1:1 minerals are essentially of a restricted range of composition.

A special characteristic of the 2:1 minerals is their tendency in soil clay systems to have a mixture of the different layer compositions within the same crystallite forming what is called a mixed-layered mineral. This reflects still another type of elemental substitution within the clay silicate structures.

Recent advances in the interpretation of X-ray diffraction spectra of clay mineral materials allows us to identify with numerical precision the proportions of these different substitutional types, which has proven to be very useful for attempts to follow changes in mineralogy brought about by changes in soil chemistry due to plants and agricultural practices. Such changes in mineralogy, and elemental composition, can be observed to occur within several years' time or even within one growing season.

Some silicate material formed by interaction of rock material with surface atmospheric chemical components (CO_2-saturated water) forms new phases that are not crystalline. This material, allophane and imogolite, is much more difficult to assess in its specific chemical behaviour, but it is none the less often of great importance to the properties of some soils.

Thus, we propose to use the terms 2:1 and 1:1 for crystalline phyllosilicate soil minerals instead of the multiple terminology most often used in published studies. Furthermore, because one of our objectives is to enable the use of soil clays to help predict soil properties, it is most useful to include other fine, hence reactive secondary phases, such as oxides, hydroxides and oxyhydroxides, along with fine phyllosilicates in the definition of clays. We therefore adopt the definition of soil clays arrived at by Churchman (2010b, p. 939) as "secondary inorganic compounds of clay (<2 μm) size in soils, regardless of their crystalline or nanocrystalline order, or their degree of disorder".

1.8 SOIL CLAY MINERALOGY

Mineralogy as a subdiscipline of soil science has been in decline for some considerable time (Hartemink et al., 2001). Its earlier results have been seen as disappointing in relation to some of the promises it offered, e.g. for predicting the potassium-supplying power of soils from their clay mineralogical composition (Loveland et al., 1999). Furthermore, fundamental studies of soil clays are unlikely to receive support, while money and effort are expended on pressing soil-related problems, such as those posed by the sequestration of carbon, involving the bonding between minerals and organic matter. Soil mineralogy, along with other soil studies, had a stronger profile as a stand-alone topic for research in the freer economic times of the mid-20th century, which were more generous to studies with probable longer-term payoffs, and as a consequence of the strong government support given to soil studies following the Dust Bowl years in the United States (see Warkentin, 2006, for examples).

New methods of curve decomposition of XRD diagrams, the predominant method of identification of soil clays, now permit a much more detailed identification and quantitative estimation of the mineral forms present and their chemical compositions. This leads to numerical estimations of the change in presence of certain cations held within the clay structures. It is possible now, for instance, to estimate the amount of potassium held in clay structures to within several tenths of a per cent using these numerical tools. These methods, along with statistical treatments of whole-profile XRD data (see Chapter 12, Section 12.7) open new areas of investigation concerning the effects of agriculture and other treatments of soil materials, which were totally impossible in the recent past (Velde and Barré, 2010).

In addition, we can draw an analogy from the example of Dokuchaev, among other examples from the history of soil science, when it comes to the search for information and principles about soil clays. Accordingly, we set out to glean information and principles concerning soil clays per se from the considerable body of mostly very recent work on carbon sequestration, in which soil clays are integral players. We also sourced recent studies devoted to other applications, such as those on soil pollution. Furthermore, we carried out these examinations of the literature armed with the knowledge that quantification of the minerals present and an understanding of their compositions is made possible by the new methods of X-ray curve decomposition.

It is our view that the very nature of clay minerals occurring in soils is often in strong contrast to that of clay minerals occurring in other, strictly geological environments and that a recognition of this contrast should enable closer predictions of soil properties from their clay mineralogies than has been the case in the past. Above all, soil clay minerals deserve their own definition. A typical geological definition of a mineral as "an element or compound that is normally crystalline and has formed as a result of geological processes" (Gaines et al., 1997) meets with difficulties when applied to clay minerals in soils. Clay minerals more generally have been variously defined by both their mineralogy, e.g., as hydrous Al/Mg/Fe layer silicates, and their size (<2 μm), which is a combination of traits for a definition that is unique among minerals (Weaver, 1989; Churchman, 2010b). However, following Churchman (2010b, p. 939), they are most usefully described as "secondary inorganic compounds of clay (<2 μm) size in soils, regardless of their crystalline or nanocrystalline order, or their degree of disorder".

Loveland et al.'s (1999) concern about the apparent irrelevance of the knowledge of a soil's mineralogy in relation to the supply of potassium may demand reconsideration as a result of this broader definition of soil clay minerals (or 'soil clays'). Indeed, we contend that knowledge of the real nature of soil clays should improve our ability to predict and explain not only the abilities of soils to supply nutrients and water to plants but also their role in promoting the physical quality of soils. This governs the ease of transport of air, water and nutrients to plant roots and, furthermore, the stability of soils against mechanical and other erosive forces. In short, a proper understanding of soil clays should clarify the key role that their constituent clay minerals play in enabling soils to meet the three global challenges of feeding the world, sustaining themselves and providing a sink for excess atmospheric carbon. It is the objective of this book to identify the unique features of clays in soils and to explain the origin of these features.

We believe that soil clays are best understood within the context of their formation. They are each products of the environment in which they have formed and developed. This approach is in contrast to the traditional approach based on their structures. Essentially, the traditional approach has treated soil clays as materials whose properties we would like to have set out in neat tables under just a small number of names. Instead, we have come to the view that they can only be understood as materials and their properties appreciated when it is realised that each is a product, maybe even a unique product, of their particular environments of formation and occurrence, i.e. of their history.

Soil clays are chemically differently reactive depending upon the types of ions held in ionically bonded sites within the clay structures. The 2:1 clays (see Chapters 2 and 8) retain ions useful to plant growth and tend to form water-retaining aggregates in associations with organic matter in the soil. Other soil clay minerals do not have these properties to as great an extent; they develop charge through interaction with H^+ and OH^- in soil solutions, and they often form aggregates through associations with iron and aluminium ions, also from solution. The nature of these minerals is a function or rock type or starting materials, climate, and topography of the soil. We will consider these parameters further on. Most important is to realise that different minerals have different effects on the chemical and physical properties of soils.

1.9 A NEW APPROACH TO THE STUDY OF SOIL CLAYS

Until relatively late in the 20th century, upon the publication of the first edition of the comprehensive volume *Minerals in Soil Environments* (Dixon and Weed, 1977), there was no widely accepted textbook on soil clay minerals. First reports on the application of X-ray diffraction to soil clays by Hendricks and Fry (1930) and by Kelley et al. (1931), which revealed their crystalline character, marked the beginning of soil clay mineralogy. The essentially crystalline nature of clay-size minerals found in soils was emphasised through the common employment of methods for the isolation of clays from soils that involved prior removal of non-crystalline components such as organic matter and oxides, hydroxides and oxyhydroxides of Fe, Al and Mn (e.g. Jackson, 1956, 1958). These procedures, essentially involving both oxidation, to decompose organic matter, and reduction, to solubilise metal (hydr)oxides, both disaggregated associations involving clays to enable their suspension as fine material and removed surface coatings from clay mineral particles to enhance their reflection of X-rays. Hence, with the crystalline character of soil clays given prime exposure, the burgeoning number of practitioners and students of soil clay mineralogy generally had their needs for information and understanding of clay minerals in soils catered for by such textbooks and reference books on clay minerals and clay mineralogy per se as Grim (1968) and Millot (1970).

Not only were the structures of soil clay minerals generally equated with those of crystalline minerals in other earth environments, but also their conditions of formation were considered to be comparable. In particular, they were considered to occur as the products of water–rock interactions which have either reached a thermodynamic equilibrium or which show a trend towards attainment of such an equilibrium state. The biological and organic inputs to formation were ignored in this approach. In the opening chapter of Dixon and Weed's (1977) book, J.A. Kittrick expressed an optimistic view that, in spite of complications among soil minerals such as those from ion exchange and crystallinity, "there do not appear to be any fundamental limitations to a marked improvement in our understanding of the soil mineral system through its (the thermodynamic approach's) use" (Kittrick, 1977, p. 23). However, the soil mineral system has proven to be extremely complex, so that modelling it upon the inorganic process of dissolution of highly crystalline minerals in contact with a large excess of water for a limited period of time has proven to be unrealistic and therefore largely unhelpful.

1.10 SOIL CLASSIFICATION

Soil classification is not a concern addressed in this book. Apart from the profile on the cover photograph, it is only applied in, e.g. Section 9.2 (Chapter 9) and Sections 10.3 and 10.5 (Chapter 10) at the level of Orders for Soil Taxonomy (Ahrens and Arnold, 2012) and, in one case (Section 10.3, a Luvisol), that of a Major Soil Unit in the World Reference Base for Soil Resources (WRB) (Michéli and Spaargaren, 2012). These broad classes are used here to identify soils.

REFERENCES

Adamo, P., and P. Violante. 2000. Weathering of rocks and neogenesis of minerals associated with lichen activity. *Applied Clay Science* 16: 229–256.

Ahrens, R.J., and R.W. Arnold. 2012. Soil taxonomy, p. 31-1–31-13. In: P.M. Huang, Y. Li, and M.E. Sumner (eds.), *Handbook of Soil Sciences: Properties and Processes*, 2nd edn. CRC Press, Boca Raton, Florida.

Bergaya, F., K. Beneke, R.W. Berry, G. Lagaly, and K.B. Tankersley. 2013. Clay science: A young discipline and a great perspective, p. 819–855. In: F. Bergaya, and G. Lagaly (eds.), *Handbook of Clay Science*. Developments in Clay Science, vol. 5. Elsevier, Amsterdam.

Bergaya, F., and G. Lagaly. 2013. General introduction, p. 1–19. In: F. Bergaya, and G. Lagaly (eds.), *Handbook of Clay Science*. Developments in Clay Science, vol. 5. Elsevier, Amsterdam.

Birkeland, P.W. 1974. *Pedology, Weathering and Geomorphological Research*. Oxford University Press, London.

Bowen, N.L. 1922. The reaction principle in petrogenesis. *The Journal of Geology* 30: 177–198.

Brady, N.C., and R.R. Weil. 2008. *The Nature and Properties of Soils*, 14th edn. Pearson Prentice Hall, Upper Saddle River, New Jersey.

Calvaruso, C., L. Mareschal, M.-P. Turpault, and F. LeClerc. 2009. Rapid clay weathering in the rhizosphere of Norway spruce and oak in an acid forest system. *Soil Science Society of America Journal* 73: 331–338.

Canarache, A., I.I. Vintila, and I. Munteanu 2006. *Elsevier's Dictionary of Soil Science*. Elsevier, Amsterdam.

Carter, D.W., and J.M. Arocena. 2000. Soil formation under two moss species in sandy materials of central British Columbia (Canada). *Geoderma* 98: 157–176.

Certini, G., and F.C. Ugolini. 2013. An updated, expanded, universal definition of soil. *Geoderma* 192: 378–379.

Chesworth, W. 2008. *Encyclopedia of Soil Science*. Springer, The Netherlands.

Churchman, G.J. 2010a. The philosophical status of soil science. *Geoderma* 157: 214–221.

Churchman, G.J. 2010b. Is the geological concept of clay minerals appropriate for soil science? A literature-based and philosophical analysis. *Physics and Chemistry of the Earth, Parts A/B/C* 35: 927–940.
Churchman, G.J., and D.J. Lowe. 2012. Alteration formation and occurrence of minerals in soils, p. 20.1–20.72. In: P.M. Huang, Y. Li, and M.E. Sumner (eds.), *Handbook of Soil Sciences: Properties and Processes*, 2nd edn. CRC Press, Boca Raton, Florida.
Dixon, J.B., and S.B. Weed (eds.). 1977. *Minerals in Soil Environments.* Soil Science Society of America, Madison, Wisconsin.
Fanning, D.S., and M.C.B. Fanning. 1989. *Soil Morphology, Genesis and Classification.* John Wiley & Sons, New York.
FAO. 2009. FAOSTAT. Food balance sheets. Food and Agricultural Organization, United Nations. http://faostat.fao.org (accessed January 2013).
Gaines, R.V., H.W. Skinner, E.F. Foord, B. Mason, and A. Rozensweig. 1997. *Dana's New Mineralogy.* John Wiley & Sons, New York.
Goldich, S.S. 1938. A study in rock weathering. *The Journal of Geology* 46: 17–58.
Grim, R.E. 1968. *Clay Mineralogy.* McGraw-Hill, New York.
Hall, S.J. 2014. Soils and the future of food: Challenges and opportunities for feeding nine billion people, p. 17–35. In: G.J. Churchman, and E.R. Landa (eds.), *The Soil Underfoot: Infinite Possibilities for a Finite Resource.* CRC Press/Taylor & Francis Group, Boca Raton, Florida.
Hartemink, A.E., A.B. McBratney, J.A. Cattle. 2001. Developments and trends in soil science: 100 Volumes of Geoderma (1967–2001). *Geoderma* 100: 217–268.
Hazen, R.M., and J.M. Ferry. 2010. Mineral evolution: Mineralogy in the fourth dimension. *Elements* 6: 9–12.
Hendricks, S.B., and W.H. Fry. 1930. The results of X-ray and microscopical examinations of soil colloids. *Soil Science* 29: 457–480.
Hillel, D. 2008. *Soil in the Environment: Crucible of Terrestrial Life.* Elsevier, Burlington, Massachusetts.
Hinsinger, P., A.G. Bengough, D. Vetterlein, and I.M. Young. 2009. Rhizosphere: Biophysics, biogeochemistry and ecological relevance. *Plant and Soil* 321: 117–152.
Jackson, M.L. 1956. *Soil Chemical Analysis – Advanced Course.* Soil Science Department, University of Wisconsin, Madison.
Jackson, M.L. 1958. *Soil Chemistry Analysis.* Prentice-Hall, Engelwood Cliffs, New Jersey.
Jenny, H. 1994. *Factors of Soil Formation: A System of Quantitative Pedology.* Dover, New York.
Kelley, W.P., W.H. Dore, and S.M. Brown. 1931. The nature of the base-exchange material of bentonite, soils, and zeolites, as revealed by chemical investigation and x-ray analysis. *Soil Science* 31: 25–56.
Kittrick, J.A. 1967. Gibbsite-kaolinite equilibria. *Soil Science Society of America Journal* 31: 314–316.
Kittrick, J.A. 1977. Mineral equilibria and the soil system, p. 1–25. In: J.B. Dixon, and S.B. Weed (eds.), *Minerals in Soil Environments*, 1st edn. Soil Science Society of America, Madison, Wisconsin.
Knoll, M.A., and W.C. James. 1987. Effect of the advent and diversification of vascular land plants on mineral weathering through geologic time. *Geology* 15: 1099–1102.
Krupenikov, I.A. 1993. *History of Soil Science: From Its Inception to the Present.* A.A. Balkema, Rotterdam.
Lambers, H., C. Mougel, B. Jaillard, and P. Hinsinger. 2009. Plant-microbe-soil interactions in the rhizosphere: An evolutionary perspective. *Plant and Soil* 321: 83–115.
Loveland, P.J., I.G. Wood, and A.H. Weir. 1999. Clay mineralogy at Rothamsted: 1934–1988. *Clay Minerals* 34: 165–183.
Lowe, D.J. 1995. Teaching clays: From ashes to allophane, p. 19–23. In: G.J. Churchman, R.W. Fitzpatrick, and R.A. Eggleton (eds.), *Clays: Controlling the Environment.* Proceedings of the 10th International Clay Conference, Adelaide, Australia (1993). CSIRO Publishing, Melbourne.
Michéli, E., and O.C. Spaargaren. 2012. Other systems of soil classification, p. 32-1–32-34. In: P.M. Huang, Y. Li, and M.E. Sumner (eds.), *Handbook of Soil Sciences: Properties and Processes*, 2nd edn. CRC Press, Boca Raton, Florida.
Millot, G. 1970. *Geology of Clays.* Springer-Verlag, New York.
Schaetzl, R., and S. Anderson 2005. *Soils – Genesis and Geomorphology.* Cambridge University Press, Cambridge.
Sposito, G., and R.J. Regnato (eds.), 1992. *Opportunities in Basic Soil Science Research.* Soil Science Society of America, Madison, Wisconsin.
Sverjensky, D.A., and N. Lee. 2010. The great oxidation event and mineral diversification. *Elements* 6: 31–36.
Velde, B., and P. Barré. 2010. *Soils, Plants and Clay Minerals: Mineral and Biologic Interactions.* Springer-Verlag, Berlin.

Verboom, W.H., and J.S. Pate. 2006. Bioengineering of soil profiles in semiarid ecosystems: The 'phytotarium' concept. A review. *Plant and Soil* 289: 71–102.
Warkentin, B.P. (ed.). 2006. *Footprints in the Soil: People and Ideas in Soil History.* Elsevier, Amsterdam.
Weaver, C.E. 1989. *Clays, Muds and Shales: Developments in Sedimentology*, vol. 44. Elsevier, Amsterdam.
White, R.E. 2006. *Principles and Practice of Soil Science: The Soil as a Natural Resource*, 4th edn. Blackwell, Malden, Massachusetts.
Wolff-Boenisch, D., S.R. Gislason, E.H. Oelkers, and C.V. Putnis. 2004. The dissolution rates of natural glasses as a function of their composition at pH 4 and 10.6, and temperatures from 25 to 74°C. *Geochimica et Cosmochimica Acta* 68: 4843–4858.

2 Soil Clays
Mineralogy

A case could be made for mineral weathering to be considered as the most important process in the geological cycle as it most directly affects the living world in general; and the life of man in particular.

M.J. Wilson (*Clay Minerals* 39: 233, 2004)

2.1 BASIC STRUCTURES OF PHYLLOSILICATES

Clay mineralogy is littered with mineral names, most based upon pure mineralogical (monophase) material which is generally not from surface alteration nor of soil origin. Their names are based upon chemical properties and crystallographic characteristics. The mineral names are or have been employed to describe minerals in natural soils formed under surface conditions. However, viewing the basic mineralogy and functioning of the minerals in soils greatly enables simplification of the terminology, providing a much clearer vision of the material present. Therefore, we propose here a very simple nomenclature system. It is based upon the initial structural configuration of the silica and oxygen ion covalent associations into layers of coordination sites called phyllosilicates based upon the two-dimensional character of their crystal structures (they are also called sheet silicates) and the chemical components present in the structures.

The basic structure is that of Si–O covalent coordinations into what is a tetrahedral configuration. The silica cations are surrounded by six oxygen anions and coordinated with them. The oxygen ions are six in number and hence have a total negative charge of 12, while the silicon ion has a total positive charge of 4. This is outlined in Figure 2.1.

In order for charge neutrality to occur, the oxygen ions must be shared in their covalent coordinations between several silicon cations hence reducing the charge per cation. The structure is essentially planar where levels of silicon cations are coordinated with oxygen anions in planes above and below the silicon ions. The basic geometry of these sheets is one of a tetrahedron on oxygen ions surrounding the silicon cation. The different planes of silicon cations are joined by a plane of oxygen anions to a plane of cations below composed of different elements. The Si–O covalently coordinated planes of ions are not found isolated in nature but are associated with another plane of cations coordinated with the apical oxygens of the tetrahedral sheets of Si–O ions. Figure 2.2 shows the joining of two tetrahedral layers by Al cations which are coordinated with the apical oxygens of the tetrahedral layers and which incorporate some hydroxyl units. The interlinking ions are bonded to six oxygens (or OH units) forming an octahedral structure of anions.

The plane of cations joining the tetrahedral sheets has a different number of coordinated anions: eight oxygens and hydroxyls. These form an octahedral-shaped oxygen network around the cations that join the tetrahedral sheets. There are two hydroxyl ions present in these sites. The two sets of ionic coordinated layers share oxygen between them and the OH units are shared with the octahedrally coordinated ions only. Octahedral site occupancy can be of two or three cations with a total of six positive charges giving the nomenclature of dioctahedral and trioctahedral minerals. The range of site occupancy is near two in the dioctahedral types and between 2.5 and 3 for the trioctahedral types.

Thus, there is an alternation of tetrahedral, silica-rich cation planes and octahedral cation planes of different cations that are interlinked by the surrounding oxygen anions in strongly covalent

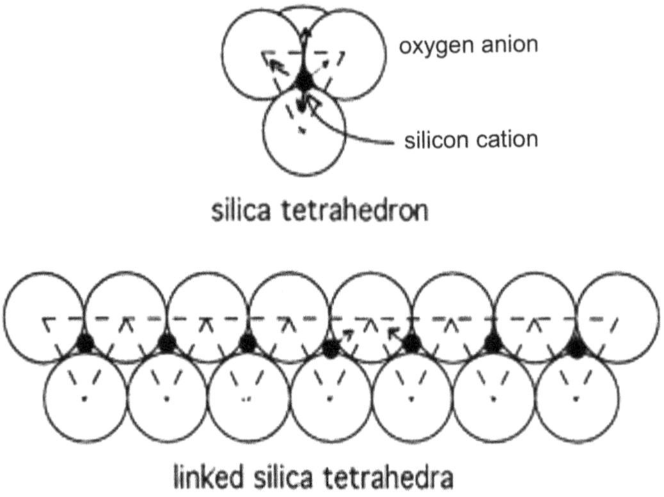

FIGURE 2.1 Schematic illustration of cation–oxygen covalent linkages in silicate clay structures. The basic geometric structure is the association of four oxygen ions with a cation in covalent bonding. The tetrahedra are linked one to another through shared oxygens into a sheet-like structure that forms the basic chemical and geometric structure of the layered silicates (phyllosilicate clay minerals).

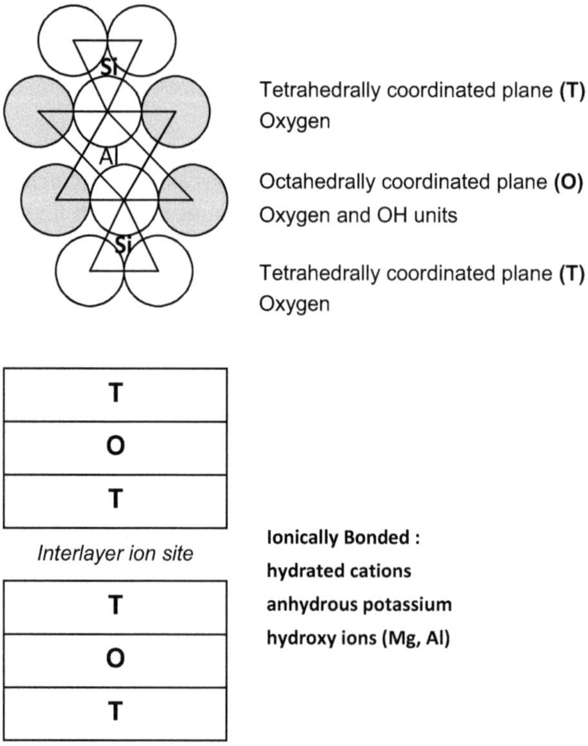

FIGURE 2.2 Illustration of the interlinkage between tetrahedra by association with a cation that is covalently bonded to oxygen ions in the tetrahedrally linked structure. The interlinking cations are coordinated with six oxygen and OH units forming an octahedron around the cation.

linkages with several cations in the silicon-based phyllosilicate minerals. There is often a three-cation plane structure of two tetrahedra (silicon-rich) and one octahedrally coordinated sheets. This is called the 2:1 structure for the existence of two silica-rich sheets and another sheet in between.

A second type is made of just one silica-rich sheet and an octahedrally coordinated sheet. It is called the 1:1 structural type (Figure 2.3). The surfaces of the sheets that are not interlinked are initially essentially neutral in charge when the cations present give the required charge per sheet, 6 for an octahedral sheet and 8 for the silicon-based sheet. It is common to abbreviate the structural terms to Tet for tetrahedral and Oct for octahedrally linked cation–oxygen sheets. Of course, this leads to the abbreviation TOT and TO for the three-sheet and two-sheet structures.

Occupancy of the octahedrally linked ion sheet is determined by a total charge of six positive charges per unit cell where ions can be either divalent or trivalent leading to occupancy by either 2 or 3 to 2.5 ions and giving a nomenclature of dioctahedral (2 ions in the octahedral sites) and trioctahedral types (nearly 3 ions in the octahedral sites).

The octahedrally coordinated ions can be either divalent or trivalent. Since the total charge balance is the same in the tetrahedra the number of ions present will vary with the ionic charge, being two trivalent ions and three or less divalent and trivalent ions. This leads to a nomenclature of dioctahedral and trioctahedral site occupancy for a basic unit structure comprised of ten oxygens and two hydroxyl units (TOT three-sheet structure) and five oxygens and four hydroxyl units (TO two-sheet unit). The three octahedrally coordinated sites compensate a negative charge of 6 and can satisfy this charge by either three divalent ions or two trivalent ions or a suitable combination of both types. These relations are shown diagrammatically in Figure 2.4a and Figure 2.4b.

2.2 LATTICE SUBSTITUTIONS

Ionic substitutions within the 2:1 structures of cations of different charge create positive charge deficiencies giving a diffuse negative charge on the oxygen surfaces of the tetrahedra in the TOT, 2:1 structures. Also, a vacancy in the octahedral sites can give the same effect of diffuse negative

FIGURE 2.3 Illustration of a two-sheet (TO) 1:1 structure of linked oxygen ions.

2:1 TOT

$Si_4Al_2O_{10}(OH)_2$	dioctahedral
$Si_4M^{2+}{}_3\,O_{10}(OH)_2$	trioctahedral
$Si_4M^{3+}{}_2\,O_{10}(OH)_2$	dioctahedral

1:1 TO

$Si_2M^{3+}{}_2O_5(OH)_4$	dioctahedral
$Si_2M^{2+}{}_3\,O_5(OH)_4$	trioctahedral

FIGURE 2.4 (a) Basic unit structures of 2:1 TOT minerals. (b) Basic unit structures of 1:1 TO minerals.

charges on the oxygens of the tetrahedral surfaces. Substitutions creating charge deficiencies on the layered structure are almost non-existent in the TO structures.

The 1:1 TO minerals, showing little cation substitution, are distinguished largely on the basis of their crystal morphology which can cause significant differences in physical and chemical properties in the soils. Chemical activity (surface properties) is largely determined by crystal morphology.

The TO, 1:1 minerals are almost exclusively dioctahedral in soils with aluminium dominant. Some cases of trivalent iron are known and, in some instances, trioctahedral minerals are present with magnesium present. These minerals do not seem to be very stable in alteration sequences, however, even though they can be produced in the early stages of water–rock interaction reactions.

Phyllosilicate clay minerals can then be considered to be of two types: those where there is a two-sheet structure of cation–oxygen complexes that are strongly covalently bonded and those with a three-sheet complex. In surface and soil clays the cations present are usually dominantly silicon, usually followed in abundance by trivalent ions such as aluminium or ferric iron. The dominant characteristic of surface clay minerals is the variable ion substitution in the structural cation sites. Figure 2.5 shows an outline of the effects of changing one cation for another.

Within the 2:1 structure built essentially of Si and M^{3+}, M^{2+}–O covalently shared ions there are various types of elemental substitutions of different amounts leaving some sites only partially filled (vacancy) (Figure 2.6).

	Elements	Cations	Charge +
Tet	Si, Al – O	2	8
Oct	M^{2+}, M^{3+} – O,OH	2 or 3 – 2.5	6
Tet	Si, Al – O	2	8

FIGURE 2.5 Illustration of the effects of changing one cation for another in a TOT structure.

Ionic Substitutions	charge change
Si – Al	−1
M^{3+}, Al – M^{2+}	−1
M^{2+} – □ vacancy	−2
M^{3+} – □ vacancy	−3

Structural site substitutions		charge change
Tetrahedral	Si_4 – Si_3Al	−1
Octahedral	$3M^{2+}$ – $2M^{2+}M^{3+}$	+1
Octahedral	$2M^{3+}$ – $M^{3+}M^{2+}$	−1

FIGURE 2.6 Illustration of effect on charge and vacancies of the exchange of one cation for another in TOT structures.

2.3 OXIDATION-REDUCTION EFFECTS

Change of oxidation state of cations, especially iron, can lead to a change in the overall charge balance on a silicate mineral structure such as

$$Fe^{2+}\left(\text{octahedral site}\right) \rightarrow Fe^{3+} = \text{Reduction of negative interlayer charge}$$

$$Fe^{3+}\left(\text{octahedral or tetrahedral site}\right) \rightarrow Fe^{2+} = \text{Increase of negative charge}$$

Since oxidation is a common factor in materials brought to the surface from rock-making environments, this can play an important role in mineral stability of phases that arrive at the surface. However, the action of bacteria on organic matter can create a reducing environment in the soil zones or certain surface sedimentary deposits. Change in oxidation state can lead to a loss of structural coherence in a mineral and hasten its dissolution, or it can contribute to a migration of some ions into or out of the structure in order to maintain a charge balance.

2.4 RESIDUAL CHARGE ON 2:1 STRUCTURE DUE TO IONIC SUBSTITUTIONS AND SITE OCCUPATION

In the 2:1 minerals there are numerous substitutions of cations one for the other. These substitutions can be of very different amounts in a structure, creating a range of chemical variability and physical–chemical properties. This is true for a mineral structure but the substitutions can also be variable from one layer of the 2:1 structure to another within a crystallite. The range of substitution amount can affect the physico-chemical properties of the mineral. This is the most important effect in 2:1 clays at the surface. The range of substitutions in the mineral leading to an overall charge imbalance or deficiency which leads to a negative surface charge on the oxygen layers determines the type and amount of interlayer cation complexes that will be present in the interlayer site. The amount of negative charge on the surface oxygen ions of the 2:1 structure is a function of the amount of ionic substitutions made in the various cationic sites of the structure. In some cases, a substitution creating a negative charge on the structure (tetrahedral Al for Si substitution for example) can be compensated by an equivalent substitution in another site (octahedral M^{3+} for M^{2+} ion) leaving the structure without a residual surface charge on the external oxygen ions. However, most often there is some substitution which gives a slightly negative charge on the structure, spread over the surface oxygens (i.e. not site specific). This residual negative charge ranges from zero to one per unit cell of 22 anionic charges ($O_{10}OH_2$). The negative charge is compensated by cations which migrate between the charged oxygen layers to bring about overall charge neutrality. The cations which compensate the residual interlayer charge are called interlayer ions and are for the most part surrounded by water molecules. The negative charges on the oxygen ions of the surface layers are not affected by pH, i.e. positively charged hydrogen ions.

The maximum residual negative charge is one electrostatic unit per unit cell of $O_{10}(OH)_2$. This leads to a mica structure when potassium is present in the interlayer site. Charge substitutions can be lower until there is no residual charge. The traditional nomenclature indicates high charge (mica-like unit of 1) to medium charge to low charge of near 0.3 charges per unit cell of $O_{10}(OH)_2$. High interlayer charge leads normally to the inclusion of an anhydrous potassium (without an H_2O complex surrounding the cation) or ammonium ion, which gives a structural interlayer spacing of near 10 Å.

The characteristic of 2:1 TOT minerals is to attract cations, usually hydrated, into the interlayer sites where the diffuse charges reside. This leads to the possibility of ion exchange or, in some cases, fixation of potassium, a more or less permanent occupation by a compensating ion in an anhydrous state. Also, the charge deficiency in these interlayer sites can be compensated by hydroxyl cation complexes, forming a structured layer of coordinated cations, either Mg or Al. The hydrated cations are relatively easily exchanged by other ions of greater abundance in environmental fluids (cation exchange ions), while the hydroxyl coordinated ions are much more stable and less liable to ion exchange.

Interlayer ion compensations due to cationic substitutions are as follows:

$$4Si \rightarrow Si_3Al \text{ and } M^{1+} \text{ interlayer or } Si_3Al \text{ and } 0.5M^{2+} \text{ interlayer}$$

$$2M^{3+} \rightarrow M^{3+}M^{2+} \text{ and } M^{1+} \text{interlayer}$$

$$3M^{2+} \rightarrow 2.5M^{2+} \text{ and } 0.5\ M^{1+} \text{ interlayer}$$

M^{2+} cations octahedra = Mg, Fe^{2+}
M^{3+} cations tetrahedra or octahedra = Fe^{3+}
M^{2+} interlayer = Ca, Mg
M^{+} interlayer = Na, K, NH_4

The types of interlayer ion compensations attracted to the diffuse charge on the layer surfaces are cations of various hydration states, i.e. hydrated with multiple water ions or anhydrous, and those that are associated with hydroxyl ions (Figure 2.7).

2.4.1 Names for 2:1 Structure Minerals in Soils

A hydrated ion occupation gives the mineral name of *smectite* or *vermiculite.* The hydroxyl complex gives the name of *hydroxyl-interlayered (HI) structure* with two types, one of very low exchangeability called *soil vermiculite* and the other of more easily exchanged cation–hydroxyl complexes called *hydroxyl-interlayered smectite.* Vermiculites have a higher charge than smectites (but not

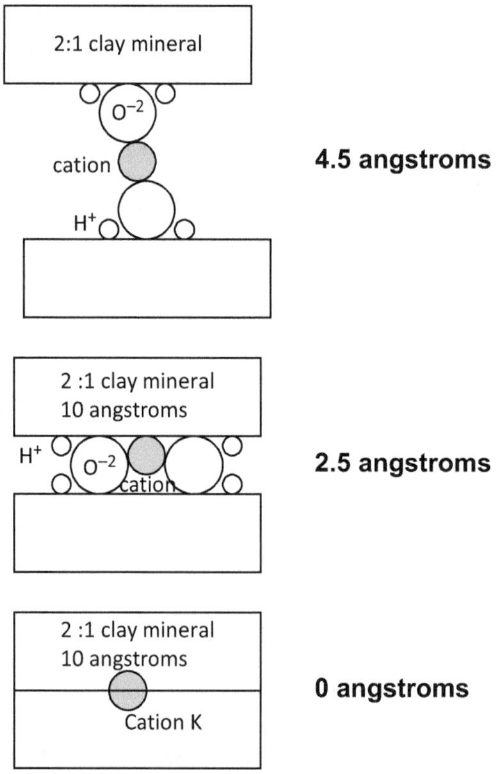

FIGURE 2.7 Locations of cations and interlayer water in relation to aluminosilicate layers in 2:1 minerals.

as high as micas or illites). When potassium is present and interlayer charge sufficiently high, the cations are anhydrous and the mineral is called *illite*. This is probably also the case for ammonium ions. In the laboratory a polar organic molecule (such as ethylene glycol) can be substituted for the water molecules that hydrate the cations in smectites and in doing so there is a change in crystal dimension from swelling which identifies the exchangeable ion characteristic. Vermiculites, when hydrated, and illites do not show swelling with polar organic molecules in the laboratory.

The 2:1 TOT minerals are then similar in structure but different in interlayer cation site occupations. Righi and Velde (unpublished) proposed that "the series of soil minerals smectite, HI and illite might well be part of the same series of continuous solid solution minerals and that the expression of the mineralogy, observed by X-ray diffraction, depends on the interlayer ion population". Basically, the same 2:1 layer structure can have different interlayer ion occupations which give it different mineral names but, more important, different chemical properties. The interlayer site ion population is largely determined by the chemistry of the soil and soil solutions. Initial formation of clays under water–rock alteration regimes gives minerals and ion site occupancies largely determined very locally in the altering material. However, upon progress towards the surface of an alteration sequence the 2:1 materials become more homogeneous and hence the chemistry of the moving aqueous solutions present determines the mineralogy to a larger extent.

Traditionally X-ray diffraction has been used to identify and characterise clay minerals of the 2:1 structural type. The basic trait used is the distance between the covalently coordinated layers in the structures (Figure 2.7). These dimensions are largely controlled by the interlayer ions present (see also Figure 2.8). The effects of interlayer ions on these dimensions are:

1. A 2:1 structure without any ions present will give an average spacing of 9.6 Å. Such material is not found in surface clay assemblages but is characteristic of metamorphic materials. The mineral names are *talc* for the trioctahedrally filled 2:1 structure and *pyrophyllite* for the dioctahedrally filled octahedral layer structure.
2. If the interlayer ion is potassium without water molecules present (anhydrous) the spacing is 10 Å. The mineral is called *illite*, which is dioctahedral and aluminous. Iron-rich varieties can be found in sedimentary material but they are not stable in alteration environments, due to the oxidation of the iron in the mineral. This mineral is *glauconite*.
3. If the interlayer ion is hydrated the interlayer spacing is near 15.2 Å and it is called *vermiculite* or *smectite*. The hydrated ions can be Mg, Ca, Na or K where two layers of water molecules are present. Other polar molecules, such as ethylene glycol or other organic molecules, can be substituted for water molecules changing the interlayer spacing of smectite but not vermiculite. However, natural organic matter rarely is found in the interlayer sites. By way of an exception, natural organic matter was found in the interlayers of smectitic minerals in a site that is very acid, probably preventing microbial breakdown of the organic matter (Theng et al., 1986). The organic interlayers, identified by ^{13}C-NMR, were resistant to oxidation and thermally stable to quite high temperatures. Mass spectrometry

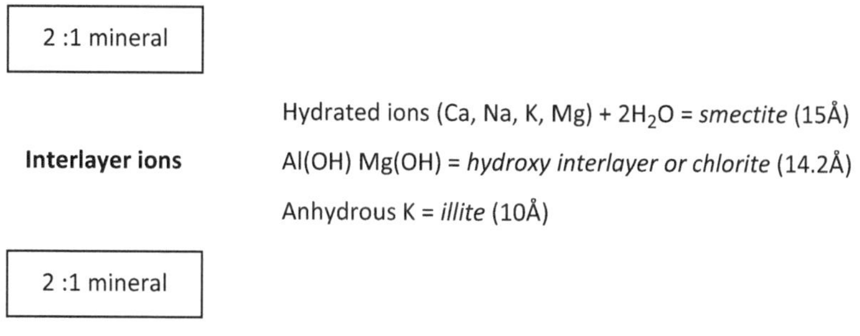

FIGURE 2.8 Names of 2:1 minerals in relation to cations in interlayer spaces.

established that the organic matter in these interlayers was approximately 7000 years old (Theng et al., 1992).

4. If the interlayer ion is associated with hydroxyl ions instead of water molecules the mineral is called *hydroxyl-interlayered mineral*, *soil vermiculite* or *soil chlorite*, and the spacing is 14.2 Å. The hydroxyl interlayered minerals such as vermiculite show a strong first-order X-ray diffraction reflection (near 14.2 Å) and a very low intensity (002) reflection near 7 Å. Chlorite shows a medium intensity reflection at 14.2 Å and a much more intense reflection at 7 Å.

Therefore, for the most part the 2:1 minerals show interlayer spacings of 14.2–15.2 Å (hydrated interlayer ions), 14.2 Å (chlorite type Al-Mg hydroxyl interlayering) and 10 Å anhydrous interlayer ions when in an untreated state at room temperature and intermediate humidity conditions.

Both sides of the 2:1 layer structure are charged negatively so that the external surface, without a contiguous layer, is still charged and can attract ions to its surface (see Figure 2.11). These ions are more open to chemical equilibrium with the surrounding solutions and hence more "exchangeable". Also, these surfaces can attract organic molecules, whereas the interlayer sites rarely contain organic matter.

2.5 CHARGE INTENSITY AND INTERLAYER CATION COMPLEX TYPES

The ionic substitutions within the silicon-oxygen networks lead to variable charges on the surface oxygen sheets of the three sheet phyllosilicate units (TOT). Two-sheet (TO) structures do not show such substitutions and hence have no interlayer ions present. Depending upon the residual charge on the oxygens between the layers of the TOT crystallites, different cation complexes will be present. If the residual charge is relatively low, i.e. less than one half charge unit per unit cell, hydrated cation complexes will be present. If the charge is high, approaching one unit of charge per unit cell, the ion complexes will be anhydrous, so are without water molecules associated with the cation or hydroxyl complexes, i.e. cations associated with OH units. The presence of these types of cation complexes can be identified using the unit cell dimensions (c-dimension), which is easily determined by X-ray diffraction methods. The interlayer dimension is a function of the ion complexes found between the TOT layers of the clay mineral (see also Figure 2.7).

Interlayer cations hydrated with

$2H_2O$/cation basal spacing = 15.4 Å
$1H_2O$/cation basal spacing = 12.5 Å
0 H_2O/cation (anhydrous) spacing = 10 Å
$Al(OH)^+_2$ or $Mg(OH)^+$ complex spacing = 14.2 Å

The chemical activity of the interlayer ion complexes is a function of their water content, i.e. whether in a hydrated ion complex state. If the ions are fully hydrated, they are relatively weakly held to the crystal interlayer site (because of overall low charge on the oxygens of the crystal) and their presence will reflect the chemical forces active in the surrounding aqueous solution of the soil clays. Ions held with a more intense electrostatic charge but anhydrous are less likely to be affected by cation activity in the surrounding aqueous solutions, and they tend to remain within the structure between the TOT layers.

2.6 CATION EXCHANGE IN INTERLAYER SITES

The exchangeable cations found in the interlayer sites where the overall charge on the layer is relatively small (0.5 to 0.2 units per unit cell compared to the maximum of 1.0 units for a mica) can be exchanged one for the other depending upon the overall chemical activity of the different ions in solution. Their presence depends upon the laws of mass action in aqueous fluids. This is indicated by the diagram in Figure 2.9.

Here the law of mass action determines which ions are present in the clay layers. Solution concentrations of different ions will determine the ions present in the solids. Thus the ratio of one ion in solution compared to another will be the determining factor of its presence. However other factors enter into the relative ionic distribution in the solids. For example the overall solution concentration of all dissolved ions can change the relative abundances of ions held in the solids. This is illustrated in Figure 2.10. One possibly important consequence is that in areas of high rainfall one ion will be favoured to remain in the soil clays, while under conditions of low rainfall another will be favoured. Also, in temperate-climate soils, rainfall is not continuous and the amount of water in soils varies, its concentration in dissolved elements changes and, as a consequence, cation selectivity will change as a function of season or water availability.

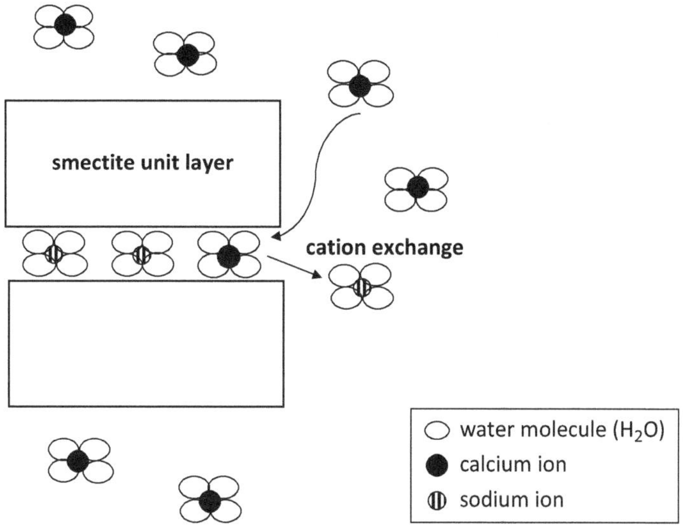

FIGURE 2.9 Illustration of the exchange of ions between solution and the interlayer sites of a 2:1 clay mineral where the preference is for certain ions (black dots) in the interlayer sites.

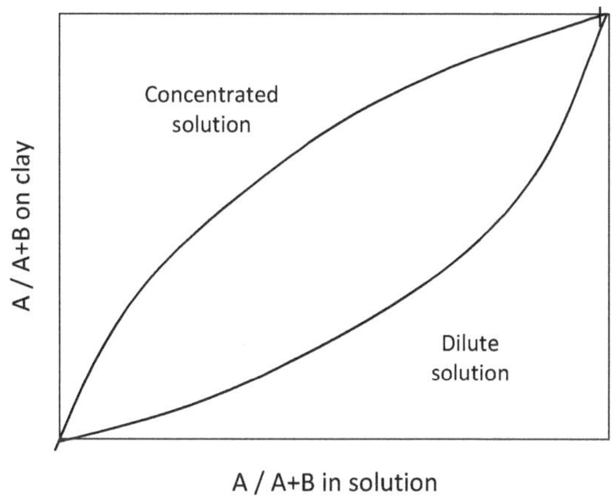

FIGURE 2.10 Illustration of possible effects of different overall solution concentration of exchange ions on the ionic selectivity of the interlayer ion sites. The concentration of A is highly favoured over that of B by the dilute solution compared to the more concentrated solution.

McBride (1979, 1980), and Xu and Harsh (1992), among others, have investigated the chemical and structural reasons for selectivity of cations from solution as they are incorporated into the clay 2:1 structures. Factors such as solution concentration, as seen earlier, can be important, but also the assemblage of ions present, di- or tri-valent and so forth. It is important to note that these effects are variable according to the solution concentration and hence climate factors.

The dynamics of diffusion of ions in the interlayer site has been studied in detail by Tertre et al. (2015) and shown to obey classical laws of atomic chemical attractions. The variables of diffusion are the chemical activity of the cations in the hydrated complex in the aqueous solution and in the interlayer sites. Since the different cations susceptible to be exchanged include monovalent and divalent cations of different hydration potential, there will be a selectivity between the ions in solution and the sites between the clay layers which will influence the retention or dilution characteristics of the elements in the dynamics of soils at the surface, which include rainfall intensity and soil structures.

It has been often noted that the potassium ions in solution enter the interlayer sites in various states of hydration. When the layers have a low charge (0.3 compared to 1.0 for micas) the potassium ions have a two-water layer structure giving a 15.4 angstrom spacing. If the charge is higher, towards 0.6, the potassium ions tend to become anhydrous and the layer dimension is that of a mica, 10 angstroms. The 10 angstrom spacing structure tends to produce a layer of potassium ions of low chemical reactivity and little exchange with ions in solution occurs. Since ammonium ions produce the same structure in laboratory experiments, they can be expected to be present in natural soil clays of high charge in the anhydrous state as are potassium ions. It is important to realise that these non-exchangeable ions can be extracted and replaced in the mica-like minerals under plant or bacterial action in soils (Adamo et al., 2015). Other hydrous cations are rare in high-charge clay layers except for hydroxyl aluminium ion complexes forming chloritic-type layers called *soil vermiculite*, *soil* or *pedogenic chlorite* or *hydroxyl-interlayered minerals*. Here we see that the amount of charge on the interlayers is a factor in ionic species selection in 2:1 soil clays.

2.7 CATION EXCHANGE SITES ON CLAY EDGES

The clay structures as described earlier are essentially of low charge intensity on the external oxygen ions of the clay layer surfaces. These sites are of high bonding energy in that each is due to an uncompensated charge on a specific ion, cation or anion. Anions are essentially oxygen ions that share electrons with other atoms but at the edge of the crystal the next atom is missing. This is indicated in Figure 2.11a. Cations or anions from the aqueous solution are fixed and bonded directly on the edge sites of the crystals. This bonding is in direct contact with the aqueous solution in contrast to the interlayer ion sites that are structured in layers of cation-water or hydroxyl complexes. The same situation or at least a similar one occurs for attractions to organic matter in soils (Figure 2.11b).

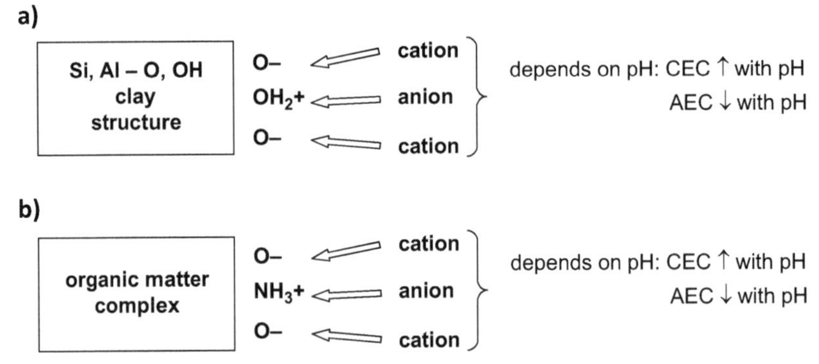

FIGURE 2.11 (a) Edge site attractions for phyllosilicate structures. (b) Edge site attractions for organic material. CEC, cation exchange capacity; AEC, anion exchange capacity.

The amount of edge site attraction is roughly a function of mineral grain size in that the smaller the grain the more edge is present compared to internal sites. This is seen in the data presented in Figure 2.12 for different grain sizes in a kaolinite sample.

The exchange dynamics for edge sites combined with internal sites is quite different from the interlayer sites alone (Robin et al., 2015). Interlayer exchange is effected between the surface oxygen ions which have a diffuse negative charge and hydrated cations which are associated by diffuse ionic bonding of hydrogen ions in the water molecules that are bonded to the cations. By contrast the edge site bonding is between a specific ionic site and a cation (i.e. covalent bonding; Figure 2.13). Thus, the chemical bonding relations are different in the two types of cation associations in 2:1 minerals with interlayer ions. One extremely important difference in bonding behaviour in the alteration and surface environments is the relation to change in hydrogen ion concentration in the surrounding aqueous solutions. The pH is not an important factor for ion exchange in the case of the hydrated interlayer ions. However, the edge site bonding is strongly influenced by solution pH. In general, the lower pH solutions (below 6 or so) show an exchange of hydrogen ions for adsorbed cations. The amount at these pH values can be affected by crystal shape (Aung et al., 2015) suggesting that crystal edges do not all present the same exchange characteristics. The region of pH which shows the strong influence for the adsorption of a cation is of course dependent upon the chemical potentials of the hydrated cation (see Bauer and Velde, 2014, for a more detailed discussion). However, in general, most adsorbed cations are lost to edge sites at pH below 5.

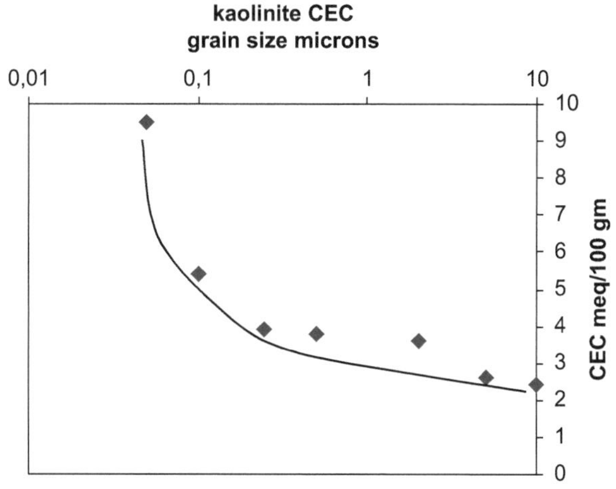

FIGURE 2.12 CEC and kaolinite grain sizes. (Data from Grim, R.E., 1953, *Clay Mineralogy*, 1st edn., McGraw-Hill, New York.)

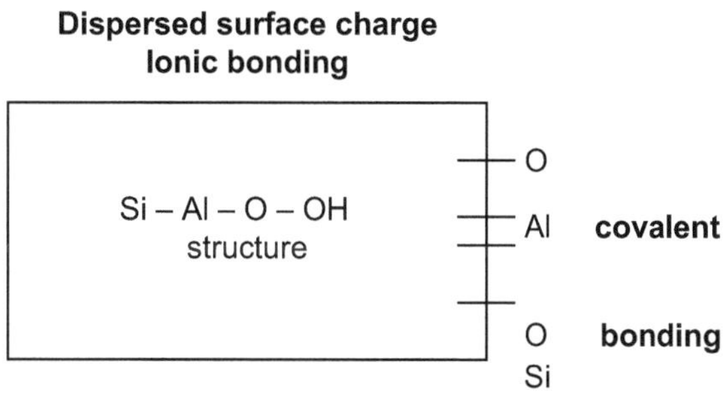

FIGURE 2.13 Nature of the bonding in the interlayer and the edges of 2:1 minerals.

Organic matter very often plays a major role in cation retention in surface materials. The major ionic interaction is between hydrogen ions on the surface sites of organic molecules (Sposito, 1989).

The reverse relations can be observed for anions, which naturally are attracted to surface sites when pH is low and hydrogen activity high. At pH values above 7 anions become less abundant on edge sites in soil materials.

As a point of comparison the hydrated absorbed interlayer ions found on a smectite (fully expandable and hydrated) is of the order of 130 centimoles of positive charge per gram, $cmol_{(+)}g^{-1}$ (or milliequivalents of charge per 100 g) of clay mineral, whereas the edge- and surface-adsorbed cations are in the range of 10 to 15 $cmol_{(+)}g^{-1}$ (or meq/100g) of charge. Thus, the mineral type present (expandable 2:1 mineral or 1:1 nonexpanding, oxide or organic material) is very important for the capacity of the materials to capture and exchange cation elements from the aqueous solutions in the soils and alterite materials of the surface. The pH of the aqueous solutions in soils is a function of the rock type, which creates the soil clay minerals and the vegetation type present and are discussed further in Chapter 7.

2.8 PERTINENT PRINCIPLES OF CATION EXCHANGE

The following is a recapitulation of the major chemical effects governing cation exchange (see Bauer and Velde, 2014).

2.8.1 Cation–Water Interaction

Cations in aqueous solution are normally surrounded by water molecules which are bound at various intensities to the cation through oxygen–electron interactions. Cations are essentially of an electron structure of ionic interaction, while oxygen is essentially of covalent electronic interaction. Thus, the bonding is relatively weak but governed by the charge density on the surface of the cation. The charge density is governed by the ionic charge of the cation and the ionic diameter, which gives a charge per surface area coefficient. Thus, water molecules are reattracted to the surface of ions as a function of the charge density on the surface of the ion. The same is true of the interaction of the cations in solution with charged solids. Most crystals and in fact non-crystalline materials that have cations on their surface that are not fully electronically compensated due to ruptures in the regularity of the crystal structure. The interaction of cations with charged (negatively) surfaces is a function of the surface density of charge on the cations (Figure 2.14).

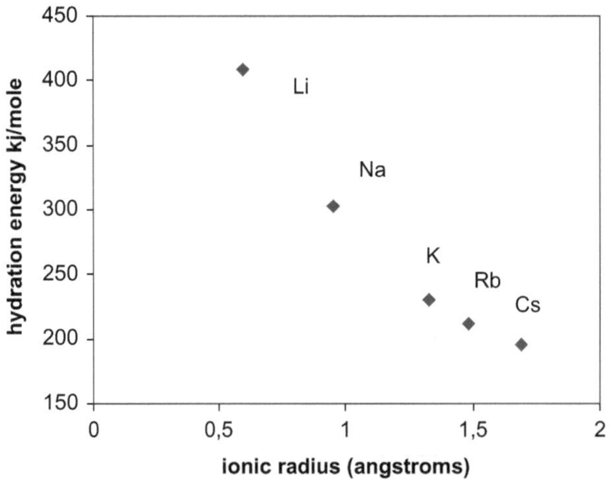

FIGURE 2.14 Hydration energy of monovalent cations in relation to their radii.

The interaction of cations with aqueous solutions is then largely governed by the charge on the ion and its diameter or more precisely its surface area.

These relations then determine the observed variables in laboratory tests of

1. Dissociation constants of compounds soluble in water
2. Hydration energy of cations (relative attraction to charged surface)

The charge–surface area relations determine the relative chemical activity of the ions in solution and the selectivity of fixation of ions on charged surfaces. A highly charged ion will be fixed before a lower charged ion, etc. In some cases, the site of fixation has some geometric constraints which will favour one ion over another; one case is the interlayer potassium fixation on highly charged 2:1 minerals discussed later. The charge intensity can be important in the selection of cations as shown in Figure 2.15 where an increase in interlayer charge on the 2:1 layers (diffuse permanent charge) is a factor of ionic selection.

The relative activity of ions in solution is also related to ionic concentration of dissolved species in solution, temperature and the different types of ions present (see Figure 2.14). Thus, the selectivity of ionic species in solution to be fixed on surfaces of solids is quite complicated and difficult to appreciate in a complex system such as a soil where many species of ions are present, and the concentration of dissolved species can vary as a function of rainfall or water percolation as can the temperature. Further, the type of charge on the solids in the soil can determine ionic preferences also. Hence a soil is a complex system when considering cation exchange.

2.8.1.1 Cation–Water Interaction: Summary

Factors which influence the interaction of cations with water in the environment of the surfaces of solids are

1. Dissociation constants of compounds soluble in water
2. Hydration energy of cations (relative attraction to charged surface)
3. Charge–surface ratios or relative charge intensity
4. Concentration–exchange relations

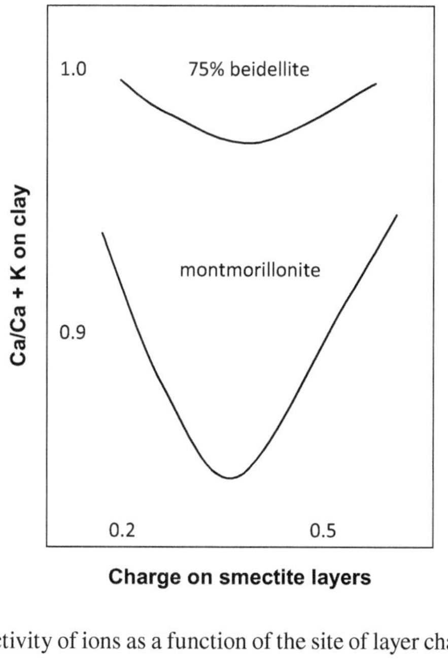

FIGURE 2.15 Change in selectivity of ions as a function of the site of layer charge in a 2:1 mineral (smectite).

2.8.1.2 Selectivity among Ions in Solution: Summary

Factors which can affect the selectivity of ions in solution to mineral or organic surfaces are

1. Relative activity in solution (function of solution concentration)
2. Ionic charge and charge density (function of charge and ionic diameter)
3. Thermal effect on selectivity
4. Amount of charge on 2:1 clay structure layers

2.8.1.3 Relations of Preference for Cations or Hydrogen Ions (Cations): Summary

There is a competition between cations and hydrogen ions in solution for charged sites on solids in soil systems depending upon

1. pH and cation attraction (seen as concentration of cations on solids vs pH)
2. Oxygen–element bond energy

2.8.1.4 Types of Exchanged Cation in Layer Silicate Interlayers: Summary

The following are the types of cations involved in exchange reactions, and their exchangeability:

1. Anhydrous ions (K), low exchangeability
2. Hydrous one layer, exchangeable
3. Hydrous two layers, easily exchangeable

In general, the cations (in various states of hydration) absorbed into the 2:1 mineral structures are stable under a large range of pH conditions (Bauer and Velde, 2014).

2.8.1.5 Cation Exchange on Edge Sites: Summary

Edge sites or charge variable sites are affected by

1. Factors of hydration energy of ions (Teppen and Miller, 2006)
2. pH and competition of cations with hydrogen ions for available ionic bonding sites (Bauer and Velde, 2014)
3. Effect of ion selectivity as a function of solution concentration, temperature or charge on substrate, among others (Xu and Harsh, 1992)

The selectivity coefficients determined in the laboratory on pure phases depend largely upon the material present, which is quite variable in soil systems. Clay minerals are rarely those used in laboratory experiments and organic materials are quite variable, and they evolve rapidly in soils as a function of time such that their surface chemical characteristics change continually and are also as a function of the type of plants that produced the material in the soils. Hence it is difficult to predict the selectivity of one ion over another for a specific soil situation. Nevertheless, certain controls override these variations.

However, it is clear that the situation of soil solutions as a function of time and place in the soil profile changes in such a manner that the selectivity of one ion over another will vary over short periods of time. The effect of solution concentration on selectivity can be expected to be important as a soil system loses water content after a rain or during evaporative periods. As plants absorb one type of ion compared to another (K vs Ca for example) the solution concentration can vary temporarily. Restoration of equilibrium conditions can be expected but as temperature and solution concentration change these restorations will be less evident.

2.8.2 Overall Effect of pH on Exchange (Capture or Loss) of Cations

Charge variable edge sites exhibit (Figure 2.16)

1. Fixation in basic solutions of cations
2. Loss of cations in acidic solutions

Silicate minerals are only one part of the solids in soils. Oxide phases, usually poorly crystallised and thus characterised as charge-variable exchange materials, are present in many cases. These materials act as ion exchangers of the crystal edge type. Further, organic material can be fixed on the charge variable exchange sites with the result that the surfaces of this organic material furnish even more charge-variable exchange sites for cations in solution. Organic material is extremely variable chemically, concerning the surface molecular and ion functions present. Thus, it is difficult to predict its effect and further, given that organic material evolves relatively rapidly in soils, measurements at a given time and place are at risk of being different with changes in the site and period of observations. These materials respond to pH as absorbed material on charge-variable sites.

The overall effect can be seen in Figure 2.17 where base exchange capacity and occupancy (exchangeable ions) have been measured for prairie and forest soils in the United States (data from

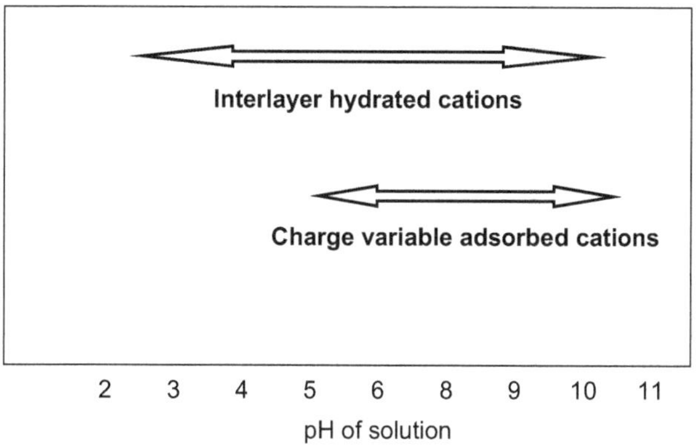

FIGURE 2.16 Schematic representation of pH effect on cations in absorbed interlayer ion sites and adsorbed surface sites (charge variable) for various pH values (see Bauer and Velde, 2014).

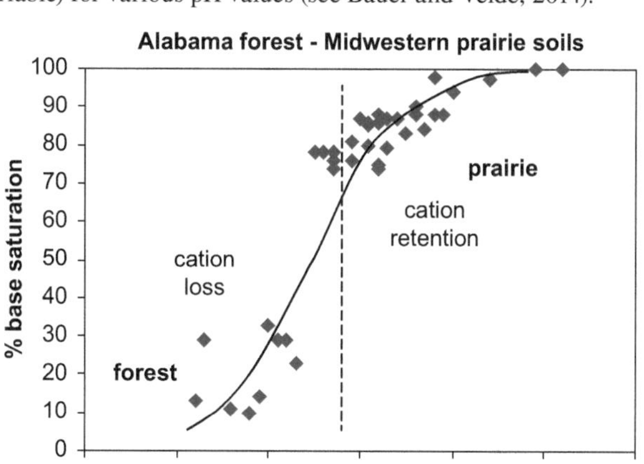

FIGURE 2.17 Indications of exchange cations (% base saturation) as a function of pH in soils. (Data from Bryant, J.P., and J.B. Dixon, 1963, *Clays and Clay Minerals* 12: 509–521; Ruhe, R.V., 1984, *Geoderma* 34: 201–219.)

Jenny, 1994). The charge-variable sites attract most of the easily exchangeable materials on the surfaces of the soil material.

Cations then can be absorbed into clays (2:1 mineral exchange sites) or adsorbed on crystal edges and surfaces of oxides and organic matter in the soil. The absorbed, internal sites are the most stable and less likely to be displaced by chemical or physical changes. The preferences for one over another ion are a function of many variables (solution concentration, hydration energy, charge on the attracting site, temperature, etc.) and hence it is difficult to predict which ions will be present in or on the soil materials. However, organic materials presenting charge-variable sites are often predominant in cation exchange in soil.

2.9 EFFECT OF CLIMATE

The concentration of ions in solution is a major variable in the process of aqueous–solid interactions. The more water present, and the greater the renewal rate, the more ions will enter the solution from the solids but in a diluted state. For example, in temperate climates rainfall is periodic and the amount of water in a soil varies with time. The relations of cations fixed on the clay materials and those in solution will then change with time (solution concentration); in a more humid climate the tendency will be for ions to enter into solution and not be fixed on clay materials due to the low average concentration of ions in the abundant water solutions. Hence the ions held on clay surfaces in a temperate climate will be most likely different from those on these surfaces in a tropical climate.

Further, the charge variable sites on clay surfaces are the most affected by the chemistry (pH and solution concentration of ions). The pH is largely determined by the vegetative cover in the soil area. Conifers and some tropical forest cover engender high hydrogen activity in the soil solutions. The interlayer sites in clays (between the 2:1 layers) are much less subject to changes in solution chemistry and are more constant in their chemical composition. This is especially true for pH effects where the cations are lost through hydrogen ion exchange on charge-variable sites. Climate is a very important variable in these relations, determining the amount of water present and the type of vegetation, which affects the local chemical configuration of dissolved materials.

2.10 MIXED-LAYERED CLAYS

Mixed-layered clays mostly comprise combinations of different types of 2:1 minerals (combinations of 2:1 and 1:1 layers also occur; see also Chapters 7 and 8). Interlayered minerals are most common in the soils zone and rather rare in the water–rock alteration zone. Mixed-layered minerals generally show layers of fixed dimension material (illite or hydroxyl interlayered layers) and vermiculite or smectite layers, where the polar molecules can be changed but are usually water molecules in soils. There are two basic types of interlayer succession in mixed-layer minerals: one where the succession is in a random order and another where there is a regular succession of the layers, called ordered mixed layering (Figure 2.18).

The proportions of the different chemical components appear to be a function of the overall chemical activities of the dissolved elements in solution, especially silica, which induces the formation of smectite that contains more silica than illite mica-like minerals. It is the most silica-rich of the clay mineral types.

Thus the 2:1 structure can lead to different mineral types according to classical mineral classifications depending upon the interlayer ion occupancy. Within the structure, if the octahedral sites are dioctahedral and contain ferric iron the mineral name is *nontronite*. Further names have been given to minerals depending upon the site in the structure where ionic substitutions lead to a residual charge on the structure. If charge is induced by substitutions in the octahedral site of dioctahedral minerals, they are *montmorillonites*, while those with substitutions in the tetrahedral site (Al for Si substitution) are called *beidellite*. Those with further substitution of magnesium in the octahedral site are called *saponite*. In *hectorite*, the octahedral cations are lithium and some other types are

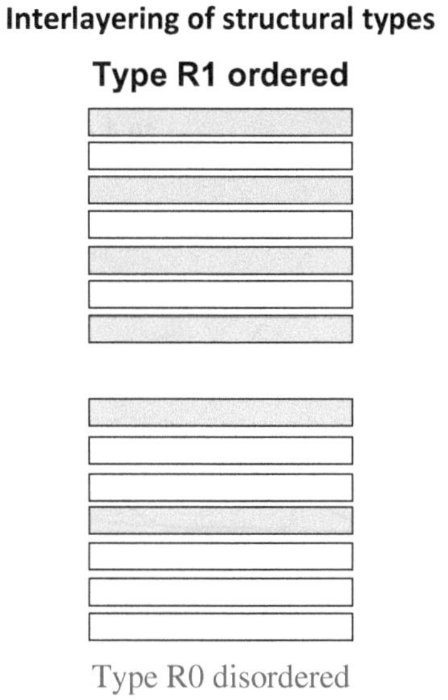

FIGURE 2.18 Examples of two types of interlayering depending upon the succession of the different layer types.

possible (see e.g. Kodama, 2012). What is most important to soil mineralogy is the interlayer ion site occupancy and the presence of iron in different sites. Iron can lead to oxidoreduction effects that change the clay mineral properties, while interlayer ion site occupancy can lead to ionic exchange, which will affect the availability of ions to plants growing in the soils. Interlayer ion site occupancy can also determine important physical properties of minerals and hence soils. If interlayer ions include sodium, in proportions as low as 5% of all interlayer cations, soils containing these clay minerals will swell and/or disperse when in contact with water.

The 1:1 TO minerals, especially *kaolinite*, can at times be found to be interstratified with 2:1 TOT minerals either smectite or at times illite. These minerals are found to form in soils by crystallisation and occur in the fine fraction of the clay minerals (Hubert et al., 2012; Viennet et al., 2014). They can become the dominant minerals at times in semi-arid soils (Vigniani et al., 2004).

2.11 IDENTIFICATION OF LAYER SILICATE CLAYS (2:1 AND 1:1 STRUCTURES) BY X-RAY DIFFRACTION

The principal method for the identification of silicate clay minerals is by X-ray diffraction techniques. The method depends on the fact that these minerals are mostly sheet shaped and when deposited on a flat slide they form parallel layers that enhance the diagnostic basal plane reflections or X-rays created by the atomic layers in the crystals. This method has been used and refined for decades. However, the X-ray diffraction intensities often overlap on one another making identification of individual components difficult, if not impossible. Nonetheless, recent progress has been made (see Lanson, 1997) using mathematical curve composition models which simulate the complex patterns occurring in a diffraction spectrum. Figure 2.19 shows the frequent aspect of a soil clay spectrum. Two specific diffraction maxima are identified, with large undefined diffraction areas between them. Using curve decomposition modelling for the large bands, three more bands which make up the major part of the diffracting material are identified as mixed-layered minerals. These bands and the material causing them are less distinct in their interaction with X-rays because the grain size is small and much of the material (the two wide bands) is of a mixed-layer type. However, the simulations give us a means of identifying the mixed-layer components and by using

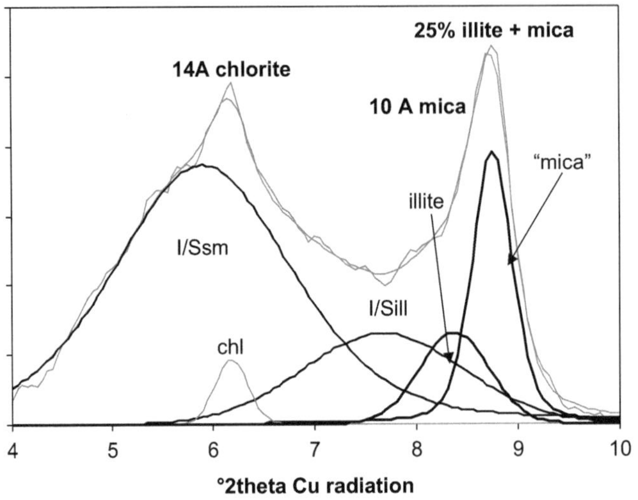

FIGURE 2.19 X-ray diffraction spectrum of clay fraction of a Chinese Vertisol. We initially see peak maxima at 14 and 10A (chlorite and mica) but the chlorite peak is in fact small, and below and around it is a strong band of mixed-layer mineral diffraction maximum (I/Ssm or smectitic illite/smectite mineral). The smaller intensity illite-smectite mixed-layered mineral (I/Sill) is not apparent in the untreated spectrum.

the peak position maxima we can estimate the proportions of the components present. Relations between the peak width and peak height give the relative peak area for a given diffraction band. If we compare these one to another, we can estimate the relative amount of a phase present in the clay fraction. Further, given that the peak position indicates the proportions of the constituent layers, it is possible to estimate the amount of a given layer type from its peak position and the relative abundance of the phase in the assemblage. Using the peak positions and the relative surface areas we can determine in many cases the chemical composition of the materials for certain elements such as potassium found in 2:1 structure 10 angstrom clay minerals such as illite (see Barré et al., 2008).

Further, profile fitting methods using appropriate computer software (Reynolds and Reynolds, 1996) enable the identification of complex minerals, in particular those with mixed layers, which are not identified by visual analysis of XRD patterns (see Chapter 8, Section 8.4). This procedure gives quantitative measures of mixed-layered minerals.

Thus, recent methods of interpretation of X-ray diffraction diagrams enable a much clearer and at times quantitative idea of clay mineral content in soils, which are almost always multi-phase with overlapping X-ray diffraction peaks and can comprise a number of mixed layers.

2.12 ALLOPHANE AND IMOGOLITE

Allophane consists of hollow, approximately spherical crystals with diameters of 3.5 to 5.0 nm (Wada, 1989). Its structure, based on that of imogolite, has a defect gibbsite Al octahedral sheet framework. Si in tetrahedra is attached to this framework by sharing three of its O atoms with an octahedrally coordinated Al on the gibbsite sheet, while the remaining O atom acquires an H atom to form an Si–OH bond pointing outwards. The Al octahedral sheet occupies the outside of the sphere, while the Si tetrahedra sheet is found on its inside (Wada, 1989; Lowe, 1995; Churchman and Lowe, 2012; Harsh, 2012). Imogolite, with tubes that show a more regular structure, gives a more distinct pattern in XRD than allophane, which gives either a poor pattern with a few broad peaks, or no distinct peaks at all (Wada, 1989). Its pattern is especially indecipherable when in a mixture with other, more crystalline minerals.

Both allophane and imogolite have high, specific surface areas (700–1500 m^2g^{-1} for allophane, 1500 m^2g^{-1} for imogolite; Churchman and Lowe, 2012). Their points of zero charge vary with their

Si:Al ratio: pH ~8 for imogolite and 1:2 Si:Al allophane, ~6–7 for 1:1.6 allophane, and ~4–5 for 1:1.2 allophane (Harsh, 2012). They also exhibit selective chemisorption (or inner-sphere sorption) of anions such as fluoride and phosphate, and also of heavy metal cations such as Pb, Cu, Zn, Co and Cd.

2.13 IRON OXIDES, HYDROXIDES AND OXYHYDROXIDES

The basic structure of iron oxides (including oxyhydroxides and hydroxides) comprises close-packed arrays of O and OH ions, with octahedral cavities occupied to different extents by Fe(III) or Fe(II) (Bigham et al., 2002; Kämpf et al., 2012). When structures are based on hexagonal close packing, they are referred to as an α-phase, and when based on cubic close packing they are denoted as a β-phase.

Goethite (α-FeOOH) comprises double chains of edge-shared octahedra linked together by sharing corners and also by hydrogen bonds. Fe(III) occupies one-half of the octahedral positions. Crystals in soils and sediments usually have rough, irregular surfaces, unlike synthetic crystals, which are typically needle shaped, although some crystals in soils and sediments show diamond shapes in their cross-sections.

Hematite (α-Fe_2O_3) is made up of sheets of edge-shared octahedra. Fe(III) occupies two-thirds of the available octahedral sites. Each FeO_6 octahedron shares edges with three neighbouring octahedra in the same plane, and a face is shared with an octahedron in an adjacent plane. Hematite crystals in soils and sediments are generally lacking in a regular shape, but synthetic hematites have a variety of shapes, such as cubes, discs, rhombohedra, spheres and ellipsoids (Bigham et al., 2002).

Lepidocrocite (γ-FeOOH) comprises double chains of edge-shared octahedra, giving corrugated layers. Well-crystalline versions have a lath- or raft-type morphology.

Magnetite (Fe_3O_4) has a unit cell that comprises 32 close-packed O atoms, with 8 tetrahedral and 16 octahedral sites occupied by cations. The tetrahedral sites contain Fe(III), while 8 of the octahedral sites contain Fe (III), and Fe (II) occupies the remaining 8 octahedral sites. This is an inverse spinel structure. They tend to have an octahedral morphology (Bigham et al., 2002; Kämpf et al., 2012).

Maghemite (γ-Fe_2O_3) has the same crystal structure as magnetite but a different chemical composition. It contains vacancies in its octahedral sites and may contain some Fe(II), although dominated by Fe(III) (Bigham et al., 2002; Kämpf et al., 2012).

Ferrihydrite ($Fe_5HO_8.4H_2O$) appears as highly aggregated spherical particles. The size of domains varies from 2 to 6 nm. Anions on the particle surface are represented by OH groups and H_2O molecules bound to them; this changes the O:OH:H_2O ratio depending on the particle volume, therefore different chemical formulae have been suggested by different authors. Its structure appears to be a defect hematite structure with both edge and face-sharing octahedra in the main, but controversy (e.g., Eggleton and Fitzpatrick, 1988; Manceau et al., 1990; Drits et al., 1993; Manceau and Gates, 1997) surrounds the possibility that some Fe(III) ions are tetrahedrally coordinated and recent evidence (Pokrovski et al., 2003; Peak and Regier, 2012) points in this direction. Fe(III) ions are randomly distributed over the octahedral (or tetrahedral) interstices and many sites are vacant. It contains more OH^- and H_2O and less Fe (III) than hematite.

Akaganéite (β-FeOOH), while rare, comprises double chains that share corners to give tunnels that are 0.5 nm in cross section. These tunnels contain Cl^- ions.

Schwertmannite ($Fe_8O_8(OH)_{8-2x}(SO_4)_x$) ($1 \leq x \leq 1.75$), with a molar ratio for Fe/S of between 4.6 and 8.0 and a mass percentage for $SO4^{2-}$ $1 \leq x \leq 1.75$) between 12.5% and 20.5% when crystal water is not considered. However, this value can range widely from 5.3% to 32% (R.W. Fitzpatrick, personal communication; see also Fitzpatrick et al., 2017). It has a similar structure to akaganéite, but with SO_4^{2-} in the tunnels rather than Cl^-.

Feroxyhite (δ-FeOOH) comprises sheets of edge-sharing octahedra as well as some face-sharing octahedra.

Green rusts ($FeOH_2$) comprise stacked sheets with this formula. They are blue-green and dominantly Fe(II) in octahedra in which some of the Fe(II) is oxidised to create a positive layer charge. The charge is balanced by interlayer anions, including Cl^-, SO_4^{2-} and CO_3^{2-}. In soils, only one variant, *fougerite* $(Fe^{2+}Mg)_6Fe_2^{3+}(OH)_{18}.4H_2O)$ has been found (Kämpf et al., 2012).

2.13.1 Surface Reactions of Fe Oxides

As with other oxides in soils, the charge on Fe oxides (including oxyhydroxides and hydroxides) is determined by H^+ and OH^- in the soil solution contacting the oxides. They each have a point of zero charge, a pH value below which they are net negatively charged, so they attract anions, and above which they are positively charged, so they attract cations. Their point of zero charge (PZC) generally falls between pH 7 and 9 (Schwertmann and Taylor, 1989; Kampf et al., 2012). While their charge is pH-variable and pH governs the uptake of many ionic species by outer-sphere bonding, they also display specific adsorption of many ions, which therefore bind by inner-sphere interactions, i.e. chemisorption. Chemisorption occurs with anions of weak acids (e.g. phosphate, fluoride, silicate, molybdate, arsenate, and also sulphate and many organic anions) and it also occurs with heavy metals (e.g. Zn, Cd, Cu, Co, Pb, also Li and Mg) (Schwertmann and Taylor, 1989; Kämpf et al., 2012). In general, the reactivity of the oxides depend on the number of sorption sites per unit mass, which is related to their specific surface area (SSA) (Bigham et al., 2002). Due largely to their small size, this can be very high (~70–250 m^2g^{-1} for crystalline Fe oxides, 200–500 for ferrihydrite) (Schwertmann and Taylor, 1989; Churchman and Lowe, 2012) so the oxides can achieve considerable adsorption of both anions and cations.

2.14 ALUMINIUM OXIDES, OXYHYDROXIDES AND HYDROXIDES

Gibbsite ($Al(OH_3)$) comprises edge-shared stacks of octahedral $Al(OH)_3$ sheets. It appears as thick, relatively rectangular crystals, or as nodules, and has also been found as pseudomorphic after feldspar (Huang et al., 2002).

Boehmite (γ-AlOOH) comprises double chains of AlO_6 octahedra that each share two edges with adjacent octahedra and two edges with an adjacent double chain to give a zigzag arrangement.

Corundum (α-Al_2O_3) is rare but has been identified in soils following bush fires where it is supposed to have derived from the dehydroxylation of Al-goethite. It is found to be isostructural with hematite on which it is templated (Huang et al., 2002).

Surface charge and interactions are similar in type to those of Fe oxides (see earlier) (Hsu, 1989). The Al oxides, in general, have SSAs of 100–220 m^2g^{-1} and a PZC between pH 5 and 9 (Kämpf et al., 2012).

2.15 MANGANESE OXIDES

In manganese oxides, Mn occurs in octahedral coordination surrounded by O atoms. They are subdivided into phyllomanganites, in which the octahedra have shared edges, and tectomanganites, in which the corners of tetrahedra are shared to form tunnels (Dixon and White, 2002). Phyllomanganites include *birnessite* ($Na_{0.7}Ca_{0.3}Mn_7O_{14}{\cdot}2{\cdot}8H_2O$), which can appear in crumpled sheets (Dixon and White, 2002). They also occur as *lithiophorite* ($Al_{0.7}Li_{0.3}Mn^{4+}{}_{0.7}Mn^{3+}{}_{0.3}O_2(OH)_2$) as nodules with a well-developed platy morphology (Dixon and White, 2002). Tectomanganites include *todorokite* ($(Na, Ca, K)_{0.3-0.5}(Mn^{4+}, Mn^{3+})_6O_{12}{\cdot}3{\cdot}5H_2O$) found in nodules in soils in a well-developed platy morphology (Dixon and White, 2002; Churchman and Lowe, 2012; Kämpf et al., 2012). *Vernadite* (δ-MnO_2) appears to have both edge-shared octahedra as in phyllomanganites and also corner-shared octahedra as in tectomanganites (Dixon and White, 2002).

Surface charge and interactions are similar in type to those of Fe oxides (McKenzie, 1989). Mn oxides generally have low PZCs (from pHs 1.5 to 7.3, depending on the species) and their SSAs also encompass a wide range (5–360 m^2g^{-1}) (Kämpf et al., 2012).

2.16 SILICON OXIDES

The most abundant of the Si oxides is *quartz*, SiO_2. It has paired helical chains of corner sharing SiO_4 tetrahedra that spiral along the Z-axis (Kämpf et al., 2012). *Tridymite* and *cristobalite* are polymorphs of quartz, with the same formula, differing among these forms of silica in the arrangement of the silica tetrahedra. They both have a sheet containing six-member rings of tetrahedra, with the tetrahedra alternately pointing above and below the plane of the basal oxygen atoms (Kämpf et al., 2012). In tridymite, the tetrahedral sheets are stacked to lie directly over each other, but in cristobalite alternate sheets are translated relative to each other so that they do not lie directly over each other.

Opal may be *opal-C* (SiO_2), *opal-CT* (SiO2) or *opal-A* ($SiO_2{\cdot}nH_2O$). In opal-C, the structure is predominantly cristobalitic (Kämpf et al., 2012). Opal-CT has random arrangements of cristobalite and tridymite arrangements. Opal-A is non-crystalline and hydrous.

2.17 TITANIUM OXIDES

Titanium as an oxide occurs widely in soils, but in small amounts, as the minerals *rutile*, *anatase* and *brookite* (all TiO_2). Rutile has single chains of edge-sharing octahedra, while the octahedra in anatase share four O–O edges. Brookite consists of deformed TiO_2 tetrahedra, which share three O–O edges. Illmenite ($FeTiO_4$) is almost isostructural with hematite (see earlier), with half of the Fe atoms replaced by Ti (Kämpf et al., 2012). Other oxides containing Fe and Ti which may be found in soils are *pseudobrookite* (Fe_2TiO_5), *pseudorutile* ($Fe_2Ti_3O_8(OH)_2$) and *ulvöspinel* (Fe_2TiO_4); see Fitzpatrick and Chittleborough (2002) and Kämpf et al. (2012) for more details.

2.18 ZIRCONIUM MINERALS

Zircon ($ZrSiO_4$) is an orthosilicate in which Zr occurs in eightfold coordination with O. Zirconium may also occur as *baddeleyite* (ZrO_2) and some other, rarer oxides (Fitzpatrick and Chittleborough, 2002).

REFERENCES

Adamo, O., P. Barré, C. Vincenza, D. Vincenzo, and B. Velde. 2015. Short term clay mineral release and re-capture of potassium in a *zea mais* field experiment. *Geoderma* 264: 54–60.

Aung, L.L., E. Tertre, and S. Petit. 2015. Effect of the morphology of synthetic kaolinites on their sorption of properties. *Journal of Colloid and Interface Science* 443: 177–186.

Barré, P., B. Velde, C. Fontaine, N. Catel, and L. Abbadie. 2008. Which 2:1 clay minerals are involved in the soil potassium reservoir? Insights from potassium addition and removal experiments on three temperate grassland soil clay assemblages. *Geoderma* 146: 216–223.

Bauer, A., and B. Velde. 2014. *Geochemistry at the Earth's Surface*. Springer, Amsterdam.

Bigham, J.M., R.W. Fitzpatrick, and D.G. Schulze. 2002. Iron oxides, p. 323–366. In: J.B. Dixon and D.G. Schulze (eds.), *Soil Mineralogy with Environmental Applications*. Soil Science Society of America, Madison, Wisconsin.

Bryant, J.P., and J.B. Dixon. 1963. Clay mineralogy and weathering of red-yellow podzolic soil from quartz mica schist in the Alabama Piedmont. *Clays and Clay Minerals* 12: 509–521.

Churchman, G.J., and D.J. Lowe. 2012. Alteration formation and occurrence of minerals in soils, p. 20.1–20.72. In: P.M. Huang, Y. Li, and M.E. Sumner (eds.), *Handbook of Soil Sciences: Properties and Processes*, 2nd edn. CRC Press/Taylor & Francis Group, Boca Raton, Florida.

Dixon, J.B., and G.N. White. 2002. Manganese oxides, p. 367–388. In: J.B. Dixon and D.G. Schulze (eds.), *Soil Mineralogy with Environmental Applications*. Soil Science Society of America, Madison, Wisconsin.

Drits, V.A., B.A. Sakharov, A.L. Salyn, and A. Manceau. 1993. Structural model for ferrihydrite. *Clay Minerals* 28: 185–207.

Eggleton, R.A., and R.W. Fitzpatrick. 1988. New data and a revised structural model for ferrihydrite. *Clays and Clay Minerals* 36: 111–124.

Fitzpatrick, R.W., and D.J. Chittleborough. 2002. Titanium and zirconium minerals, p. 667–690. In: J.B. Dixon and D.G. Schulze (eds.), *Soil Mineralogy with Environmental Applications*. Soil Science Society of America, Madison, Wisconsin.

Fitzpatrick, R.W., L.M. Mosley, M.D. Raven, and P. Shand. 2017. Schwertmannite formation and properties in acidic drain environments following exposure and oxidation of acid sulfate soils in irrigation areas during extreme drought. *Geoderma* 308: 235–251.

Grim, R.E. 1953. *Clay Mineralogy*, 1st edn. McGraw-Hill, New York.

Harsh, J. 2012. Poorly crystalline aluminosilicate clay minerals, p. 23-1–23-13. In: P.M. Huang, Y. Li, and M.E. Sumner (eds.), *Handbook of Soil Sciences: Properties and Processes*, 2nd edn. CRC Press/Taylor & Francis Group, Boca Raton, Florida.

Hsu, P.H. 1989. Aluminium hydroxides and oxyhydroxides, p. 331–378. In: J.B. Dixon and S.B. Weed (eds.), *Minerals in Soil Environments*, 2nd edn. Soil Science Society of America, Madison, Wisconsin.

Huang, P.M., M.K. Wang, N. Kämpf, and D.G. Schulze. 2002. Aluminum hydroxides, p. 261–290. In: J.B. Dixon and D.G. Schulze (eds.), *Soil Mineralogy with Environmental Applications*. Soil Science Society of America, Madison, Wisconsin.

Hubert, F., L. Caner, A. Meunier, and E. Ferrage. 2012. Unravelling complex <2 um clay mineralogy from soils using X-ray diffraction profile modelling on particle size sub-fractions: Implications for soil pedogenesis and reactivity. *American Mineralogist* 97: 384–398.

Jenny, H. 1994. *Factors of Soil Formation: A System of Quantitative Pedology*. Courier Corporation, North Chelmsford, Massachusetts.

Kämpf, N., A.C. Scheinost, and D.G. Schulze. 2012. Oxide minerals in soils, p. 22-1–22-34. In: P.M. Huang, Y. Li, and M.E. Sumner (eds.), *Handbook of Soil Sciences: Properties and Processes*, 2nd edn. CRC Press/Taylor & Francis Group, Boca Raton, Florida.

Kodama, H. 2012. Phyllosilicates, p. 21-1–21-49. In: P.M. Huang, Y. Li, and M.E. Sumner (eds.), *Handbook of Soil Sciences: Properties and Processes*, 2nd edn. CRC Press/Taylor & Francis Group, Boca Raton, Florida.

Lanson, B. 1997. Decomposition of experimental X-ray diffraction patterns (profile fitting): A convenient way to study clays. *Clays and Clay Minerals* 45: 132–146.

Lowe, D.J. 1995. Teaching clays: From ashes to allophane, p. 19–23. In: G.J. Churchman, R.W. Fitzpatrick, and R.A. Eggleton (eds.), *Clays: Controlling the Environment*. Proceedings of the 10th International Clay Conference, Adelaide, Australia (1993). CSIRO Publishing, Melbourne.

Manceau, A., J.M. Combes, and G. Calas. 1990. New data and a revised structural model for ferrihydrite: Comment. *Clays and Clay Minerals* 38: 331–334.

Manceau, A., and W.P. Gates. 1997. Surface structural model for ferrihydrite. *Clays and Clay Minerals* 45: 448–460.

McBride, M.B. 1979. An interpretation of cation selectivity variations in M+-M+ exchange on clays. *Clays and Clay Minerals* 27: 417–422.

McBride, M.B. 1980. Interpretation of the variability of selectivity coefficients for exchange ions between ions of unequal charge on smectites. *Clays and Clay Minerals* 28: 255–261.

McKenzie, R.M. 1989. Manganese oxides and hydroxides, p. 439–465. In: J.B. Dixon and S.B. Weed (eds.), *Minerals in Soil Environments*, 2nd edn. Soil Science Society of America, Madison, Wisconsin.

Peak, D., and T.Z. Regier. 2012. Direct observation of tetrahedrally coordinated Fe(III) in ferrihydrite. *Environmental Science and Technology* 46: 3163–3168.

Pokrovski, G.S., J. Schott, F. Farges, and J.-L. Hazemann. 2003. Iron (III)-silica interaction in aqueous solution: Insights from X-ray absorption fine structure spectroscopy. *Geochimica et Cosmochimica. Acta* 67: 3559–3573.

Reynolds, R.C. Jr., and R.C. Reynolds III. 1996. *NEWMOD II, A Computer Program for the Calculation of the Basal Diffraction Intensities of Mixed-Layer Clay Minerals*. R.C. Reynolds, Hanover, New Hampshire.

Robin, V., E. Tertre, D. Beaufort, O. Regnault, P. Sardini, and M. Descostes. 2015. Ion exchange reactions of major cations on beidellite: Experimental results and new thermodynamic database. Toward a better prediction of contaminant mobility in natural environments. *Applied Geochemistry* 59: 74–84.

Ruhe, R.V. 1984. Soil-climate system across the prairies in Midwestern USA. *Geoderma* 34: 201–219.

Schwertmann, U., and R.M. Taylor. 1989. Iron oxides, p. 379–438. In: J.B. Dixon and S.B. Weed (eds.), *Minerals in Soil Environments*, 2nd edn. Soil Science Society of America, Madison, Wisconsin.

Sposito, G. 1989. *The Chemistry of Soils*. Oxford University Press, New York.

Teppen, B.J., and D.M. Miller. 2006. Hydration energy determines isovalent cation exchange selectivity by clay minerals. *Soil Science Society of America Journal* 70: 31–40.

Tertre, E., A. Delville, D. Prêt, F. Hubert, and E. Ferrage. 2015. Cation diffusion in the interlayer space of swelling clay minerals – A combined macroscopic and microscopic study. *Geochimica et Cosmochimica Acta* 149: 251–267.

Theng, B.K.G., G.J. Churchman, and R.H. Newman. 1986. The occurrence of interlayer clay-organic complexes in two New Zealand soils. *Soil Science* 142: 262–266.

Theng, B.K.G., K.R. Tate, and P. Becker-Heidmann. 1992. Towards establishing the age, location, and identity of the inert soil organic matter of a spodosol. *Zeitschrift für Pflanzenernährung und Bodenkunde* 155: 181–184.

Viennet, J.C., F. Hubert, E. Ferrage, E. Tertre, A. Legout, and M.-P. Turpault. 2014. Investigation of clay mineralogy in temperate acidic soils of a forest using X-ray diffraction profile modeling: Beyond the HIS and HIV description. *Geoderma* 241–242: 75–86.

Vingiani, S., O. Righi, S. Petit, and F. Terribile. 2004. Mixed-layer kaolinite-smectite minerals in red-black soil sequence from basalt in Sardinia (Italy). *Clays and Clay Minerals* 52: 473–483.

Wada, K. 1989. Allophane and imogolite, p. 1051–1087. In: J.B. Dixon and S.B. Weed (eds.), *Minerals in Soil Environments*, 2nd edn. Soil Science Society of America, Madison, Wisconsin.

Xu, S., and J. Harsh. 1992. Alkali cation selectivity and surface charge of 2:1 clay minerals. *Clays and Clay Minerals* 40: 567–574.

3 Geology
Defining the Starting Point for Soil and Clay Formation

Never take rocks for granite.

Anonymous

3.1 THE GEOLOGICAL CYCLE

Basically, there are two origins for rocks at the surface of the earth: magmatic action by fusion of rock material, and sedimentation and transformation of sedimentary materials by changes in pressure and temperature (metamorphism). To a large extent, magmatic materials have been formed and chemically differentiated at high temperatures with minerals forming at later stages. This gives granites, basalts and other compositions of magmatic rocks. The minerals have been formed at high temperatures and are for the most part not stable at earth surface conditions in the presence of water. Initially, continents were composed of these rocks of magmatic origin and altered by water–rock interaction as described in Chapter 4. Here, loss of silica is the most important action followed by that of alkalis and alkaline earths such as Ca. Oxidation is important also, forming insoluble oxide phases. With the introduction of plants on continents, rocks altered in a different manner, forming more silica- and alkali-rich clay minerals (phyllosilicates) especially of the 2:1 structure type in the soil zone. The erosion and transport of these minerals to form sediments created materials of a different chemical and mineral nature, being richer in silica and potassium and calcium plus some magnesium. Sodium was essentially lost to the sea, and hence sodium chloride became concentrated to form sea salt. These materials, when subjected to the geological cycles of burial and tectonic movement, changed in mineralogy but maintained their mineral type – phyllosilicates – to a large extent. Thus, a predominance of muscovite, chlorite and biotite in metamorphic rocks and in clay minerals in sedimentary rocks is quite apparent.

If we consider the physico-chemical characteristics of magmatic and metamorphic–sedimentary rocks, the magmatic types will show high chemical instability at the surface, forming entirely new phases, while metamorphic and sedimentary rocks will form new phases but will be dominated by minerals from the transformation of the initial phyllosilicate mineralogy.

3.2 GEOLOGY OF THE CONTINENTAL SURFACES

Most of the outcropping, and hence soil-forming, rocks on continents are of sedimentary origin and have retained much of the mineralogical characteristics of the materials that formed them. Three general categories can be defined: carbonate-dominated rocks, sandstones and shales, or soil clay-based sediments. Of course, mixtures of these components can be found in sedimentary rocks in layers or in mixed proportions in a given layer. The more flat-lying rocks are more propitious for agriculture and the installation of human activity. Such rocks tend to be made of sediments that have not been subjected to major tectonic events and hence where the original mineralogy has not been changed to a great extent. The range of relatively low temperatures experienced during burial, usually below 200°C, transforms the sedimentary minerals to a certain extent but leaves them essentially in the same mineral structural categories, i.e. 2:1, 1:1 and carbonate types.

A second category of rock found on continents is of higher temperature origin, magmatic or metamorphic, where the mineralogy is strongly out of equilibrium with surface continental conditions, i.e. slightly acidic rainwater and oxidising air. This material is typically found in mountain chains but not always.

A third category of material found on continental surfaces is sedimentary and unconsolidated, i.e. inherited. Riverbeds, internal freshwater lake deposits and continental edge salt marsh materials are largely composed of transported soil materials. Aerial deposits of fine-grained minerals called loess, composed of materials from arid zones are frequent, especially in Northern Hemisphere areas. They also occur throughout New Zealand, mainly as a result of materials from riverbeds that were windblown during glacial and periglacial periods (several papers in Eden and Furkert, 1988). They can be very important components of soil zones.

Thus, the materials susceptible to form a soil can be of varied origin, from high-temperature minerals to those formed at the surface and either slightly transformed or not transformed at all. Contact with plants stabilises the fine-grained material, produces new minerals and promotes the longer residence time of rainwater which interacts with the high-temperature minerals in the rocks.

3.3 PRIMARY MINERALS IN ROCKS: RAW MATERIAL FOR ALTERATION

3.3.1 Silicates

The most common broad types of silicates in rocks are given in Table 3.1. These are the raw materials for the most abundant minerals in most soils, whether through their alteration or as residual fractions of the primary minerals themselves.

Other silicates may occur in soil-forming rocks either as major components but only occasionally. They may also be widespread but in minor amounts, or else they may occur only rarely. Chlorites, which are phyllosilicates that contain Mg, Fe and/or Al in their interlayers, can be major components of low-grade schists and can also occur as early alteration products of other primary minerals, such as pyroxenes, amphiboles and biotite. They are easily altered by weathering. Some

TABLE 3.1
Nature and Relative Strengths of Bonding in Different Types of Silicates Common in Soil-Forming Rocks (in Order of Increasing Stability)

			Silicate Unit		
Structural Type	**Group Name**	**Formula, Structure**	**Shared O per Si**	**Weakest Bonds**	**Examples**
Isolated	Nesosilicates	SiO_4^{2-}	0	With divalent cations	Olivines
Double chains	Inosilicates	$Si_4O_{11}^{6-}$, with Al substitution	2	With divalent, and other cations	Amphiboles
Single chains	Inosilicates	SiO_3^{2-}, with Al substitution	2.5	With divalent, and other cations	Pyroxenes
Layer	Phyllosilicates	$Si_2O_5^{2-}$, with Al substitution, as a sheet joined to Al–, Mg–, Fe–OH sheet	3	With interlayer cations, usually K^+	Micas
Framework	Tectosilicates	SiO_2, with Al substitution	4	With cations (K^+, Na^+, Ca^{2+})	Feldspars
	Tectosilicates	SiO_2	4	Si–O	Quartz

rocks also contain zeolites, while others harbour a variety of magnesium-rich serpentines, among them chrysotile, antigorite, lizardite, amesite and berthierine. Zircon $ZrSiO_4$ is widespread in rocks but mostly in low abundance.

Silicate minerals (those containing Si and other elements) are susceptible to chemical destabilisation. This instability is roughly inversely proportional to their temperature of formation. High-temperature minerals tend to have less Si–O–Al bonding, which is highly covalent and more stable under surface conditions. The clay minerals are of different types depending upon the proportions of Si and Al present. In soil systems when rainfall is moderate Si-Al minerals are present with Si dominant over Al. These are the 2:1 minerals. As rainfall and chemical undersaturation of aqueous solutions are greater, the minerals are less siliceous, i.e. 1:1 minerals of the kaolinite type where Al = Si in content. As rainfall and solution undersaturation increases, silica is lost and Al remains as a hydroxide. Eventually, in extreme cases of intense rainfall, even Al is lost and oxides of iron remain.

Thus, the relative undersaturation of altering aqueous solutions determines the relative Si content of the minerals present in alteration sequences. Minerals of low silica content can be found in the early stages of rock alteration (see Velde and Meunier, 2008, for details) in the presence of minerals of high silica content and high-temperature origin. Mineralogy in the early stages of rock alteration can be highly varied due to the differences in the flow rate of altering solutions on a micro to macro scale. As alteration of minerals becomes more prevalent, in soils the homogenisation of the structure of the altered materials tends to make water and silicate-oxide systems more homogeneous, and the mineralogy becomes more consistent and follows the relative rainfall intensity of the alteration system.

Overall, the minerals of highest temperature origin (volcanic and igneous rocks) have the least Si–Al content and are most easily transformed by surface solutions, while silicates such as feldspars with more Al present are more stable and resistant to weathering (aqueous solution interaction) processes. One exception to this sequence is that of quartz, which is formed at high temperatures but is very little affected by surface aqueous fluid interactions, which gives the high silica content beach sands under highly varying climatic conditions. Quartz is considered to be metastable under surface conditions.

The change in silica content of soil clays is a very significant aspect of alteration as a function of climate. Plants tend to counteract this trend as they can, by moving silica from depth to the surface by root action as phytolitic deposits (solid amorphous silica), which encourages the presence of 2:1 minerals in soils.

When the soil-forming rocks are sedimentary, many of the raw materials for alteration may already be secondary minerals, having been formed from primary minerals by alteration in one or more previous weathering cycles.

The change in charge of cationic ions in the phyllosilicate mineral structures is extremely important in the formation of clay minerals in soils. Most schematic descriptions of soil clay mineral formation begin with the interaction of oxygen saturated rainwater and silicate minerals in rocks of high-temperature origin, either igneous or metamorphic. In these rocks, the minerals are far out of chemical equilibrium with surface conditions and they most often do not contain many phyllosilicate minerals. They tend to be dissolved and recrystallised. In these cases, there is a great difference in thermodynamic stability between the low-temperature aqueous environment and the silicate minerals in the rocks. However, this is far from the only origin of soil clay phyllosilicate minerals.

At the continental surfaces, one finds large surface areas of rocks that are the result of low temperature transformations of soil clay minerals themselves: sedimentary rocks. On many continents, these rocks formed as sedimentary basins and buried to depths of only several kilometres in moderate temperatures of a hundred degrees or so, are flat-lying, and contain minerals that are transformed from sedimentary clays but which still maintain the basic phyllosilicate structures (2:1 and 1:1 types) found to be stable in soil environments, especially those in temperate climate situations. North America and Europe are examples of this situation. The areas of higher relief are usually due to the upward movement of rocks that have experienced higher temperatures and greater mineral transformations. The flat-lying sedimentary rocks are well suited to be transformed by the chemical and physical forces or the surface into mineral zones that are of the soil clay mineral type without significant erosion and loss of material. The relatively flat sedimentary basin rock structures form soils that are conducive to the formation of prairie biospheres under temperate and subdesertic climates. In sedimentary rock materials, the main mineralogy is illite-chlorite and illite-smectite mixed-layered minerals. The expandable clays (mixed-layered minerals) are transformed to illite and chlorite with increasing diagenetic conditions (time and temperature). These minerals, when brought to the soil zones, tend to be slightly oxidised and form more expandable minerals, especially from the chlorite fractions. The basic 2:1 mineral structures are conserved and hence the minerals can be called products of transformation forces. Since these materials are very often found in flat-lying rock landscapes, they are the basis of agricultural resources. Hence much of farmed land is formed from previous soil minerals and is immediately useful for agriculture.

Returning to Table 3.1, the alteration of a particular mineral begins with the disruption of its weakest bond. Bonds in all silicates are based on silica tetrahedra. Their strength depends on (1) the nature of the links between tetrahedra; (2) the extent of substitution of four-valency Si within tetrahedra by three-valency Al, and hence gain of negative charge (isomorphous substitution); and (3) the extent of incorporation of charge-balancing cations, and their location, in the structure. While Table 3.1 generally holds for major silicate phases of interest in soil-forming rocks, the devil can sometimes lie in the detail. Following Loughnan (1969), for example, zircon $ZrSiO_4$ and andalusite Al_2SiO_5 are both neosilicates like olivine and have $(Si{:}O)_{molar}$ ratios of 0.25 and 0.2, respectively, that are equal or less than those of olivine and yet are both very stable (i.e. resistant) to breakdown by weathering, in contrast to olivine, which is notably unstable according to Goldich's (1938) series (see Chapter 1, Figure 1.1).

The generalisations about relative rates of breakdown of primary phases that are shown in both Table 3.1 and Figure 1.1 may provide useful guidelines. Even so, they are subject to specific influences of such environmental factors as the particular composition of the solutions bathing the minerals (often these are acidic), the dynamics of redox conditions and the rate of throughflow of water past the minerals. These can each play a decisive role in the kinetics, course and ultimate products of their alteration.

3.3.2 Non-Silicates

For the most part, non-silicates are resistant to change at the surface of the earth. They tend to make up a portion of sand in streams and rivers and on lake or ocean shores. These minerals are often mined for their rare metal or element (e.g. rare earths, zirconium and so forth) content, which is concentrated through the action of water transport of these insoluble minerals. In general, these minerals are of low abundance in sediments as they were in their host rocks. However, they are significantly concentrated by the fact of their insolubility and high density compared to clay minerals.

Carbonates are, for the most part, formed through the interaction of atmospheric CO_2 with aqueous solutions. One concentration mechanism is by organic precipitation as shell material for aqueous species. However, some carbonates are formed when isolated units of seawater are evaporated, and the concentrate forms a carbonate matrix. Under dry climates, capillary water movement to the surface can bring carbonate-rich solutions to the surface to form precipitates in the alteration sequence upon evaporation of the aqueous solution. These carbonates do not form at the surface but several tens or more centimetres in depth. They are not in the plant zone of soil formation, which is too acidic for their chemical stability.

Phosphates, particularly apatite $Ca_5(PO_4)_3(OH, F, Cl)$ but sometimes also fluorapatite, hydroxyapatite, variscite, wavellite and monazite, occur as accessory minerals in some igneous and metamorphic rocks. As magnetite Fe_3O_4 and also the related titanomagnetite, iron also occurs as an oxide in some rocks, particularly those with a volcanic origin. These minerals are resistant to chemical action, occur at the surface of the earth and persist in sediments through sedimentary transport.

Non-silicates may also occur in soil-forming rocks either as major components but only occasionally, as minor components that are widespread, or also as rare components. Carbonates, whether calcite $CaCO_3$, dolomite $CaMg(CO_3)_2$ or a combination of these, are often dominant in sedimentary limestones. Other carbonates, including aragonite, siderite $FeCO_3$ and ankerite $Ca(Mg, Fe, Mn)(CO_3)_2$, could also occur in limestones, usually in lesser amounts. As a rutile, titanium oxide TiO_2 is widespread in minor amounts in soil-forming rocks. It may also occur as brookite or the related mineral sphene $CaTiSiO_5$.

3.3.3 The Initial Production of Clays in Weathering

The fundamental principle proposed by Dokuchaev in 1883 of soil being the product of a number of different environmental factors (Krupenikov, 1993) was formulated into an equation-like relationship by Jenny (1941). This relates soil and its defining properties (*S*) to the so-called soil-forming factors of time (*t*), climate (*cl*), organisms (*o*), relief (*r*) and parent material (*p*):

$$S = f(t, cl, o, r, p)$$

If all of these factors, except time, were constant across a series of soils, formed on the same parent materials, in the same climate, under the same plants and other organisms and under similar conditions of relief such as aspect and slope, then these soils would constitute a chronosequence, and many such sequences have been studied (e.g. Bockheim, 1980, 1990). Even so, chronosequences, as with other sequences in which one soil-forming factor varies while the others remain invariant (including climosequences, vegesequences, geosequences; see Churchman, 1980, 2000, for instances), are generally rank approximations in the multifaceted real world (see e.g. Crocker, 1952). This means that any attempt to trace the effect of time on the development of soils or, in our sphere of interest, clays within soils is most likely complicated by accompanying changes in other soil-forming factors over time which confound the effect of time alone.

Some indications of the early course of clay production come with caution from chronosequences for these reasons, but some come simply from studies carried out when rocks or, more strictly, parent materials of soils, first became exposed to the forces of weathering. This has occurred with the retreat of glaciers and in the aftermath of volcanic eruptions. Examples where glacier retreat under current global warming has helped understand early soil development come from the European Alps (Egli et al., 2001; Mavris et al., 2011; Dümig et al., 2012). The 1980 eruption of Mount Saint Helens in the Western United States provided fresh materials for early weathering studies by Dahlgren et al. (1997), among others. Early soil development has also occurred under different conditions for the other soil-forming factors, most notably climate and parent materials, but also vegetation and relief, each incidentally not always independent of climate, parent materials or, indeed, time. Early stages of weathering may persist for considerable periods of time when the driving forces for weathering are weak, especially in cold and/or dry climates. The dry valleys of Antarctica have enabled studies of changes under both cold and dry conditions (Bockheim, 1990; Salvatore et al., 2013), while the effects of higher rainfall under cold conditions have also been studied in maritime Antarctica (Simas et al., 2006); in the cold Northern climes of Spitsbergen, Svalbard archipelago at 77° (Kabala and Zapart, 2012); and in Alaska (Burt and Alexander, 1996). They also show when weathering has occurred in climates that are merely cool, especially in mountainous areas (e.g. Mokma et al., 1973, and Churchman, 1980, both in New Zealand and Bain, 1977, in Scotland). Early stages of weathering can also be shown in weathering rinds on exposed rocks even in tropical climates like that of Costa Rica (Navarre-Sitchler et al., 2013) and in soils that are newly initiated as a result of the erosive removal of the previous soil at that site (Churchman, 1980, 2000, and Tonkin and Basher, 2001, all studying soils in New Zealand). Overall, the various rocks studied in these situations have encompassed igneous, sedimentary and metamorphic types and their physical states prior to weathering have ranged from solid, through comminuted rocks to newly erupted volcanic ejecta.

At the earliest stages of weathering, soil formation is a largely inorganic process, albeit that biological influences often take very little time to exert themselves. Virtually regardless of temperature, and also of type of parent rock, the initial products of weathering of igneous, metamorphic and sedimentary rocks are quite similar. The rates of breakdown of primary minerals and of production of secondary products are affected by temperature, as expected from the well-known accelerating effect of temperature on chemical changes generally. It is probably only in the very early stages of weathering that other environmental factors, besides time but including climate (both temperature and rainfall), geology, vegetation, other organisms and relief, can easily be ignored in comparing between sites in vastly different environments worldwide where soils are under development.

Initial changes on weathering tend to include:

1. Loss of the alkali and alkaline earth cations, Na^+, K^+, Ca^{++} and Mg^{++}, and also Si (as silicic acid, H_4SiO_4 or $Si(OH)_4$).
2. Oxidation of iron from Fe^{2+} to Fe^{3+}. The products of oxidation are oxides, hydroxides and oxyhydroxides. These are characterised (and quantified) together as the fraction of the soil extractable by dithionite, although aluminium oxides, oxyhydroxides and hydroxides may also be removed by this treatment (see Chapter 12, Section 12.2 herein). After only a little time for soil development, there is usually an increase in the proportion of total Fe that is extractable by oxalate in relation to other forms of Fe (see also Section 12.2). Hence nanocrystalline forms of Fe, principally ferrihydrite, constitute common early products of weathering in a wide range of soils. The dependence of the effects of oxidation on rock type is illustrated in Figure 3.1.
3. Al is often also released as nanocrystalline forms that are soluble in oxalate. Principal among these is likely to be allophane.
4. Build-up of organic matter can occur as soon as vegetation and other soil organisms become established, and this plays a role in enhancing the breakdown of primary minerals

FIGURE 3.1 **(See colour insert.)** Alteration of rocks in a wall of the Abbaye de Beauport, Brittany, France, showing differences in reaction rates between granite blocks and meta-basaltic rocks in a church wall 400 years old. Dissolution of the iron-rich basaltic minerals in dark-coloured rocks is far greater than that of granites (grey blocks to the left). Basalt mineral alteration is largely due to the effects of oxidation of iron in the basalt minerals. Reddish meta-sandstone (centre, right) is resistant to alteration.

and solubilisation of metals that are released as a result. It is evidenced by a rise in forms of Fe and Al that are extractable by pyrophosphate and the effect is confirmed when organisms are present concurrent with, or prior to, rock weathering, as occurs in bird colonies.

5. Oxidation of manganese and the formation of oxyhydroxides of Mn may also occur.
6. It appears that the transformation of chlorite and/or mica, especially their trioctahedral forms, almost always constitutes the first process of alteration of primary minerals that results in crystalline secondary products. The transformation process is considered to occur largely intracrystalline in the solid state, albeit in contact with solution.
7. Chlorites can be transformed into smectites. Micas are commonly transformed to interstratified product phases. These include both randomly and regularly interstratified mica (or illite) with either vermiculite or smectite. The crystalline products of these processes may be partially formed, so that, for instance, a swelling chlorite results rather than a smectite per se, at least in the initial stages of transformation of chlorite. Interlayering of the products of transformation of chlorites and/or micas, whether smectites or interstratified mica-vermiculites (or illite-vermiculites) or mica-smectites (or illite-smectites) with hydroxyl forms of metals, often Al, commonly occurs.
8. The extent of alteration observable or the extent which has occurred in a given time likely depends on the availability of surfaces exposed to solutions and to the atmosphere, and hence on physical factors such as the particle size and the degree of fracturing of parent rock material, as well as rock type.

3.3.4 Geological Deposits, Rock Types and Clay Minerals

3.3.4.1 Rocks

On the continents as we see them today, the largest part of exposed surfaces is composed of sedimentary rock material. Areas of eruptive or high-grade metamorphic materials are relatively rare. Metamorphic materials and intrusive volcanic materials tend to form topographic high relief, while sedimentary rocks form the flatter areas on many continents. The sedimentary rocks and low-grade metamorphic rocks have a dominant silicate mineralogy of potassic mica (various types of illite and muscovite) and chlorite. Most of the iron content of the rocks is in the chlorite minerals. The chlorite is susceptible to chemical alteration (oxidation of the iron ions), which leads to the formation of expanding smectite-like minerals of various types and interlayer charge. The micas or illite in the rocks are little affected by surface alteration, at least in the early stages of interaction with water at the surface. Hence soils and alterites of these materials will tend to have smectites derived from chlorite (trioctahedral in nature) and mica-type clays. These minerals are of the transformation type.

The higher-grade metamorphic materials can contain aluminous micas, but the chlorite tends to be destabilised in favour of other non-phyllosilicate minerals such as garnet. And a portion of the muscovite changes into feldspars. The non-phyllosilicate minerals formed by metamorphism are eventually transformed into different types of smectite minerals or kaolinite under weathering conditions. Acidic volcanic or intrusive rocks follow similar alteration patterns. Basaltic extrusive materials eventually form smectites and kaolinites under conditions of surface alteration. These minerals are formed largely by crystallisation processes and the smectites are dominantly dioctahedral with either Al^{3+} or Fe^{3+} in the octahedral layer sites.

The transformation of different rocks into alterites and eventually the soil zone is a function of the lithology of the rock as illustrated by Oh and Richter (1995) in altered rocks from the Southeastern United States (Figure 3.2).

3.3.4.2 Sediments

A significant portion of soils utilised by man is composed of recently deposited altered material: river sediments and coastal deposits as salt marshes and others. These materials form soils and can be affected by plant action at the surface just as those formed from rocks. Interactions of organics in

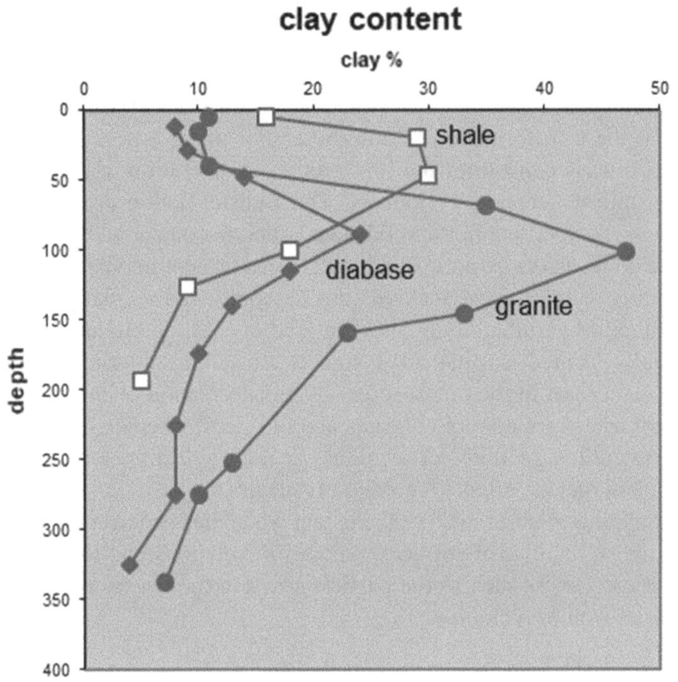

FIGURE 3.2 Clay content for alterites on different rock types in the Eastern United States. Depth in centimetres. (Data from Oh, N.-H., and D.D. Richter, 2005, *Geoderma* 126: 5–25.)

the soils and the clays can be very important concerning their aggregation properties for example. Also, it is frequent that the new plant interactions modify the clays in the sediments. This is seen in Delaware Bay salt marsh sediments after several tens of years of plant growth (Figure 3.3).

Similar changes can occur in river sediments in and along valleys as shown in Bain et al. (1993). These areas were the first used by man for agriculture in the Neolithic period of the Northern Hemisphere. Then pericontinental areas of the salt marsh type were colonised in the Southern Hemisphere. The advantage of such deposits is that the proportions and mineral types were quite similar to those in soils where plants grew.

3.3.4.3 Loess and Dust

Loess is well known as a surface deposit of recent origin (Quaternary in age) which is present in some of the most fertile areas of agricultural practice. The major distinction is the slightly reddish colour due to iron oxide formation and a rather homogeneous fine-grained matrix. A very important aspect of loess soils is that they are or were renewed periodically with new, unaltered material from wind transportation. Loess soils are found in most Western Hemisphere temperate climates. They can form very thick layers (several meters) of material. The important factor is that loess materials come from non-soil sources and are thus very reactive to soil (organically controlled) chemistry when plant communities are established.

One of the major sources of loess materials was generated by the transport of little-altered rocks via glacial action and eventually wind transportation of glacial deposits after their deposition at the glacier front. For the most part, initial loess materials are of the chlorite-illite type, with relatively little altered material (i.e. complex 2:1 minerals with varying interlayer combinations). However glacial deposits as glacial drift can sometimes contain altered high-temperature minerals of the soils type (i.e. mixed-layered 2:1 minerals). The transport and accumulation of loessic materials dominated with illite-chlorite continues today. In Central China almost 2 mm of this loess is deposited each year in certain places, which renews the soil zone clays each 100 years.

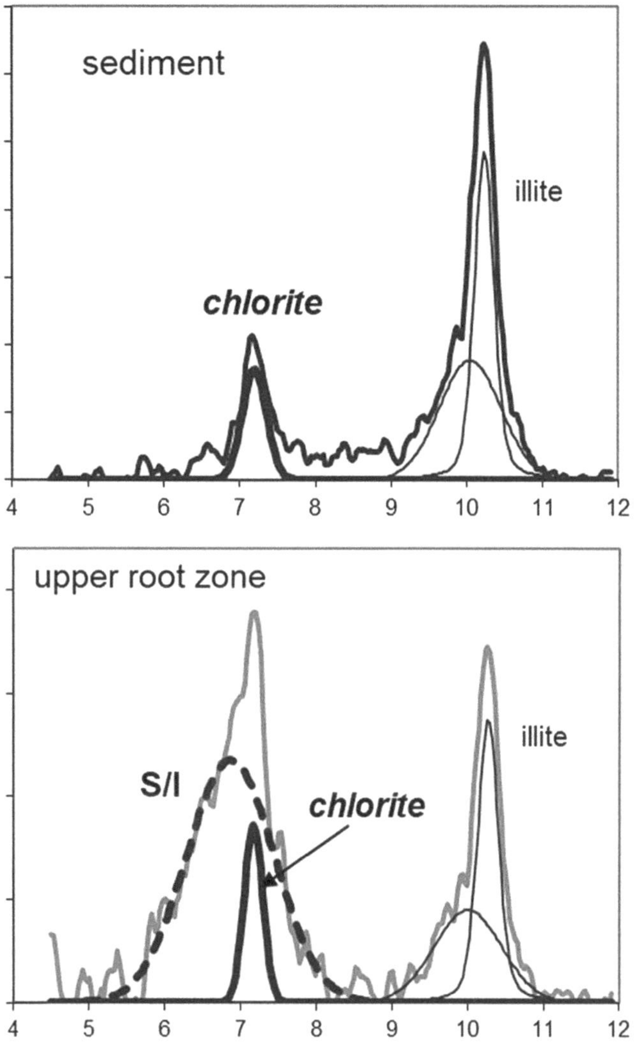

FIGURE 3.3 Decomposed X-ray diffraction pattern of clay fractions of Delaware Bay soils developed on salt marsh sediments. Vertical axis indicates relative intensity of signal and horizontal axis position in degrees two theta copper radiation. S/I indicates the presence of an interstratified smectite-illite mineral.

The movement of clay fraction materials by high-altitude wind currents over great distances is a rather special situation. The clay materials must be fine in size and if possible of sheet structure, which keeps them from drifting out of the atmosphere. In fact, much of the material is formed from isolated grains. This means that there is very little, if any, interparticle junction that is affected by clay organic interaction. Such a situation suggests an origin in a plant-free zone such as a desert or arid climate area. Hence, if there is little plant interaction, we can expect to find unaltered material, such as chlorite-illite material, present. Thus, the initial loessic material is little altered, having been formed largely by mechanical disaggregation of the initial rock materials (glacial grinding). Local transportation from the glacier deposition front by rainwater action forms deposits such as shallow lake bottom materials that have a large surface area that, when dry, are susceptible to displacement by wind action. This type of transportation is still prevalent in China. Old Quaternary deposits can be reactivated and are still active in certain regions (Dodonov and Zhour, 2008).

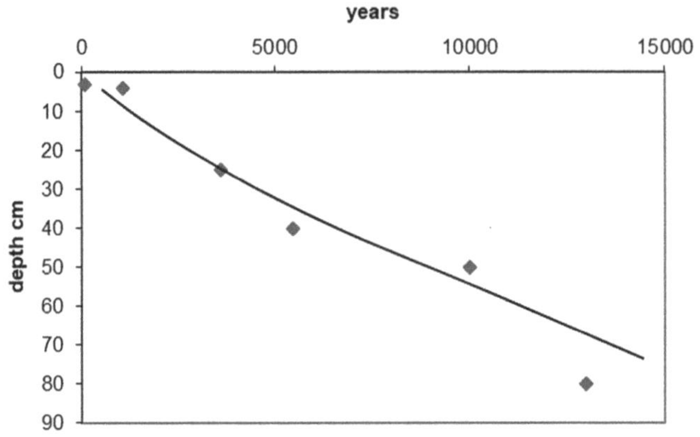

FIGURE 3.4 Depth limit of alteration of chlorite as a function of time in river sediments in Scotland. (From Bain, D.C., et al., 1993, *Geoderma* 57: 275–293.)

The presence of illite and chlorite presents materials which can form 2:1 charge variable expanding materials that can fix and release nutrient elements such as potassium, which is found in the illite fraction. The magnesium of the chlorites is also useful for the chlorophyll of plants. The 2:1 structures fix organics to form aggregates whose microporosity acts as a reservoir of water for the plants during dry seasons. Thus illite-chlorite loessic materials become fertile soils for plant growth and sustenance. The data given by Bain et al. (1993) (Figure 3.4) indicate that the conversion of illite and chlorite into 2:1 expanding mixed-layered clays in the soil zone of river sediments can occur over periods of 4000 years.

Other dry climate soils can furnish dust (clay-sized or fine-silt-sized particles) for wind transportation. Some is moved from continent to continent, and some is deposited closer to its source. All depends on the grain size, wind currents and movement into the stratosphere. Climate and air mass movement determine the presence and transportation of dust materials from soils and exposed sediment surfaces (Rex et al., 1969; Churchman et al., 1976; Collyer et al., 1984). Many desert areas form shallow deposits of clay-sized materials that are moved by wind currents to various distances from their source. In these areas, newly formed clay and other minerals occur due to evaporation of the transporting water. Here silicate minerals that are stable under alkaline conditions (the clay silicates palygorskite or sepiolite for example) are found, but they do not resist the chemistry generated by plant-dominated soils and they disappear from the mineral assemblages (Bigham et al., 1980).

However, dry soils (from the plant interaction zone), especially those worked by present-day farmers, are often appropriate for wind transportation over shorter distances. The finer aggregates or non-aggregated clay materials are moved by strong wind currents and deposited at some distance from their original site of deposition. Here there is the entire assemblage of soil materials in different proportions depending upon wind current intensity and the availability to air mass movements of the material during dry periods. A classic example of this effect is the period of the Dust Bowl in the Central Western United States during the 1930s where the movement of newly denuded soils was so intense as to significantly lower the fertility of the land and force farmers to migrate to other regions. In these areas, there was no plant cover in the fall and winter periods leaving the soils totally denuded and thus susceptible to wind erosion. The fertile soil zone was highly disturbed and plant productivity greatly reduced.

Some areas experience massive movements of soil materials, such as Eastern Australia where large amounts of material are moved by wind and deposited over several-hundred-mile distances. These have been traced to New Zealand, about 4000 km away (Glasby, 1971; Collyer et al., 1984). In these cases, the variation of climate effects (rainfall and dry periods) determines the possibility of movement of surface materials where no plants are present.

REFERENCES

Bain, D.C. 1977. The weathering of chloritic minerals in some Scottish soils. *Journal of Soil Science* 28: 144–164.

Bain, D.C., A. Mellor, M.S.E. Robertson-Rintoul, and S.T. Buckland. 1993. Variations in weathering processes and rates with time in a chronosequence of soils from Glen Feshie, Scotland. *Geoderma* 57: 275–293.

Bigham, J.M., W.F. Jaynes, and B.L. Allen. 1980. Pedogenic degradation of sepiolite and palygorskite on the Texas High Plains. *Soil Science Society of America Journal* 44: 159–167.

Bockheim, J.G. 1980. Properties and classification of some desert soils in coarse-textured glacial drift in the Arctic and Antarctic. *Geoderma* 24: 45–69.

Bockheim, J.G. 1990. Soil development rates in the Transantarctic Mountains. *Geoderma* 47: 59–77.

Burt, R., and E.B. Alexander. 1996. Soil development on moraines of Mendenhall Glacier, southeast Alaska. 2. Chemical transformations and soil micromorphology. *Geoderma* 72: 19–36.

Churchman, G.J. 1980. Clay minerals formed from micas and chlorites in some New Zealand soils. *Clay Minerals* 15: 59–76.

Churchman, G.J. 2000. The alteration and formation of soil minerals by weathering, p. F3–F76. In: M.E. Sumner (ed.), *Handbook of Soil Science*. CRC Press, Boca Raton, Florida.

Churchman, G.J., R.N. Clayton, K. Sridhar, and M.L. Jackson. 1976. Oxygen isotope composition of aerosol size quartz in shales. *Journal of Geophysical Research* 81: 381–386.

Collyer, F.X., B.G. Barnes, G.J. Churchman, T.S. Clarkson, and J.T. Steiner. 1984. A trans-Tasman dust transport event. *Weather and Climate* 4: 42–46.

Crocker, R.L. 1952. Soil genesis and the pedogenic factors. *The Quarterly Review of Biology* 27: 139–168.

Dahlgren, R.A., J.P. Dragoo, and F.C. Ugolini. 1997. Weathering of Mt. St. Helens tephra under a cryic-udic climatic regime. *Soil Science Society of America Journal* 61: 1519–1525.

Dodonov, A., and L. Zhour. 2008. Loess deposition in Asia: Its initiation and development before and during the Quaternary. *Episodes* 13: 222–224.

Dümig, A., W. Häusler, M. Steffens, and I. Kögel-Knabner. 2012. Clay fractions from a soil chronosequence after glacier retreat reveal the initial evolution of organo-mineral associations. *Geochimica et Cosmochimica Acta* 85: 1–18.

Eden, D.N., and R.J. Furkert (eds.). 1988. *Loess: Its distribution, geology and soils: Proceedings of an International Symposium on Loess, New Zealand, 14–21 February 1987.* A.A. Balkema, Rotterdam.

Egli, M., A. Mirabella, and P. Fitze. 2001. Clay mineral formation of two different chronosequences in the Swiss Alps. *Geoderma* 104: 145–175.

Glasby, G.P. 1971. The influence of aeolian transport on marine sedimentation in the South-west Pacific. *Journal of the Royal Society of New Zealand* 1: 285–300.

Goldich, S.S. 1938. A study in rock weathering. *The Journal of Geology* 46: 17–58.

Jenny, H. 1941. *Factors of Soil Formation: A System of Quantitative Pedology*. McGraw-Hill, New York.

Kabala, C., and J. Zapart. 2012. Initial soil development and C accumulation on moraines of the rapidly retreating Werenskiold Glacier, S.W. Spitsbergen, Svalbard archipelago. *Geoderma* 175–176: 9–20.

Krupenikov, I.A. 1993. *History of Soil Science: From Its Inception to the Present.* A.A. Balkema, Rotterdam.

Loughnan, F.C. 1969. *Chemical Weathering of the Silicate Minerals.* Elsevier, New York.

Mavris, C., M. Plötze, A. Mirabella, D. Giaccai, G. Valboa, and M. Egli. 2011. Clay mineral evolution along a soil chronosequence in an Alpine proglacial area. *Geoderma* 165: 106–117.

Mokma, D.L., M.L. Jackson, J.K. Syers, and P.R. Stevens. 1973. Mineralogy of a chronosequence of soils from greywacke and mica-schist alluvium, Westland, New Zealand. *New Zealand and Journal of Science* 16: 769–797.

Navarre-Sitchler, A.K., D.R. Cole, G. Rother, L. Jin, H.L. Buss, and S.L. Brantley. 2013. Porosity and surface area evolution during weathering of two igneous rocks. *Geochimica et Cosmochimica Acta* 109: 400–413.

Oh, N.-H., and D.D. Richter. 1995. Elemental translocation and loss from three highly weathered soil-bed rock profile in South-East United States. *Geoderma* 126: 5–25.

Rex, R.W., J.K. Syers, M.L. Jackson, and R.N. Clayton. 1969. Eolian origin of quartz in soils of Hawaiian Islands and in Pacific pelagic sediments. *Science* 163: 277–279.

Salvatore, M.R., J.F. Mustard, J.W. Head, R.F. Cooper, D.R. Marchant, and M.B. Wyatt. 2013. Development of alteration rinds by oxidative weathering processes in Beacon Valley, Antarctica, and implications for Mars. *Geochimica et Cosmochimica Acta* 115: 137–161.

Simas, F.N.B., C.E.G.R. Schaefer, V.F. Melo, M.B.B. Guerra, M. Saunders, and R.J. Gilkes. 2006. Clay-sized minerals in permafrost-affected soils (Cryosols) from King George Island, Antarctica. *Clays and Clay Minerals* 54: 721–736.

Tonkin, P.J., and L.R. Basher. 2001. Soil chronosequences in subalpine superhumid Cropp Basin, western Southern Alps, New Zealand. *New Zealand Journal of Geology and Geophysics* 44: 37–45.

Velde, B., and A. Meunier 2008. *The Origin of Clay Minerals in Soils and Weathered Rocks*. Springer-Verlag, Berlin.

4 Primary Minerals and Their Alteration by Weathering

In the confrontation between the stream and the rock, the stream always wins. Not through strength but through persistence.

Anonymous

4.1 PRIMARY MINERALS AND THEIR WEATHERING PRODUCTS

4.1.1 Amphiboles, Pyroxenes and Olivines

The ferromagnesian minerals amphiboles, pyroxenes and olivines each tend to give rise to *trioctahedral smectites* at early stages of their alteration (Velde and Meunier, 2008). However, the nature of these, and the subsequent products formed from them, vary between these three easily weatherable types of primary minerals. Microsystems occupying relatively closed pores lead to different types of initial products, even between different types of amphiboles. Thus, *hornblende* and *actinolite*, respectively rich in Ca and Mg, lead in turn to saponite + *trioctahedral vermiculite*, and talc + *hectorite-like smectite* (Proust, 1982, 1985; Velde and Meunier, 2008). Differences between the various types of amphiboles disappear when microsystems become more open, with first a tendency to form *saponite*, then, in fully open systems, *kaolinite*, *gibbsite* and *Fe-oxyhydroxides* result (Velde and Meunier, 2008). Pyroxenes also give rise first to trioctahedral smectites and these are easily replaced by their dioctahedral counterparts, with the type formed depending on availability of Al versus Fe, hence *montmorillonite* versus *Fe-montmorillonite* (Velde and Meunier, 2008). Olivines typically give rise first to mixtures of poorly crystallised minerals that are known as *iddingsite* or *bowlingite*, and are variable in composition. *Fe-Mg smectites* like *nontronite* or *Fe-saponite* result in due course, but themselves are unstable, and *halloysite* and *goethite* have been observed to form in their place. Since olivine contains no Al, the formation of 1:1 halloysite is enabled by the dissolution of other components such as feldspars and volcanic glass (Velde and Meunier, 2008).

4.1.2 Serpentinites

As for the ferromagnesian minerals, amphiboles, pyroxenes and olivines, the alteration of serpentines depends on extent of leaching, hence drainage, as well as to a degree on the composition of the originating serpentine (Velde and Meunier, 2008; Churchman and Lowe, 2012). Where leaching is restricted, as in some microsystems, serpentines tend to weather to *Mg-rich trioctahedral smectites* such as *saponite* and also *talc*, along with *Fe-oxyhydroxides*. In more open systems, the products from serpentine alteration tend to be *Fe-rich dioctahedral smectites* such as *Fe-montmorillonite* or *nontronite*, along with Fe-oxyhydroxides. In general, products of serpentine weathering tend to be heterogeneous (Caillaud et al., 2004; Hseu et al., 2007). They can include *vermiculite*, *chlorite-vermiculite*, and, eventually, *kaolinite* and *quartz*.

4.1.3 Volcanic Glass

Volcanic glass is easily dissolved. It is an amorphous solid comprising cations and occurring within loosely linked SiO_4 tetrahedra (Churchman and Lowe, 2012). Its breakdown occurs through (1) loss of divalent and monovalent cations via their exchange with protons, (2) loss of Al also by exchange with protons, and (3) dissolution of silica. Glasses with the lowest silica contents are most easily dissolved. Hence basaltic glasses are less stable than those from andesites, and those from rhyolites are the most stable. However, all have high surface areas and porosities, so they break down easily.

4.1.4 Feldspars

These can give rise to a wide variety of minerals on weathering, among them *smectites, illites* (the so-called sericites), *gibbsite*, *quartz*, and, most commonly, *kaolinite* and *halloysite.* As tectosilicates, they must first go into solution before recrystallising out as phyllosilicates, oxides or hydroxides. It appears that weathering first produces etch-pits, occurring along cracks and other points of weakness in the mineral structure (Wilson, 2004). Often, an amorphous phase has been identified prior to formation of a crystalline silicate and/or oxides or hydroxides. In the lower part of a weathering profile (the 'alterite'), the aqueous solution surrounding feldspars and other primary minerals may be in equilibrium with primary minerals such as feldspars, which dissolve congruently. By contrast, the soil solution in the upper part of the profile (the soil proper) is dynamic in composition and in concentrations of both inorganic and organic reagents, and the soil product phase tends to be out of equilibrium with its bathing solution (Wilson, 2004). Laboratory studies of the weathering of feldspars have usually identified a leached layer forming on the primary mineral, but this does not seem to form under field conditions (Wilson, 2004; Churchman and Lowe, 2012).

4.1.5 Micas

These are unlike other primary minerals (except chlorites) insofar that they are already (2:1) phyllosilicates and so their transformation – so-called because they involve changes within the structure, actually in their interlayer regions – may occur in order to produce other, clay-size 2:1 phyllosilicates. In particular, most micas contain potassium ions in their interlayer region to balance the charge on their layers. The potassium ions are replaced by other, hydrated cations, often Ca or Mg in the soil environment, to give rise to *vermiculite. Smectite* results from some structural changes in the aluminosilicate layers, most often also oxidation of Fe in the octahedral sheet, and its ejection from the layers (Wilson, 2004). A complete replacement of potassium ions would lead to a vermiculite as an expanded phase, while only partial replacement leads to an *illite.* Replacement of approximately one-half of the K^+ leads to an interstratified *illite-vermiculite.* Together with considerable changes occurring in the layers, most likely due to hydrogen ions displacing structural ions such as those of iron, *illite-smectite* may form, and a discrete smectite phase forms when all potassium ions are replaced and the layer charge is decreased due to these structural changes. Among the micas, biotite is much more prone to break down by transformation than muscovite, which can undergo transformation nonetheless. When alteration of micas proceeds to the interstratified phases, either *illite-vermiculite* (sometimes given as *mica-vermiculite*, or from biotite, *hydrobiotite*) or *illite-smectite* (sometimes *mica-smectite*), the interstratification may be either random or regular. Regular interstratification has been identified as occurring in soils formed in cold climates, often in mountainous zones (Churchman and Lowe, 2012). It appears that displacement of potassium from one interlayer region enhances the strength of binding interaction of K^+ with the adjacent region and therefore further K^+ is displaced only from every other interlayer region (Churchman and Lowe, 2012).

4.1.6 Chlorites

Chlorites are also phyllosilicates and are subject to transformations as in micas, with the interlayer hydroxy species (most often based on Mg, but also on Al and/or Fe) being prone to replacement by hydrated cations as with micas. They tend to be more unstable than muscovites (Churchman, 1980), and alter to a randomly interstatified phase, and to *chlorite-vermiculite.* A semi-regular chlorite–vermiculite has been identified and its regularity attributed to the same alternate interlayer displacement mechanism as for regularly interstratified illite–vermiculite or illite–smectite phases (Wilson, 2004). *Vermiculite* and *smectite* can also be formed. It has also been observed that chlorites occasionally disappear as crystalline phases, giving amorphous and soluble products (Churchman and Lowe, 2012) and also, on occasions, *kaolinite* and *halloysite* (Wilson, 2004).

In general, a feature of the alteration of rock minerals (with the possible exception of volcanic glass) is that it is heterogeneous (Velde and Meunier, 2008). Rarely do particular rock minerals give rise to specific secondary clay minerals as products (Churchman and Lowe, 2012). The chemical composition of solutions resulting from the breakdown of rocks will generally comprise contributions from more than one mineral, whether primary minerals or else secondary minerals resulting from the early stages of breakdown of an assemblage of primary minerals.

4.1.7 Heterogeneity of Products

Heterogeneity is often also shown in the composition of minerals within a soil profile. Soil profiles comprise a lower portion and an upper portion. The lower portion, or the C horizon, often overlies the saprolite or saprock, or disintegrated rock, and comprises some secondary minerals that resemble the secondary minerals there but mostly derive from the breakdown of their primary minerals. By contrast, the upper portion, and especially the uppermost A horizon, is home for plants and their associated species, including bacteria and fungi, and the secondary minerals found there may reflect the influence of the plants and their activities, as we have seen for transformation products of micas.

As well as their effect on mineralogy resulting from the cycling of nutrients, such as potassium, silica, and nitrogen as ammonium ions, plants influence A horizon mineralogy because they generally require an aerated zone for their growth and survival, so minerals there become oxidised. This promotes the de-stabilisation of minerals that carry iron in its divalent state. Trioctahedral minerals are often unstable in upper horizons of soils as a result.

4.2 MECHANISMS OF ALTERATION OF PRIMARY MINERALS

4.2.1 Oxidation

Among the most weatherable primary minerals in soil-forming rocks, several contain significant concentrations of structural iron. These include biotite $K(MgFe^{II})_3AlSi_3O_{10}(OH,F)_2$; amphiboles, such as hornblende $(Ca, Na)_{2.3}(Mg, Fe^{II})_4(Al, Fe^{III})(Si, Al)_8O_{22}(OH, F)_2$; pyroxenes, including augite $(Ca, Mg, Al, Fe)_2(Si, Al)_2O_6$; and chlorite $(Fe^{II}, Mg)_{10}Al_2(Si, Al)_8O_{20}(OH,F)_{16}$. Fe generally occurs in its reduced form Fe(II) in primary minerals and is easily converted to its oxidised form Fe(III) simply through the drying of soils. It has been claimed (Millot, 1970) that the most important weathering reactions involve iron. The generally less common manganese (Mn) likewise usually occurs as Mn(II) in primary minerals. It is also easily converted to its oxidised form Mn(IV). These changes in valency result in a charge imbalance in the appropriate minerals, leading to a loss of other ions from their structures, which can destabilise the minerals, enhancing their further breakdown by hydrolysis, assisted by the hydrogen ions produced during oxidation.

The oxidised forms of Fe and Mn occur as oxides and oxyhydroxides (in the case of Fe), and those of iron, in particular, are almost ubiquitous in soils (Churchman, 2010; Churchman and Lowe, 2012). Mineralogical compositions have been summarised by Churchman (2010) for soils with particular dominant secondary aluminosilicates (comprising kaolinite, halloysite, illite, vermiculite, smectite or allophane) that are either at early stages of weathering (from geologically active New Zealand) or are mostly highly weathered (from ancient Australian land surfaces), and also some from high pH, dry environments (in Iran), which are dominated by palygorskite. Together with a compilation in Churchman and Lowe (2012, Table 20.5), of mineralogical analyses of soils worldwide representing a wide range of lithologies of parent rocks, depth horizons and soil types (as given by Soil Taxonomy), these sets of data give some indication of ranges of contents of iron 'oxides' (including oxyhydroxides) that can occur in soils. Up to 15% free Fe_2O_3 (or 10.4% Fe) occurred throughout a New Zealand soil dominated by kaolinite, and between 10% and 20% of different horizons of more than one Australian soil also dominated by kaolinite comprised either or both of the iron oxides hematite and goethite (Churchman, 2010).

4.2.2 Reaction Rates and Parameters Determining Alteration as a Function of Time

As with all geological situations, change is a function of chemical parameters and time. The development of soil materials from rocks is a good case. Given the various types of processes possible in the formation of clays in surface alteration, we might expect different rates of formation depending upon the kinetics of each fundamental process that is operative (as defined by the slowest step in the process). The different processes will depend upon the relative chemical potentials in solution and the kinetic energy necessary to form the new phases, the surfaces of the individual phases exposed to solutions, and the removal rate of the solutions.

The question of clay formation, the factors that converge to form clays and their relative importance has been considered by Jenny (1994) in an attempt to numerically treat the problem. He considers climate, living organisms, topography, parent material and time. However, his conclusion is that relations in natural situations are rarely linear, presenting general trends of values which are rarely clearly defined.

Dissolution is probably the most rapid process in that, if the solution is sufficiently undersaturated, the mineral goes into the solution at the solution–solid interface, molecule by molecule. Here the energy of reaction is that of the rupture of bonds between elements engendered by the chemical undersaturation of the solution. The mechanism of ionic diffusion (substitution of hydrogen ions for cations) necessary to break Si–O–Si bonds is likely to be much slower than direct ionic dissolution in that the movement of cations is within a mineral structure. Thus, the two destabilisation processes will have different reaction kinetics. Growth of crystals from elements in solution should be relatively rapid, as is the dissolution process, in that the chemical attraction of elements necessary for crystal growth depends upon the depletion of ions from solution by crystal growth. The movement of ions is within the aqueous solution. The case of mineral growth from a gel or amorphous mass of ions will be much slower in that the energy necessary for crystal growth involves the reorganisation of covalently bonded ions with oxygen which must be reorganised to form new mineral phases. The last type of reorganisation of matter to form clays is that of the modification of pre-existing minerals. Here only a small part of the ions in the material are concerned, and the movement is by local diffusion. However, the rate of each reaction will depend upon the type of elemental bonding (covalent change being the slowest and cationic migration more rapid) and number of bonds to be modified. Brantley et al. (2008) discuss at length theoretical parameters and calculations of mineral dissolution and resultant dissolution rates. We can look at the overall results in cases of water–mineral interactions involving different types of geological materials and under different types of climate conditions where parameters of rock structure and climate (rainfall and frequency) have considerable effects not included in theoretical studies of mineral dissolution.

4.2.3 Effects of pH

One parameter, obvious from the descriptions of water–rock interactions, is the overall pH of the alteration system under consideration. Since hydrogen ions are the major vector of chemical displacement and change in silicate materials, pH will undoubtedly be primordial in the amount and rate of change. This is borne out by numerous laboratory studies of mineral solubility as a function of pH. Minerals containing magnesium such as olivine and amphibole are strongly affected by pH. Values below 5 increase dissolution by several orders of magnitude. This indicates that the availability of hydrogen ions changes the rates of dissolution and mineral reaction. For the most part pHs of altering solutions are largely controlled by the overall rock composition, basic rocks having higher solution pHs than acidic rocks (granites, etc.). The overall alterite material tends to buffer the altering solution pH. This is due to the type of dissolved ions present, Si giving a lower pH and Mg a higher pH to solutions. Thus, rock materials tend to maintain pH and hence alteration rates. Hence both basic and acidic rocks appear to alter at similar rates in nature (see discussions in Brantley et al., 2008).

However, in the soil zone (organic–alterite interaction) pH can be controlled by the biological action of the plant regime. Prairie soils are more basic in nature than forest soils. This can have a significant effect on the rate of transformation of the silicates present in these zones.

REFERENCES

Brantley, S.L., J. Bandstra, J. Moore, and A.F. White. 2008. Modelling chemical depletion profiles in regolith. *Geoderma* 145: 494–504.

Caillaud, J., D. Proust, D. Righi, and F. Martin. 2004. Fe-rich clays in a weathering profile developed from serpentinite. *Clays and Clay Minerals* 52: 779–791.

Churchman, G.J. 1980. Clay minerals formed from micas and chlorites in some New Zealand soils. *Clay Minerals* 15: 59–76.

Churchman, G.J. 2010. Is the geological concept of clay minerals appropriate for soil science? A literature-based and philosophical analysis. *Physics and Chemistry of the Earth, Parts A/B/C* 35: 927–940.

Churchman, G.J., and D.J. Lowe. 2012. Alteration, formation and occurrence of minerals in soils, p. 20.1–20.72. In: P.M. Huang, Y. Li, and M.E. Sumner (eds.), *Handbook of Soil Sciences: Properties and Processes*, 2nd edn. CRC Press, Boca Raton, Florida.

Hseu, Z.Y., H. Tsai, H.C. Hsi, and Y.C. Chen. 2007. Weathering sequences of clay minerals in soils along a serpentinitic toposequence. *Clays and Clay Minerals* 55: 389–401.

Jenny, H. 1994. *Factors of Soil Formation: A System of Quantitative Pedology*. Courier Corporation, North Chelmsford, Massachusetts.

Millot, G. 1970. *Geology of Clays*. Springer-Verlag, New York.

Proust, D. 1982. Supergene chlorite in alteration of metamorphic chlorite in an amphibolite from Massif Central, France. *Clay Minerals* 17: 159–173.

Proust, D. 1985. Amphibole weathering in a glaucophane-schist (Ile de Groix, Morbihan, France). *Clay Minerals* 20: 161–170.

Velde, B., and A. Meunier. 2008. *The Origin of Clay Minerals in Soils and Weathered Rocks*. Springer-Verlag, Berlin.

Wilson, M.J. 2004. Weathering of the primary rock-forming minerals: Processes, products and rates. *Clay Minerals* 39: 233–266.

5 Driving Forces of Alteration

The interaction between clays and organic matter is as vital to the continuation of life as, and less understood than, photosynthesis.

G.V. Jacks (*Soils & Fertilisers* 26: 147, 1963)

5.1 CLIMATE

The important climate function in reaction rates is that of the availability of water, which can bring about the reactions that change the mineral phases at the surface. If rainfall is constant, even in more or less minor quantities, strong alteration will occur, whereas if rainfall is periodic the alteration will be less, due to the smaller amount of time that the water is in contact with the altering minerals. Thus, total rainfall measurements are an indicator of alteration potential, but the number of days that the soil and alterites are water-saturated is a better measure. The movement of unsaturated solutions is the major motor of surface alteration. Solutions flowing along fractures or within a porous medium will effect more chemical change than those under slow movement. However, the type of alteration and the reaction product will not be the same. In general, the more water movement, the more silica loss and the higher the loss of soluble elements such as alkalis and alkaline earths. Concomitant with these effects of the silica and ionically bonded ions, the more dissolution, the more oxidation will occur and there will be a greater amount of accumulation of iron and manganese oxides.

In temperate climates, for example, rainfall is periodic and the amount of water in a soil varies with time. The relations of cations fixed on the clay materials and those in solution will then change with time (solution concentration). In a more humid climate the tendency will be for ions to enter into solution and not be fixed on clay materials due to the low average concentration of ions in the abundant water solutions. Hence the ions held on clay surfaces in a temperate climate will be most likely different from those on these surfaces in a tropical climate.

5.2 TOPOGRAPHY

One factor which is very important in the interaction process between rainwater and silicate rock and sedimentary material is that of topography. The slope of a surface essentially determines the rate of water flow out of the system, usually downhill towards a stream or river. An induced flow rate increases the transport of dissolved and above all particulate material which can flow through soil and alterite zones to be deposited at a lower level in the landscape. Accumulation or transport of alterite clays can change the rate of flow very locally within the alterite–soil zones and change the composition of the altering fluids (Birkeland, 1999). This is especially important for the soil zone where clays are formed and accumulated. The qualities of soils can often depend upon topographic factors such as clay content and organic matter content. For example, higher slope and thus shorter periods of water transit (contact time) lead to clays of the 1:1 type (halloysite) on granitic rocks in the northern Sierra Nevada (reported in Birkeland, 1999, p. 288), whereas in soils at the foot of the slopes, the 2:1 clays illite and smectite, which contain more silica, occur.

Of course, the more well-known factor of soil erosion (i.e. massive displacement of soil and alterite materials) as a result of topographic differences is a constant and evident factor in the evolution of materials at the surface. The more material taken away, the more chemical interaction will be important where the altering solutions are less buffered with stable minerals and come in contact with remaining unaltered high temperature minerals in the profile.

5.2.1 Interaction at the Water–Rock Interface

The combined effects of alteration parameters on the rate of alteration can be seen by looking at the water–rock interface as a function of depth and age. Here the initiation of the alteration profile has been observed in several cases (Figures 5.1 to 5.4).

Data from Evans and Cameron (1979) show that the depth of alteration or the distance to the rock–alterite interface on Baffin Island in the far north is a direct function of time (Figure 5.1). The rate of progression is of the order of one metre per 140,000 years. The rate of overall mineral change (development of new clay minerals from rock minerals) is initially rapid and slows somewhat in its overall transformation of rock to saprock. Under climatic conditions similar to those of Baffin Island on Alpine plateaus, Egli et al. (2001) found that the per cent of clays in the alterite (C horizon) was 5% after 12,000 years, but the rate of change quickly reached an apparent maximum. More

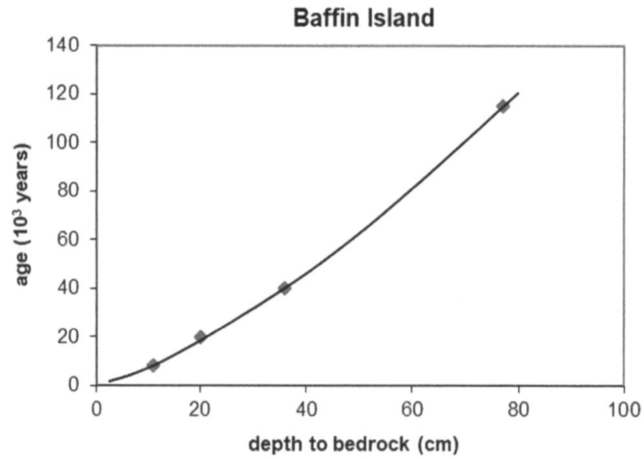

FIGURE 5.1 Plot of depth of the bedrock–alterite interface as a function of age on Baffin Island. (From Evans, L.J., and B.H. Cameron, 1979, *Canadian Journal of Soil Science* 59: 203–210.)

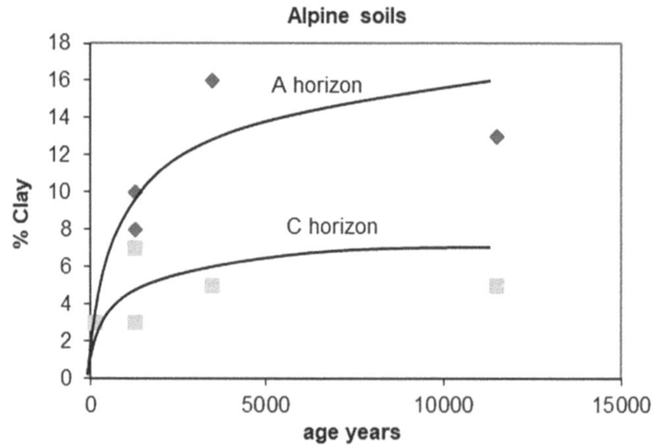

FIGURE 5.2 Plot of clay content developed under an Alpine climate for the A and C horizons as a function of age. (Data from Egli, M. et al., 2001, *Geoderma* 104: 145–175.)

clay is found in the soil clay horizons. Under more humid and warmer conditions in New Zealand, Lowe (1986) found above 40% clay in altered tephra after 200,000 years, with there being a steady increase with time.

These observations include the functioning of several alteration processes, dissolution, diffusion and mineral transformation and crystal growth. The overall result is a loss of coherence of the initial rock due to a small portion of mineral reaction. The much longer times necessary to form a total clay assemblage indicate the length of time necessary to complete the soil-forming process.

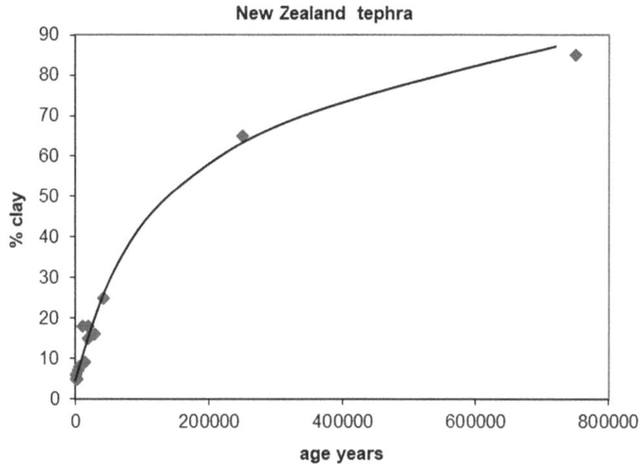

FIGURE 5.3 Plot of clay content of soils developed on volcanic tephra in New Zealand as a function of age. (Data from Lowe, D.J., 1986, Controls on the rates of weathering and clay mineral genesis in airfall tephras: A review and New Zealand case study, p. 265–330, in S.M. Colman and D.P. Dethier (eds.), *Rates of Chemical Weathering of Rocks and Minerals*, Academic Press, Orlando, Florida.)

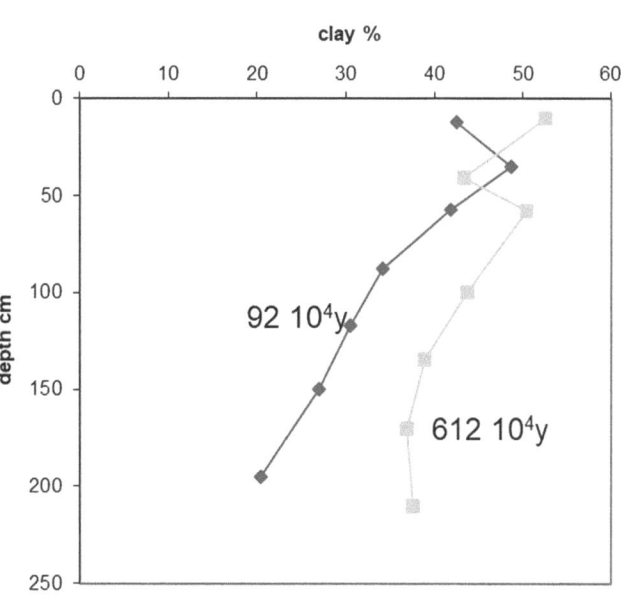

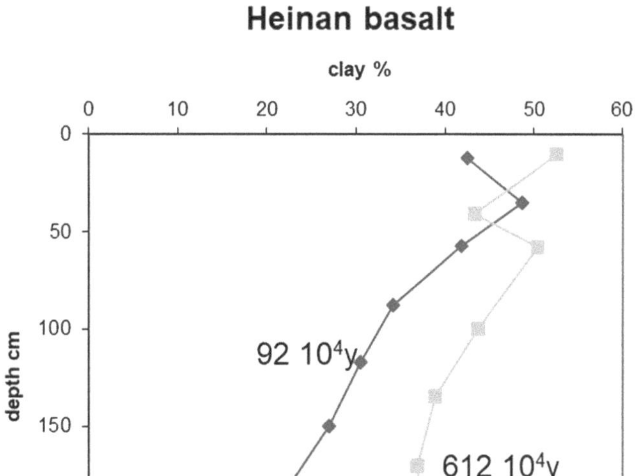

FIGURE 5.4 Clay development from basalt on Hainan Island (China) under sub-tropical climates indicate that the clay content increases downward with time but there is a convergence of clay content in the A horizon. The scale is millions of years. (Data from He, Y., et al., 2008, *Geoderma* 148: 206–212.)

The time frame for mineral transformation can be much smaller than that for mineral dissolution and crystallisation, several years as observed by Velde and Church (1999) for chlorite and muscovite in a salt marsh soil zone. In this process, change in charge balance of the structure by oxidation decreases the ionic charge holding interlayer ions which can diffuse out of the network to the altering solution. The relatively rapid change in mineralogy due to diffusion of ionically bonded ions demonstrates the importance of the strong covalent bonding of the silicate framework compared to ionically bonded ion forces. The resistance to mineral transformation at the surface depends upon the chemical forces involved and the thermodynamic–chemical characteristics of the minerals in the rocks. The mineral that is among the most resistant to alteration is the pure silica phase of quartz. Here there are no ionically bonded cations in the structure which can be displaced by diffusing hydrogen ions. Even though thermodynamically speaking, according to laboratory studies, quartz should be much more soluble, it remains as the typical resistant mineral in surface deposits where sand grains, the remains of the surface alteration process, are in the majority quartz.

Data from White et al. (2008) shows that the two feldspars present in acidic rocks being altered in terrace materials in California are altered at different rates (Figure 5.5). Potassium feldspar is lost by about the same amount in alteration profiles of 65,000 years and 226,000 years duration, while plagioclase is much more strongly affected at the longer time period. Hence different minerals can react very differently under the same physical–chemical conditions of alteration.

Further, the charge variable sites on clay surfaces are the most affected by the chemistry (pH and solution concentration of ions). The pH is largely determined by the vegetative cover in the soil area. Conifers and some tropical forest cover engender high hydrogen activity in the soil solutions. The interlayer sites in clays (between the 2:1 layers) are much less subject to changes in solution

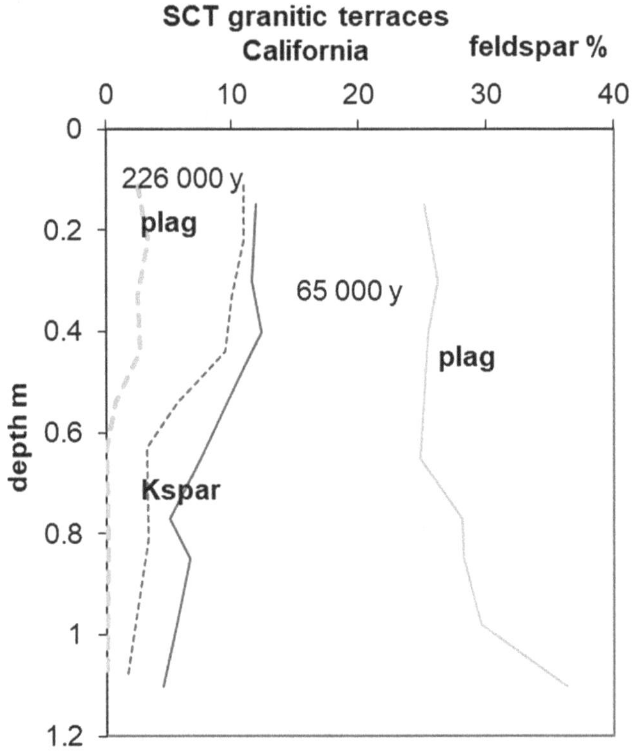

FIGURE 5.5 Comparative losses of feldspars (potasssic, Kspar; and plagioclase, plag) in granitic terrace material in California. (Data from White, A.F., et al., 2008, *Geochimica et Cosmochimica Acta* 72: 36–68.)

chemistry and are more constant in their chemical composition. This is especially true for pH effects where the cations are lost through hydrogen ion exchange on charge-variable sites.

Hence basic chemistry is the key to understanding the processes and their rapidity at the surface.

5.3 GEOLOGICAL PARAMETERS

Since the alteration reaction is between rock and water, the relative proportions of each will also determine the reaction rate. The less dissolved material in the altering solution, the greater the tendency to equilibrate the rock–solution system, and the amount of reaction will change as a consequence. The more rain, the more alteration. This is of course an obvious relationship which has been observed for quite some time. Tropical climates show greater alteration rates. However, another factor can become important in certain cases. This is the initial porosity due to fractures in the altering rock. If a rock is dense without tectonic fracturing, then there is less fluid flow initially into the rock and there will be little interaction between the rock and the altering fluids. The more initial fractures, the more interaction and more dissolution will occur along fractures. The reverse is frequently seen in the alteration of carbonates. Many high mountain massifs are formed of carbonates, which are highly soluble minerals compared to silicates. However, carbonates have a tendency to recrystallise under tectonic stress, closing tectonically induced cracks and fissures, and thus likely to form more or less impermeable surfaces to the introduction of rainwater. The high cliffs are then in fact due to mineral morphology and strongly bonded crystal faces. This is a geological dimension which should be considered in assessing reaction rates.

Further, the rock type can be a factor in alteration rates. Differences in solubility of constituent minerals from one rock type to another can influence the reaction rate and the rate of weathering. If the rock is one of metamorphic origin, composed of significant quantities of phyllosilicates such as muscovite and chlorite, the tendency to change the mineral structures will be lessened and the reaction rate will be slower. If the rock is of high temperature origin and composed of highly soluble minerals such as olivine (basalts, etc.) the reaction rate will be enhanced by the inherent tendency to come to chemical equilibrium at a greater rate. Thus, mineral type can be a factor in the rate of alteration of silicate rocks.

5.3.1 Alteration Profile

Our point of view is concerned with the existence of clay complexes (mineral and organic components) at the surface of the earth and a coherent description of the chemical and physical forces which make the soil clay mineral complex. In the soil zone, this is one of the plant–mineral interactions. It is quite clear that the present-day surface of the earth is for the most part dominated by the products of mineral–plant interactions. Roots, leaves, different substrates exuded from plant roots and biological action of living organisms interacting with the products of rock alteration in the upper part of the altered zone of the earth's surface produce what is normally called soil. Soil however is based upon the presence of a relatively high proportion of fine-grained mineral material, essentially silicate material in the form of phyllosilicates. Other materials of mineral origin such as oxides and hydroxides of iron and aluminium, and insoluble high temperature phases such as quartz and feldspars are present in various amounts. The dominant characteristic of soils can thus be considered to be the interrelation of organic material and silicate–oxide materials of mineral origin. In this sense without plant life – in the past or at present – there is no soil.

In the areas where vegetal life is almost non-existent, in deserts or on frozen alpine or arctic areas, the very fine-grained clay mineral and oxide material present is almost non-existent. In these cases a fine-grained material forms at the rock–atmosphere interface by thermal action of differential expansion and contraction where silicate minerals are separated from one another in the rock, forming small mineral grains. This is due to the fact that primary silicate minerals in rocks are highly asymmetric in their response to thermal expansion. This creates differential movements at mineral interfaces, which tend to weaken the mineral cohesion in the rock. Thus, small mineral grains, or grain fragments, are formed at the rock–atmosphere interface by thermal action. This fragmental material is not physically stable at the rock–atmosphere interface and under conditions of rare rainfall or wind action it is transported and concentrated in sedimentary-type deposits. As a result, the fine-grained material abiotically produced by fragmentation is not physically stable and will not form a significant layer of alterite material. Layers of alterite are stabilised by plant action, roots and the accumulation of the products of organo-clay associations. The first surface interaction, rock fragmentation and the liberation of individual mineral grains, produces an unstable material that does not remain in place for long without biological stabilisation.

Thus soil, fine-grained material of physical (by thermal fracturing) and chemical (by chemical change within an aqueous environment) action, is an accumulation of altered rock materials and mineral grains in a fine-grained state, which is stabilised by the interaction of organic material with the fragmented inorganic material to give some newly formed minerals that have formed from reaction with rainwater at the surface. This being the case, soil becomes the primary substrate for plant life at the surface of the earth, on continents and islands, which is maintained in place by the plant life itself. Plants stabilise the fragmented and slightly altered silicate minerals from rocks from which they can extract elements as the minerals are altered into clays and oxides.

5.3.2 Rock Alteration by Pore Water

It is clear from an analysis of the numerous reports on the structure and content of soils and the alteration profiles below them that there are several distinct zones of development under the plant-dominated surface where rocks become clay-dominated assemblages of material. The variation within profiles is due to the different stages of development of the chemical alteration processes at the surface, which is a function of climate, among other factors, especially time.

The deepest interaction is at the interface where infiltrating rainwater is in contact with rock material of high temperature origin. The thermodynamic properties of high-temperature minerals, for the most part anhydrous, in the presence of aqueous solutions at relatively low temperatures are such that chemical interaction is inevitable. This is the water–rock interface where new minerals are formed. As time goes on, the water–rock interface is found at deeper and deeper levels as the altered material is kept in place, essentially by the protection of plant systems. The new minerals produced by the interaction of slightly acidic, oxygenated rainwater are generally of small grain size (largely less than 2 micrometres, or μm) and hydrous (containing hydrogen ions) for the most part. Iron, largely ferrous in high-temperature minerals, is oxidised producing oxides or ferric clay minerals. This is the starting point for alteration and the formation of soils, generally called the saprock stage. In this material the typical structure is one of inhomogeneous materials where fractures and other passages allow preferential alteration of the solid materials as they interact at different rates with the rainwater flowing through the rock. Rock structure is largely maintained but coherence is weakened by the formation of new phases and mineral dissolution.

As alteration proceeds, producing an increasing amount of fine-grained material mixed with rock and mineral grain fragments, the material becomes more homogeneous in texture, and the variety of alteration mineralogy created by mineral transformation and fracture passageways becomes less important. This is called here the alterite material, where many but not all of the high-temperature materials have reacted, and the structures and texture of the rock have largely disappeared. The alterite zone is one of a general tendency towards a more homogeneous, mineralogical and textural structure. It tends to be porous and promotes the movement of water and dissolved or fine-grained suspension material following gravitational forces. This is the zone where the deep roots of the plant cover find elements necessary for growth by absorbing elements as ions in solution present through the dissolution of the remaining minerals in the alterite material that have not yet reacted with the rainwater that flows through it. It is noted that the term 'alterite' to designate the transformation towards the soil zone is not often used, but since the use of other terms seems not to have found a general consensus (see Taylor and Eggleton, 2001, p. 159), we will use alterite to designate the gradual transition from structured, partially transformed rock to more or less homogeneous fine-grained altered material.

Thus, at the surface, of differing thickness depending upon climate and the age of the alteration sequence, is the soil zone, where organic matter interacts with the fine-grained minerals (essentially clays). This is the focus area of our interest here. The origin of minerals in the alterite zone is the focus of another book (Velde and Meunier, 2008). As the soil zone develops, more clay minerals (silicates and oxides) appear. In many instances these fine-grained minerals are moved downward from the organic-rich soil zone by fluid transport. They are accumulated in a layer of varying thickness, which is called the B horizon, whereas the organic zone is called the A horizon. One major difference between the two zones is that the organic soil zone has a general chemistry which is controlled by plant action, whereas the B horizon has largely lost the imprint of this activity, even though the minerals present were formed or altered under the soil zone chemical conditions.

Since the physical configuration of the different reaction zones (water–rock, alterite and soil) have specific physical and chemical variations, the chemical changes and the minerals produced vary from one zone of the alteration profile to the other. The chemistry of the water–rock and alterite zones depends upon the rate at which water circulates and on its overall quantity. The more water in contact with the rock, the less dissolved material it contains and the more rock that will be incorporated in it. The amount of reaction that occurs is dependent upon (1) the contact time of a given amount of water with minerals which determines the amount of material that enters into solution and thus the relative saturation of the solutions with dissolved elements, and (2) the total amount of water that comes into contact with the reacting minerals.

The chemistry of the soil zone depends upon the plant regime present and the rainfall factors.

5.3.3 Movement of Clays

The fine-grained materials, clays, tend to be transported by moving water within the alteration profiles. This is a function of their interaction with organic matter, which tends to form clay-organic aggregates that are larger in size than the individual components and hence less likely to be moved by flowing water. Organic–mineral interactions stabilise clays protecting them from physical displacement to a certain extent. However, many soils show a significant downward movement of soil clays from the surface, plant–clay interaction zone to a level below the soil zone in the surface. The concentration of clays below the surface, organic-rich horizon (A horizon), constitutes the B horizon (Figure 5.6). In some cases the zone just below the A horizon is almost devoid of any clay material which accumulates below. This is called the E horizon, which is mainly composed of quartz grains and other relatively insoluble fine-grained materials. The grain size is well above that of the clays, <2 μm. The reasons for the loss of clays from the E horizon is that the chemistry of the plant–clay horizon, with low pH values, is such that there is little interaction between clays and organic materials. They do not aggregate significantly, and hence clays are transported to lower levels by rainwater in a very fine-grained state. Such a structure (A–E–B horizon) is typical of acid soils such as those found in conifer forests (Figure 5.7).

If the clays can move downward within the soil zone into the upper alterite zone, they can also move laterally within the soil–alterite layers when on a slope creating a clay-rich deposit at lower levels, in streams or lakes. Such movement can be very important where the clays concentrate not only below the A horizon but downslope from the soil zone in zones of accumulation where this clay material creates a deposition of clay-rich materials.

5.3.4 Geology and Alteration

The chemical interaction of rainwater (water in equilibrium with atmospheric gases) with crystalline rock materials is the classical starting point for the analysis of alteration and soil-forming processes.

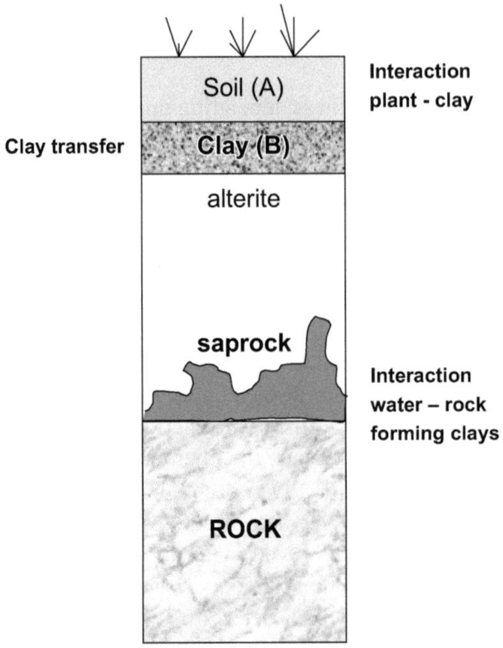

FIGURE 5.6 Schematic representation of the zones of alteration in a developed sequence of rock alteration. In young sequences, recently glaciated zones for example, the depth to the water–rock interface is not great, whereas in old alteration sequences in tropical zones the depth to rock can be tens of metres. The relative dimensions of each zone depend upon climate and the duration of the contact time between rock, alterite and rainwater.

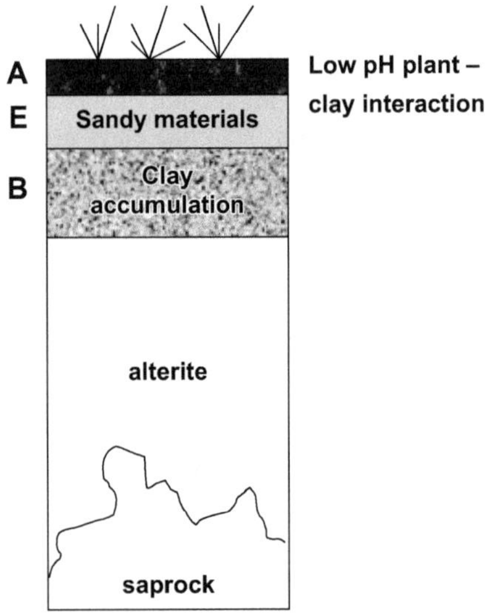

FIGURE 5.7 Soil clay movement in an acid environment where a largely clay-free zone (E horizon) is formed below the A soil horizon. Below the E horizon is accumulation of clays (B horizon).

However, within the geo-materials group of rocks, several distinct types of material occur. Rock is a descriptive term of geologists to indicate a solid (to the impact of a hammer) material as opposed to sediments and soils. Within this varied group are those which have formed from molten materials (silicates) and formed crystals or remained in an amorphous state from temperatures above 800°C or so. The phases and minerals are strongly out of thermodynamic equilibrium at the surface of the earth. Temperature is the main factor but also most of the rocks of igneous (molten) origin formed in the presence of only slight amounts of water, which puts them out of equilibrium with surface chemical conditions where water is abundant. The tendency of alteration to form more stable minerals is very strong for these rocks when they arrive at the surface.

A second category of rock materials is that of metamorphic rocks which have experienced high temperatures but have not melted entirely. Here a significant range of temperatures has caused the formation of the minerals in the different rocks from those of melted materials to the initiation of large-scale recrystallisation of the materials present. Such rocks, usually based upon sedimentary materials which were initially of various states of hydration and water content, contain minerals with hydrogen present which are of specific crystallographic types, namely, amphiboles and phyllosilicates (chlorite, muscovite and biotite). Lower temperature of formation and the presence of hydrogen in the minerals give them greater thermodynamic stability at the surface. Sedimentary rocks that are formed at still lower temperatures, towards 200°C, often remain highly hydrated. They have even less tendency to change their crystallographic identity under conditions of surface alteration. Other sedimentary rocks which are largely anhydrous, such as carbonates, tend to dissolve under surface chemistry conditions, i.e. slightly acid solutions. The resistance to alteration of quartz leads to its concentrations in sandstones, which give such sedimentary rocks a great resistance to change at the surface. Thus, the conditions of rock formation in the geological cycles give different degrees of mineral stability or instability to the rocks present at the surface.

The geologic history of rocks can influence their physical characteristics by producing cracks or broken surfaces, which are revealed as the materials come to the surface and into a low confinement environment. Tectonically fractured rocks present many passageways for surface water to penetrate into the mass and promote alteration. However, not all rocks are fractured and some tend to resist fracturing. Limestones are such a case. The carbonate material is susceptible to internal deformation and recrystallisation upon the application of asymmetric tectonic constraints. Such reactions create tightly bound crystal surfaces within the rock, which when exposed do not lead to water penetration. This is why there is the anomalous case of rather soluble carbonate materials forming great mountain massifs, while much less soluble materials such as granites tend to be altered at a greater rate. Thus, the geologic history of the rock material can influence its reaction to surface alteration processes.

However, rocks are not the only materials involved in soil formation. Sediments, made of materials either transported by water or wind, can bring silicate and oxide materials to the alteration surface which can form deposits that are the basis for soil formation. Windborne volcanic material for example brings high temperature materials, often silicate glasses, into the soil zone directly before going through the water-rock interaction zone. Continued renewal of the volcanic material can produce specific soil types and mineralogies. Wind-blown sediment, largely derived from other zones of alteration, can be deposited on soil systems as loess materials. This material is closer in chemical and mineralogical character to the soil minerals than the volcanic input. However, chemical interaction with plants and soil formation (organic-silicate, oxide interaction) are important for these materials in soil-forming processes. Here the key interactions occur in A horizons. And finally, sediments transported by rivers can form new zones of plant–clay interaction along river bottoms or on flood plains as well as along the shores of oceans.

A rarer case of clay mineral formation occurs in soil–alterite formation where periodically surface water is introduced and evaporated successively. These are evaporite deposits, which are generally salts that rapidly dissolve again. However, in some cases silicate minerals are precipitated from solution, which is the case for the magnesium silicates sepiolite and palygorskite. These chain-structured minerals (in contrast to phyllosilicate clay minerals which are mostly tabular) are not stable in the soil zone of plant–mineral interaction. They form below the soil zone in the alterite zone or in sediments below the soil zone mainly due to their pH values being lower than those of their formation.

Another aspect of geologic importance is the effect of morphology or topography. Depending upon the slope and supply of water to the soil zone, the movement of solids by water action and movement of groundwater can be of great importance for the development of soil materials. One aspect is the impact of topography on the biome present, i.e. root depth and plant type, which can influence the relative activity and chemistry of the plant–silicate and oxide material. Topography can also influence the relative content of fine-grained materials moved by water transport out of the immediate alterite and soil system. The presence and time of transit of water can also determine the amount of alteration of rock materials. These factors are controlled by topography.

Thus, the configuration of the surface layers and their origins are factors which are very important in the formation of altered materials that are the basis for soils. Topography is an important factor in that it enables movement of material within the alterite and soil zones which is accumulated elsewhere, thinning the alteration zone. This also affects the soil thickness. A second factor of great importance is that of the time span over which there has been water–rock interaction in a stable topographic configuration. The older the profile, the deeper the alterite zone. And of course, the climate is a major factor in alteration in that it determines the amount of water in contact with the silicate materials, which affects alteration; the more diluted the solutions, the more transformation will occur. One major factor in climate is not just the total amount of rainfall but also its regularity. Since contact time is a fundamental factor in silicate transformation and recrystallisation, a climate of regular and constant rainfall will produce more alteration than one with sporadic, intense rainfall events. Another factor which may be relevant is that of temperature. Following the rough guide for this, the rate of reaction of the alteration is expected to double for each 10°C rise in temperature (www.chemguide.co.uk).

REFERENCES

Birkeland, P.W. 1999. *Soils and Geomorphology*, 3rd edn. Oxford University Press, New York.

Egli, M., A. Mirabella, and P. Fitze. 2001. Clay mineral formation in soils of two different chronosequences in the Swiss Alps. *Geoderma* 104: 145–175.

Evans, L.J., and B.H. Cameron. 1979. A chronosequence of soils developed from granitic morainal material, Baffin Island, N.W.T. *Canadian Journal of Soil Science* 59: 203–210.

He, Y., D.C. Li, B. Velde, Y.F. Yang, C.M. Huang, Z.T. Gong, and G.L. Zhang. 2008. Clay minerals in a soil chronosequence derived from basalt on Hainan Island, China and its implication for pedogenesis. *Geoderma* 148: 206–212.

Lowe, D.J. 1986. Controls on the rates of weathering and clay mineral genesis in airfall tephras: A review and New Zealand case study, p. 265–330. In: S.M. Colman and D.P. Dethier (eds.), *Rates of Chemical Weathering of Rocks and Minerals*. Academic Press, Orlando, Florida.

Taylor, G., and R.A. Eggleton. 2001. *Regolith Geology and Geomorphology*. John Wiley & Sons, Chichester, United Kingdom.

Velde, B., and T. Church. 1999. Rapid clay transformations in Delaware salt marshes. *Applied Geochemistry* 14: 559–568.

Velde, B., and A. Meunier. 2008. *The Origin of Clay Minerals in Soils and Weathered Rocks*. Springer-Verlag, Berlin.

White, A.F., M.S. Schulz, D.V. Vivit, A.E. Blum, D.A. Stonestrom, and S.P. Anderson. 2008. Geochemical weathering of a marine chronosequence, Santa Cruz California: Interpreting rates and controls based upon soil concentration depth profiles. *Geochimica et Cosmochimica Acta* 72: 36–68.

6 Chemistry of Alteration by Weathering

The internal machinery of life, the chemistry of the parts, is something beautiful. And it turns out that all life is interconnected with all other life.

Richard P. Feynman (1918–1988)

6.1 ALTERATION CONTEXT

The major chemical forces acting upon solids at the surface of the Earth are those of interaction with rainwater. This aqueous fluid is slightly acidic (atmospheric CO_2 saturation) and oxidising (saturation with the Earth's air atmosphere). The chemical action of water in large quantities is that of dissolution of solids until they reach a concentration that allows some materials to remain as solids. Dissolution is of two kinds: (1) high water content and integral dissolution of the phases into the liquid, and (2) partial (incongruent) dissolution where some elements are oversaturated in the solution and combine to form new phases, whereas other elements are undersaturated and remain in solution. Some minerals have a very low effective solubility for different reasons and remain in many cases unaffected by contact with rainwater, quartz being a special case. During the dissolution process or perhaps slightly before, the oxidation powers of air-saturated water are apparent in the oxidation of divalent metal ions found in the silicate and other high-temperature phases. The formation of oxy-ions, essentially iron oxide, is a very important process in that it destabilises the initial mineral by changing the overall electrochemical balance and very often disrupts the crystalline structure of the mineral. This is not always the case, especially for the phyllosilicates biotite and chlorite, which can re-equilibrate through internal (interlayer site) ionic diffusion.

The interaction of rock materials with rainwater occurs throughout the alteration column, from the abiotic water–rock interaction zone up to the soil zone of organic and mineral interaction. Soil clay minerals are for the most part an inheritance of water–rock clays formed at some depth in the profile which have been affected by changing chemical conditions as they interact with the organic matter and plant-controlled chemistry of the surface.

One source of material in soils that at times can be very important is the deposition of solids by wind currents, loess materials (fine-grained unconsolidated sedimentary or volcanic material). In these instances, the unaltered rock type mineralogy interacts with rainwater in solutions where a large part of the chemistry is controlled by biological action.

Hence soil clays can be due to bottom-up movements or top-down interactions. The reaction rates and types of minerals formed are likely to be rather different in the two cases.

6.2 CHEMICAL FORCES

The chemistry of surface alteration can be identified with the interaction between rainwater and solids (rock or sediment). The driving mechanism is the interaction of hydrogen ions with minerals, which is therefore a function of solution pH and the interaction of electrons from dissolved oxygen ions in the aqueous solution producing oxidation effects. The amount and type of reaction is determined by the quantity of water available and its content of dissolved elements. This is the basis of surface alteration.

6.3 CHEMISTRY OF ELEMENTS AND MINERAL STABILITY

The chemistry of elements is a very well-known discipline and treated in many texts. Some points are be reiterated here as a means of putting the chemistry of elements in rocks into perspective.

The minerals in rocks that make soil clays are silicates. Silicates are, as expected, dominated by silicon and its chemistry. The most important aspect of silicon in rock minerals is the covalent bonding with oxygen ions that are shared with other silicon ions. This sharing, covalency, is the salient characteristic of minerals in silicate rocks. A second and very important aspect of silicate minerals is that they are strongly stabilised chemically when other covalent bonding cations, especially aluminium, are present. Aluminium has the same tendency to share oxygen ions in covalent bonding with other cations, especially silicon. Hence the chemical reactions between Si–O and Al are such that they tend to form solids rapidly (reach saturation at low dissolution values) under surface conditions. More 'soluble' elements, those more ionic in the character of their bonding, such as alkalis and alkaline earths, are less likely to form solid phases under conditions of surface alteration. Iron is a special case in that it is the only element of major abundance to be relatively oxidised or reduced in its oxidation state under surface conditions. The solubility of divalent iron is relatively high, while that of oxidised iron is low, so these cations tend to form insoluble oxides. However, trivalent iron can be found in silicate minerals at the surface in the usual structural sites of aluminium and hence Fe^{3+} is at times a major component of clay minerals, most notably 2:1 clays.

The general relations of the major elements in rock minerals relative to their tendency to form solid phases upon contact with large amounts of water is schematically shown next:

$$Fe^{3+}, Al > Si > Mg, Fe^{2+} > K, Na, Ca$$

Trivalent ions are the least mobile or soluble, forming oxides or hydroxides; silicon is slightly more soluble in its different combined forms, while Mg and Fe^{2+} tend to be displaced by hydrogen ions in minerals and the most labile elements are the highly ionically bonded K, Na and Ca ions. The alteration process usually starts then with the ionic exchange in the solid state of hydrogen for the cations of K, Na, Ca and eventually Mg. This leaves an electronically stable material, where loss of cations is compensated by hydrogen ions. However, the structure of the initial mineral is strongly modified, and the initial geometry is altered such that there is a strong tendency to form new solids in the form of a gel, which, in incorporating hydrogen, eventually becomes 'hydrated' clay minerals upon crystallisation. Dissolution of the material in order to attain saturation with the alteration solutions also affects silicon. The relative amounts of silicon in the gel determines the types of clay minerals that will form, those that are silica-rich, silica-poor or which eventually do not contain silica.

The interaction of silicate solids from rocks, formed at temperatures and pressures above those of the surface, is to a large part determined by the mineral structure. Tectosilicates (feldspars) containing K, Na or Ca as ionically bound cations and inosilicates (olivine) containing Fe and Mg are unstable, while chain silicates (pyroxenes and amphiboles) are more stable. Among the most stable high-temperature silicates are the phyllosilicates (micas and chlorites). The most persistent mineral, although theoretically unstable, is quartz. It would appear that the lack of ionically bound cations leads to its stability in that the process of hydrogen diffusion and cation replacement is not operative. Hydrogen will diffuse into quartz, but apparently just at the surface where it is fundamental in the dissolution process. Crystal structure is thus important as well as the type of elemental chemical attractions and bonding within the mineral structure. As usual, geological systems are complicated and difficult to define with simple formulae. Nevertheless, as a general rule, the higher the temperature of formation of a mineral, the more unstable it is in the surface chemical environment. Minerals in sedimentary rocks are the least modified by surface chemical actions in that they formed at temperatures under aqueous conditions near those of the surface.

However, the new minerals formed by alteration processes reflect the state of undersaturation with dissolved elements in the altering solutions (governed by rate of water flow) and they can display different types depending on rainfall intensity. The silicate minerals formed at the surface tend to show different silica contents and cationic contents depending upon water flow and undersaturation state. The most silica-rich, the 2:1 minerals, are formed in the initial stages of interaction, while less silica-rich minerals (1:1) are found under conditions of lower silica content of the solutions. Eventually silica is totally lost from the clays, and Al or Fe oxides or hydroxides are found. Thus, silica loss is a fundamental part of the alteration processes.

A second and extremely important surface chemical interaction is the oxidation by electron exchange of Fe and Mn ions in the high-temperature minerals. Most high-temperature minerals have Fe and Mn in the reduced divalent state, while the equilibrium with oxygen-saturated rainwater favours higher ionic states. A change in oxidation state of an ion in a mineral leads to electronic instability of the structure and eventual migration of the oxidised ion into another structural position.

The relative importance of each type of chemical force in the alteration process on alterations occurring in a rock will depend on the chemistry and physical state of the rock involved. Figure 3.1 (see Chapter 3) shows that basalts can alter more rapidly than granite, and both eruptive rocks are more liable to change than meta-sedimentary quartz-rich rocks.

The initial stages of alteration can be accomplished from millimetric to centimetric depths as indicated in the figure. The basalt is the most readily altered mostly because of oxidation processes that destabilise the ferromagnesian minerals. Ionic interdiffusion and consequent dissolution is less rapid as seen in the granite and the quartz-rich metasedimentary rocks are very stable under alteration conditions after 400 years of alteration.

6.4 MECHANISMS OF ALTERATION

Several types of interaction and resulting processes occur as high-temperature minerals come into chemical contact with altering aqueous solutions at the surface. The first leads to the chemical and structural change of pre-existing minerals in the rocks brought to the surface. The second is the reorganisation of the altered materials into new minerals, which form the basis for clays in the surface environment.

6.4.1 Mineral Change: Loss of Mineralogical Identity

Clay materials are essentially hydrated, i.e. containing hydroxyl ions or hydrogen ions, and the iron ions which they contain are in the majority in the oxidised state. This reflects the essential chemical forces of the slightly acidic oxygen-saturated water that reaches the solid–atmosphere surface. However, the clays that form indicate the approach to various states of equilibrium with dissolved ions in aqueous solution. There is a sequence of products which are produced corresponding to different concentrations of dissolved elements in the rock–water interaction sequence. This is based upon the relative solubilities or tendency to dissolve of the different minerals present in the rocks. Almost all rock minerals are unstable at surface conditions, but the tendency to react with aqueous solutions is variable. Hence the most unstable minerals, such as olivine in a basalt with a restricted silicon–oxygen network of covalently bonded ions, will rapidly release silica and magnesium into solution, and other minerals react with this material. The least soluble element, iron, tends to form a trivalent oxide phase, often forming an individual mineral compound, upon reaction to an oxidised state in contact with rainwater. The release of silica will have different effects on the other minerals in the basalt rock, such as plagioclase and pyroxene. In the end, after a sufficient amount of time has elapsed and water has flowed over the rock interface, the new mineral assemblage will converge on a stable assemblage, depending upon the total water flow (which determines its state of undersaturation with dissolving elements).

However, the new minerals reflect the state of undersaturation (rate of water flow) and can display different types depending on rainfall intensity. The silicate minerals formed at the surface tend to show different silica and cationic contents depending upon water flow and undersaturation state. The most silica-rich, the 2:1 minerals, are formed in the initial stages of interaction, while less silica-rich minerals (1:1) are found under conditions of lower silica content of the solutions. In volcanic ash, the 1:1 mineral halloysite forms when precipitation is low (≤~1200 mm y^{-1}; Parfitt et al., 1983), leading to a low water throughflow (<250 mm y^{-1}, according to Parfitt et al., 1984). However, allophane (generally as a 1:2 mineral) is favoured with higher rainfall and throughflow of water (Parfitt et al., 1984; Churchman and Lowe, 2012). Eventually, silica is totally lost from the clays, and Al or Fe oxides or hydroxides are found. Thus, if the rainfall and undersaturation of the alteration solutions with silicate material are sufficiently low, even the new clays formed can become unstable. The amount of ionic material in solution determines the minerals that will form by precipitation of these elements from solution.

These overall processes are accomplished by specific mechanisms of chemical interaction. The processes involve total mineral dissolution in the aqueous solution, an interdiffusion of ions where the ionically bonded ions are lost to solution, and a reorganisation of ions in the oxidised state forming oxides, essentially of iron composition. Some minerals are modified to a lesser extent through the processes of oxidation of iron components, which creates clay minerals from high-temperature phyllosilicates.

6.4.2 Dissolution

As rainwater interacts with solids at the surface, it is initially undersaturated with most of the elements found in the minerals and hence tends to dissolve the phases present. As interaction occurs, the water is gradually saturated with the different elements and comes into equilibrium with various mineral phases. In soils the minerals present are less likely to interact with rainwater in that they have already attained an equilibrium with aqueous solutions. Nevertheless, some dissolution will occur in the initial stages of reaction. As the rainwater descends into the alteration sequence, it contains more dissolved ions and will have less effect on the products of incongruent dissolution, which are essentially the clay minerals. The main chemical factor in the dissolution of silicates is the strength of the Si-O covalently bonded network. Carbonates and other minerals can be dissolved by exchange of the ionically bonded cations with the formation of individual oxyanion molecules. However, silicates have strongly bonded cation–anion structures even though the crystal structure might be disrupted.

The initial phase of the dissolution of silicates is accomplished by the diffusion of hydrogen ions into the crystal structure where they become associated with the oxygen ions that link the silicon ion networks. Here the hydrogen ion bonds with an oxygen, changing the electronic balance of the Si-O structure (McSween et al., 2003, p. 112). The diffusion of hydrogen ions into the network is of the order of several micrometres. The surface networks are disrupted, which weakens the interlinkage creating the possibility of formation of oxyanion cations, of Si–O–H ions, that allows silicon–oxygen ionic units to be separated from the Si-O network and to enter into the aqueous solution. Hydroxylation of the Si-O network is the major step in the process of dissolution. The Si-O oxyanions, such as $H_3SiO_4^-$ and $H_2SiO_4^{2-}$, in solution can migrate in the solutions under chemical potential differences where they are taken into growing mineral structures, or they can be taken out of the system by fluid flow.

6.4.3 Interaction by Diffusion and Ion Exchange

The two most important actions of rainwater with rock silicates are that of hydration and oxidation. Hydration is the introduction of hydrogen ions into the minerals in exchange for ionically bonded elements, leaving the covalently bonded ions in a strong oxygen network. Oxidation involves iron and other transition elements, which are for the most part divalent in the high temperature rock minerals.

Hydration and eventually oxidation are accomplished in zones of intimate contact between the minerals (usually silicates) and air-saturated aqueous solution. In the solution there are hydrogen ions produced by the interaction of CO_2 with water. These ions are highly active because of their relatively small size compared to oxygen and other cations present in silicates. The charge of plus one is highly concentrated over the surface of the hydrogen ion. This gives it a strong potential to enter the silicate mineral by diffusion and to displace larger, less densely charged cations that are ionically bonded within the Si-O network by elements that are covalently bound. This is an interdiffusion process where the charge balance is respected. An example of ion exchange by interdiffusion in the solid is shown in Figure 6.1. Once the initial cation is displaced, the geometry of the mineral structure is disturbed and it loses its chemical-thermodynamic stability, creating an amorphous state.

A demonstration of such a process in silicate material can be observed in stained glass windows where the silica-based glass interacts with slightly acidic rainwater. Figure 6.2 shows the relations of relative concentration of elements in a 13th century glass pane from the Angers Cathedral (France). The distance from the unaltered glass to the pane surface is on the order of a millimetre and a half showing preferential loss of ionically bonded elements from the amorphous glass, while the covalently bonded ions are more stable. However, silica is the least concentrated of these elements suggesting that there is some loss from the glass by water–silicate interaction.

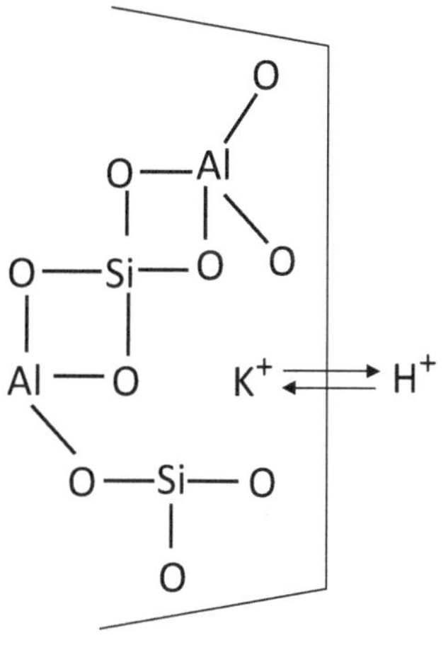

FIGURE 6.1 Schematic representation of the diffusion of a hydrogen ion into a crystal where it replaces a potassium ion which diffuses out of the crystal.

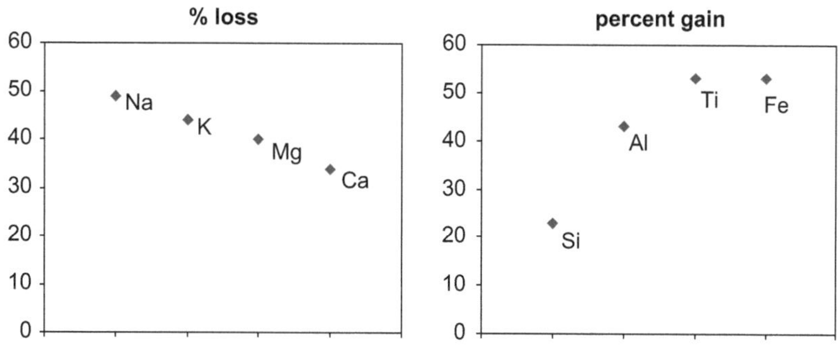

FIGURE 6.2 Relative amounts of elemental loss and gain as percentage of the altered material compared to the original glass composition of a 13th century stained glass window from the Angers Cathedral (France). Total alteration distance is near 300 μm. (Velde, unpublished data.)

The interaction by diffusion of hydrogen in silicate minerals creates an essentially amorphous material, which can be seen in a thin section of altered rock as a gel material (see Velde and Meunier, 2008). The hydrogen tends to remain as a cation bonded ionically to the overall structure as were the initial mineral cations. Along with the loss of constituent cations of the initial mineral and the creation of a gel there is a tendency for oxidation of the materials in the former mineral through the migration of dissolved oxygen ions in the aqueous solutions. Eventually the thermodynamically unstable gel material (amorphous) forms new crystals. Some of the silica is also lost during the process. The hydrogen-rich material tends to form silicates which incorporate these ions into the structure forming what are called hydrous minerals, those typical of clays. This is alteration by ion exchange and loss, and the eventual formation of new phases in the place of a pre-existing phase is called incongruent dissolution. The general relations of these processes are shown in Equations 6.1 and 6.2, which illustrate the two main interactions of rainwater with silicate rock minerals. The result of hydrogen ion diffusion into the mineral with exchange for ionically bound cations and the oxidation of iron are the two processes that dominate the interaction phenomenon.

Hydration

$$\underset{\text{Plagioclase}}{Ca\,Al_2Si_2O_8} + 2H^+ + H_2O \Rightarrow \underset{\text{kaolinite}}{Al_2Si_2O_5(OH)_4} + \underset{\text{ion}}{Ca^{2+}} \tag{6.1}$$

Oxidation

$$\underset{\text{Olivine}}{(Fe,Mg)_2\,SiO_4} + 3H_2O + 2O^- \Rightarrow \underset{\text{iron oxide}}{Fe_2O_3} + \underset{\text{solution}}{(Mg\ Si)(OH^+)_6} \tag{6.2}$$

It is notable that quartz, relatively unstable in the presence of water in the laboratory, is very resistant to alteration by rainwater in nature. Here the network of Si-O covalent bonds cannot be modified by loss of cations which are ionically bound in the crystal structure. Thus, the chemistry (chemical bonding type) in this case is key to understanding the relative stability of silicate minerals under conditions of weathering.

6.4.4 Oxidation

Oxidation is a very important process in most silicate rocks. This is the result of electronic exchange between oxygen ion associations in solution or on crystal edges and elements within the mineral structure where divalent ions (essentially Fe) are changed to trivalent ions, which changes the overall electronic balance of the initial mineral structure that can have various effects (Faure, 1991). In cases of phyllosilicate structures (chlorite and biotite) there is a tendency for the structure to remain intact releasing interlayer ionically bonded ions or hydroxyl complexes in order to establish overall electronic balance.

In cases of non-phyllosilicate structures, the results are much more important in that the Si-O networks are ruptured by the change in oxidation state or iron. The oxygen ions are linked to hydrogen ions and the smaller silicon ions. The units formed can enter into solution or form a gel-like substance. One common effect is the formation of iron oxides as independent phases, but this is not always the case in that many basic rocks form ferric phyllosilicates upon alteration by oxidation. In certain cases (Banfield et al., 1995) oxidised manganese forms new smectite minerals.

There is a strong tendency as weathering intensity increases (from time and rainfall) for the iron oxides to form mineral grains and eventually dominate the clay assemblages of alterite and soil clay assemblages. Hence oxidation can form oxide minerals or ferric silicate clay minerals (see Velde and Meunier, 2008, chapter 4 for more details).

In fact, the mechanisms of reaction are the diffusion of hydrogen ions into the silicate structures which initiates ionic interdiffusion and hydration when the hydrogen ions interact with the oxygen ions. Recrystallisation leads to hydrated minerals, i.e. those containing OH units in the crystal structure. Oxidation involves electron transfer from valence-variable ions to the activated oxygens in the aqueous solution. The result is a change in charge on the valence-variable ions, which disturbs the electronic balances in the crystal. If significant element migration has occurred, the oxy-ions form oxide phases.

6.5 FORMATION OF NEW CLAYS

6.5.1 Crystal Growth

New clay minerals can be formed by the growth of crystals using dissolved materials from the altering solutions. Here the dissolution of minerals is the source of new minerals. In order for this system to function, the solutions containing the dissolved minerals must move sufficiently slowly for the growing crystals to adequately deplete the solutions to move ions to the growth interface from sites of mineral dissolution. As the crystals grow, they deplete the solutions of dissolved ions which must migrate to the site of crystal growth. Such phenomena have been identified in the early stages of alteration at the rock–water interface (Velde and Meunier, 2008, chapter 4). Crystals growing from dissolved material in solution can fill voids which were left by the interaction of hydrogen ions with a pre-existing mineral by diffusion–precipitation mechanisms.

At times the dissolution–growth process is relatively rapid where the growing crystal is subadjacent to the dissolving phase. The new mineral follows the dissolution front into the old mineral. This produces what is called pseudomorphism (see Putnis, 2014). Here the hydrogen-ion-added, and alkali-depleted, zone, is very thin, whereas in other cases it can be much thicker in size.

Dissolution of clay phases under conditions of high rainfall can induce transformation very near the old minerals such as cases where different forms of 1:1 aluminous minerals are converted one into another (halloysite, kaolinite changes) or eventually to the silica-free gibbsite (see Churchman and Lowe, 2012). In these cases, the changes occur throughout the altering material, which is more or less homogeneous alterite.

It has been noted that in some instances biological activity (lichens and other organisms) can produce limited amounts of clays by dissolution–crystallisation processes on rock surfaces (Arocena et al., 2003), but it is unlikely that these processes result in important zones of new minerals.

6.5.2 Mineral Growth from Amorphous Materials

In the amorphous material formed by ionic diffusion of hydrogen ions and elemental loss in the high-temperature phases in rocks, the disordered materials tend to form new minerals corresponding to the overall chemistry of the amorphous material. These minerals are formed by crystal growth mechanisms from the disordered material in the gel state, which forms oxyanion units that dissociate from the gel into the aqueous solution and can diffuse to sites of crystal growth. In appearance this gives a pseudomorphism process where new multicrystalline materials replace a pre-existing mineral. These diffusion–transfer gels lead to both 2:1 and 1:1 minerals forming with oxides.

These new clays form in response to the saturation of the aqueous solutions in the amorphous mass and in the solutions surrounding them and the chemistry of these solutions. The formation of clays from amorphous materials appears to be slow, and in some cases very slow such as in the alteration of volcanic glass to structured phyllosilicate minerals. Migration of other elements from outside the gel material can occur, which produces new mineral phases.

Oxides can form as new minerals from the dispersed oxidised material of the gel mass. Some elemental migration, over short distances, is assumed in order to form these concentrations, even though the oxidised material has a low solubility in aqueous solutions.

One obvious case of the alteration of amorphous material is in the interaction of surface water with volcanic glass. Overall it appears that 2:1 clays are slow to form and instead 1:2 allophane or imogolite (poorly ordered tectosilicates) form as alteration products. Upon alteration under conditions of lower rainfall, more silica-rich minerals such as halloysite can form (Churchman and Lowe, 2012). It is possible that the formation of 2:1 minerals is more difficult than the formation of 1:1 minerals from volcanic glasses.

6.5.3 Mineral Transformation

In cases where the pre-existing minerals in the rocks that are interacting with surface aqueous solutions have a structure similar to those of the stable minerals at the surface (phyllosilicates, oxides and hydroxides), the tendency to dissolve or disrupt the pre-existing structure is not great. These minerals (biotite, chlorite and muscovite) tend to retain their initial structure, though not necessarily their crystal size. However, the high-temperature mineral chemistry is no longer stable and the minerals change their chemistry to a certain degree, largely due to oxidation of Fe ions. This can be called mineral transformation. The tendency is for the high-temperature minerals to develop a lower charge on the interlayer surfaces and then to have interlayer ions that can be exchanged for others depending upon the chemical potentials of the elements in the solutions. The mineral transformation process is to a large extent found in the soil zone where biological processes drive the reactions; however, it can be important in the initial water–rock interactions and in the alterite zone. The apparent motor of mineral change is oxidation of divalent iron ions to trivalent forms within the structure. The change in charge induced on the overall phyllosilicate structure reduces the fixation capacity of the 2:1 layers, and interlayer ions such as K in micas, or hydroxyl complexes of Mg and Fe in chlorites are released. The process is usually irregular, leaving various residual charges on the 2:1 structures and inducing a capacity for ion exchange of various amounts. Frequently these minerals show the existence together of unaffected layers and those where the exchange capacity for hydrated ions is induced. This leads to mixed-layered minerals.

Diffusion of covalently bonded ions within the 2:1 structures is not frequent but can occur in certain circumstances, glauconite alteration for example (Courbe et al., 1981), and is suggested in changes in Si–Al ratios as chlorite alters to vermiculite-smectite (Murakami et al., 1996), and possibly was observed in laboratory experiments on montmorillonite (Birgersson et al., 2014). Biotite has been seen to become oxidised with a loss of iron atoms but little else by Fordham (1990). Other examples are summarised by Churchman and Lowe (2012).

The transformation of 2:1 primary minerals into other clay mineral types reflects the relative ease in changing the composition of the interlayer ion site ions of these 2:1 minerals to stable covalent clay mineral structural complexes by changing the overall charge on the structure, whereas it appears that 1:1 minerals must recrystallise and form new minerals with more difficulty. Mineral transformation, largely induced by oxidation of iron ions in phyllosilicate structures, occurs on quite another time scale compared to mineral dissolution. Here the basic mineral structure remains much the same, with the basic cation–oxygen network remaining largely intact, while minor changes occur in the elemental composition of the mineral which nonetheless effect relative important changes in the physical and chemical properties of the phases. These much more minor changes can be expected to occur more rapidly. An example can be observed in the transformation of illite-chlorite minerals from diagenetic or low-grade metamorphic rocks into smectite minerals in the soil zone.

The basic silicate–oxygen interconnected structure is essentially left intact leaving the 2:1 structural units in the minerals formed. Such a transformation is much more rapid than the dissolution–crystallisation process necessary to form new phyllosilicates from other silicate minerals. These reactions involve 2:1 clay mineral structures where the overall ionic charge balance is slightly changed by oxidation of some ions (generally Fe) in the structure. The change in overall ionic charge on the 2:1 structure, making it more positive, is such that interlayer cations leave the structure and enter the altering solution. The amount of change in charge is dependent upon the importance of oxidation and the relative amount of Fe ions in the original structure. The physical and especially chemical properties of the transformed minerals are such that they characterise the new alterite and soil clays.

Dissolution may precede reprecipitation, including neogenesis, but may also leave a different mineral phase as a residue. Gibbsite $Al(OH)_3$ often forms under conditions of strong leaching, which selectively removes Si from aluminosilicates at the expense of Al. There may be gradations within leached profiles, with more gibbsite at the surface than in lower, less strongly drained horizons (Huang et al., 2002) and similarly within landscapes, with gibbsite in the highlands and kaolinite in the lowlands (Norfleet et al., 1993, in a temperate climate; and also Herbillon et al., 1981, in the humid tropics).

As the common structural element of aluminosilicates that is generally mobilised most by leaching at the soil surface, Si can become reprecipitated lower in soil profiles, as quartz and also cristobalite, opaline silica, and as silica cements forming hard pans in some soil profiles.

The Al oxyhydroxide, boehmite AlOOH, occurs in laterite profiles and in tropical soils, generally mostly near their surfaces and is thought to have resulted from the dehydration of gibbsite, while corundum Al_2O_3, found in at the surface of lateritic profiles in Western Australia, is thought to have resulted from further dehydration occurring in bush fires (Anand and Gilkes, 1987).

Rare phosphate minerals in soils such as plumbogummite $PbAl_3(PO_4)_2(OH)_5{\cdot}H_2O$ and related varieties based on Al (variscite), Fe (strengite), Ca (crandallite) and Ba (gorceixite) probably derive from the alteration of apatite, fluorapatite or hydroxyapatite, or else from reactions between phosphate fertilisers and soil minerals (Churchman and Lowe, 2012).

Sulphides occur in sediments as a result of bacterial reduction of sulphates in seawater. Upon contact with air on drying these sediments, sulphides, which include pyrite FeS_2, the most common sulphide, but also mackinawite and pyrrhotite, both FeS and greigite Fe_3S_4 (Burton et al., 2009) can easily be oxidised to sulphates, including sulphuric acid, hence giving rise to troublesome acid sulphate soils. These soils, labelled the 'nastiest soils in the world' (Dent and Pons, 1995, p. 263) are generally coastal but may also occur in association with dryland salinity (Fitzpatrick et al., 1996). They contain a variety of hydroxyl Fe sulphate minerals, particularly jarosite $KFe_3(OH)_6(SO_4)_2$ or 'cat clay', and also schwertmannite $Fe_8O_8(OH)_{8-2x}(SO_4)_x nH_2O$. Hydrated sulphates of Na and Fe (sideronitrile and natrojarosite), Na and Al (tamarugite), Mg and Fe (copiapite) and Pb and Fe (plumbojarosite), as well as gypsum, halite, the Fe oxyhydroxide, akaganéite and elemental sulphur (Burton et al., 2009) can also form in these soils.

Soils often contain small residues of titanium and zirconium, as oxides and other compounds, e.g. zircon $ZrSiO_4$. These are highly stable and are useful as invariant internal standards to enable assessments of extents of weathering of other components of soils. By great contrast, some highly soluble minerals such as halite NaCl and other halides, various sulphates and carbonates, nitrates, borates, chromate and perchlorate salts among others are precipitated out in soils in extremely arid environments such as in Antarctic dry valleys and in deserts.

REFERENCES

Anand, R.R., and R.J. Gilkes. 1987. The association of maghemite and corundum in Darling Range laterites, Western Australia. *Soil Research* 25: 303–311.

Arocena, J.M., L.P. Zhu, and K. Hall. 2003. Mineral accumulations induced by biological activity on granitic rocks in Qinghai Plateau, China. *Earth Surface Processes and Landforms* 28: 1429–1437.

Banfield, J.F., G.G. Ferruzzi, W.H. Casey, and H.R. Westrich. 1995. HRTEM study comparing naturally and experimentally weathered pyroxenoids. *Geochimica et Cosmochimica Acta* 59: 19–31.

Birgersson, M., O. Karnland, P. Korkeakoski, O. Leupin, U. Mäder, P. Sellin, and P. Wersin. 2014. Montmorillonite stability under nearfield conditions. Nagra NTB Report 14–12.

Burton, E.D., R.T. Bush, L.A. Sullivan, R.K. Hocking, D.R.G. Mitchell, S.G. Johnston, R.W. Fitzpatrick, M. Raven, S. McClure, and L.Y. Jang. 2009. Iron-monosulfide oxidation in natural sediments: Resolving microbially mediated S transformations using XANES, electron microscopy, and selective extractions. *Environmental Science and Technology* 43: 3128–3134.

Churchman, G.J., and D.J. Lowe. 2012. Alteration, formation and occurrence of minerals in soils, p. 20.1–20.72. In: P.M. Huang, Y. Li, and M.E. Sumner (eds.), *Handbook of Soil Sciences: Properties and Processes*, 2nd edn. CRC Press, Boca Raton, Florida.

Courbe, C.B., B. Velde, and A. Meunier. 1981. Weathering of glauconites: A reversal of the glauconitisation process in a soil profile in western France. *Clay Minerals* 16: 231–243.

Dent, D.L., and L.J. Pons. 1995. A world perspective on acid sulphate soils. *Geoderma* 67: 263–276.

Faure, G. 1991. *Principles and Applications of Inorganic Geochemistry*. MacMillan, New York.

Fitzpatrick, R.W., E. Fritsch, and P.G. Self. 1996. Interpretation of soil features produced by ancient and modern processes in degraded landscapes. V. Development of saline sulfidic features in non-tidal seepage areas. *Geoderma* 69: 1–29.

Fordham, A.W. 1990. Weathering of biotite into dioctahedral clay minerals. *Clay Minerals* 25: 51–63.

Herbillon, A.J., R. Frankart, and L. Vielvoye. 1981. An occurrence of interstratified kaolinite-smectite minerals in a red-black soil toposequence. *Clay Minerals* 16: 195–201.

Huang, P.M., M.K. Wang, N. Kämpf, and D.G. Schulze. 2002. Aluminum hydroxides, p. 261–289. In: J.B. Dixon and D.G. Schulze (eds.), *Soil Mineralogy with Environmental Applications*. Soil Science Society of America, Madison, Wisconsin.

McSween, H.S., S.M. Richardson, and M.E. Uhle. 2003. *Geochemistry: Pathways and Processes*. Columbia University Press, New York.

Murakami, T., H. Isobe, T. Sato, and T. Ohnuki. 1996. Weathering of chlorite in quartz–chlorite schist: I. Mineralogical and chemical changes. *Clays and Clay Minerals* 44: 244–256.

Norfleet, M.L., A.D. Karathanasis, and B.R. Smith. 1993. Soil solution composition relative to mineral distribution in Blue Ridge Mountain soils. *Soil Science Society of America Journal* 57: 1375–1380.

Parfitt, R.L., M. Russell, and G.E. Orbell. 1983. Weathering sequence of soils from volcanic ash involving allophane and halloysite, New Zealand. *Geoderma* 29: 41–57.

Parfitt, R.L., M. Saigusa, and J.D. Cowie. 1984. Allophane and halloysite formation in a volcanic ash bed under differing moisture conditions. *Soil Science* 138: 360–364.

Putnis, A. 2014. Why mineral interfaces matter. *Science* 343: 1441–1442.

Velde, B., and A. Meunier. 2008. *The Origin of Clay Minerals in Soils and Weathered Rocks*. Springer-Verlag, Berlin.

7 Formation of Clays in the Soil Zone of Alteration

Death is an incident producing clay.

John Gilling (1912–1984)

Initially we will look at the formation of soil clays from the standpoint of development of alteration profiles. The initial stages of alteration are the interaction between altering fluids (rainfall in various stages of saturation) and altering minerals. This is the water–rock zone of reactions. Once altered to different extents the rock loses its strength and texture, and gradually becomes more and more clay-rich. This is the alterite zone. Alteration to contain more clay mineral content continues up to the plant–soil zone. The alteration involves loss of ionically bonded elements in the silicates, oxidation of iron and manganese ions and gradual but important losses from the initial silicate rock mineral assemblages. Loss of elements from the water–rock zone is mainly by movement of fluids containing dissolved species in the solutions. As we approach the upper soil zone, we find transportation of clays in the altering fluids along with the dissolved materials. Thus, the effects of water on rocks are not only those of mineral change but also of transportation of altered products from the surface. The following is a description of the formation processes and products in the lower portion of an alteration sequence, below the zone where soil clay minerals are found, and their reactions in the plant–soil regime.

7.1 CRYSTALLISATION FROM INCONGRUENT DISSOLUTION

If the aqueous solution is left in contact with silicate solids for an adequate period of time, then there is a selective dissolution where the less soluble elements are left in a modified structural state and the more soluble elements are taken into solution. This is accomplished by the diffusion of hydrogen ions into the structure, which ejects ionically bound ions such as K, Ca or Mg in order to maintain electrostatic equilibrium. Essentially the elements which are left in a solid state are those which have a strong covalent attraction for oxygen. The solid material remaining has either an initially amorphous state (see Velde and Meunier, 2008, chapter 4) or a newly formed crystalline state. Amorphous gels can often be seen in the initial stages of silicate rock weathering of igneous rocks. The alteration process is initiated by the diffusion of hydrogen ions into the pre-existing high-temperature silicate crystals susceptible to such chemical processes such as tectosilicates (feldspars).

As hydrogen diffuses into the crystal, other positively charged ions diffuse out of it. Thus, the anhydrous silicates gain hydrogen ions and lose cations to the solution. The less soluble ions such as Al and Si remain in a complex configuration in covalently bonded associations with the hydrogen ions and oxygen ions (Churchman and Jackson, 1976; May et al., 1986). This material loses its crystal structure. It can be dissociated into oxyanion units which enter solution or form a destructured mass of ions which is called a gel. Here the loss of larger soluble ionically bonded elements disrupts the crystal structure to a point where no diffusion of ions is possible internally to maintain a crystallographic structure. New minerals form by gradual reorganisation of the materials in the gels and the formation of new crystals or by movement of oxyanions to sites of crystal growth. This operation gives a new mineral structure that is found in the old volume of the original crystal, a sort of pseudomorphism.

7.2 CRYSTAL GROWTH FROM ELEMENTS IN SOLUTION

All minerals are slightly soluble, especially in unsaturated aqueous solutions, and transport of dissolved material to a new location in an altering rock is frequent either by movement of the water or by diffusion in the solution. As the minerals dissolve, the activity of the dissolving elements is initially high. Here new minerals can form on mineral substrates growing into voids formed by the alteration processes if water flow rate is low, such as along intergranular cracks (Figure 7.1). Formation of minerals along cracks in a rock or at crystal contacts is described in Velde and Meunier (2008). Water flow must be sufficiently low so that the diffusing elements released from dissolving phases can reach the zones of lower chemical potential where new crystals are forming, while retaining a state of relatively high concentration at the solution–growing crystal interface. Here the crystallisation of new phases decreases the elemental activity in solution, which calls for more dissolution of the dissolving phase. These are very local processes, millimetres in dimension, and the new minerals then depend upon the chemistry of the adjacent dissolving phases.

7.2.1 Neogenesis

The essence of neogenesis is the dissolution of primary minerals or of unstable initial alteration products of these minerals to yield their elemental components in solution which, given time, recombine and precipitate out of solution as new secondary minerals. All primary minerals are soluble in water that is usually acidified by carbon dioxide albeit that the dissolution is often incongruent, leaving a solid residue.

In order to form clay minerals in soils, silicates in rocks may be seen as undergoing hydrolysis in acid aqueous solutions (Carroll, 1970; Chamley, 1989; Churchman and Lowe, 2012). The active agent is H^+, which exchanges for cations in the structure of minerals. The ease of removal of cations from minerals by hydrolysis is related to the electrostatic valency of the structural cation by Pauling's electrostatic valency principle (Pauling, 1929; Churchman and Lowe, 2012). The electrostatic valency is given by the ratio of the valency of the cation to its coordination number, or the number of ions of opposite charge surrounding the cation. Hence potassium – in micas and some

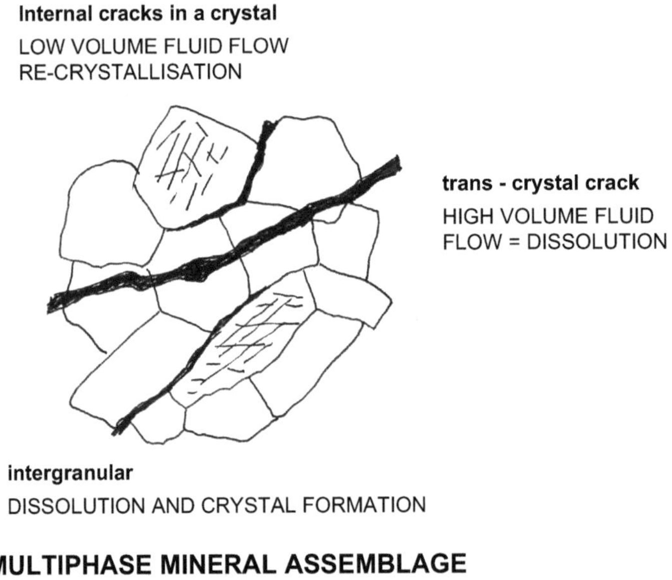

FIGURE 7.1 Illustration of different alteration sites in a rock at the surface. Small fractures between grains show little alteration, while large fissures show significant dissolution and clay phase growth.

feldspars – is more easily removed than aluminium by hydrolysis. Within groups with the same valency, ionic size has an effect, so that those with a larger size are more easily removed (Paton et al., 1995). However, organic compounds can also aid extent of dissolution, as can pH and temperature.

An important factor affecting extent of dissolution and also the course of recrystallisation that may occur is that of the nature of the microenvironment in which reactions and changes take place. Early alteration of rock minerals particularly takes place within cracks and fissures with a variety of sizes throughout the rocks and hence with a variety of extents of access to water flowing through the disintegrating rocks. The rate of access to this water dictates the chemical composition of the water following different stages of dissolution and hence the chemical potential that provides the driving force for further changes. Hence the different reactions and changes take place within a whole range of "plasmic microsystems", in Velde and Meunier's (2008) terms. In some of these, where gravity feeds water liberally down wide and open cracks and fissures, water availability and flow are no more limiting than they are above or outside the confines of the rocks. At the other extreme, in dead ends of closed cracks, water is barely available and dissolved ions and other species can build up to high concentrations. As a result, a myriad of products can form at the early stages of rock alteration.

Thus, for example, minerals generally considered as products of the later stages of weathering such as kaolinites and Fe-oxyhydroxides are found to have formed from biotite (Ahn and Peacor, 1987; Banfield and Eggleton, 1988; Velde and Meunier, 2008). Dissolution and recrystallisation has occurred in this particular microsystem. In other parts of the rocks, illites and trioctahedral vermiculites have formed, as commonly occurs by transformation reactions. Similarly, kaolinite and halloysite can form from muscovite in some microsystems (Singh and Gilkes, 1991; Velde and Meunier, 2008). For chlorites, while transformation commonly occurs to give first mixed-layer chlorite-vermiculite, then trioctahedral vermiculite per se, oxidation of Fe^{2+} and also leaching of Mg^{2+} can lead to a dioctahedral vermiculite, itself sometimes converted to an aluminium hydroxyl interlayered (Al–HI) mineral, accompanied by kaolinite and Fe-oxyhydroxides. These changes occur by dissolution–crystallisation (Velde and Meunier, 2008).

The restrictions often imposed by the geometry of pores within rocks and their effect on chemical environments for dissolution and recrystallisation may help explain why the course of formation of products from feldspars within disintegrating rocks can be vastly different from that seen in laboratory experiments on the dissolution of feldspars (e.g. Churchman and Lowe, 2012). The rate of alteration can also differ greatly between laboratory experiments and field observations, with the former generally much more rapid. One particular difference in products from the two approaches is that amorphous products appearing as gels in laboratory dissolution experiments are not found to be essential products of natural weathering and are best regarded only as metastable states (Velde and Meunier, 2008). The alteration of feldspars, including both K-feldspars and plagioclases, can lead to a variety of mineral assemblages, depending on the local physicochemical conditions in microsystems. Products can include kaolinite and/or halloysite, illite (from K-feldspar) or smectite with kaolinite, and also gibbsite with kaolinite and/or halloysite (Velde and Meunier, 2008). The rate of renewal of solutions in the microsystems helps govern the nature of the initial products (Velde and Meunier, 2008). The nature of the products formed from feldspars may also be strongly influenced by other reactive minerals that may occur in close proximity. The several observations of the formation of illite in feldspar-rich environments (Churchman and Lowe, 2012) provide common examples of this generalisation. When an illite that developed at a boundary between muscovite and orthoclase within weathering granite was found to contain more Fe and Mg than either the muscovite or orthoclase, it was concluded by Meunier and Velde (1976) that it was formed by neogenesis.

7.2.1.1 Thermodynamic Explanation of Stability of Minerals

The explanation of the stability of minerals from the composition of soil solutions has shown considerable promise (e.g. Kittrick, 1967, 1973). By this method, stability fields are established for possible minerals that are in actual or potential (given the passage of time) contact with the solution, in order to predict the equilibrium mineral composition of the solid phase from the equilibrium concentration of ions in the solution phase. Stability fields are established on axes that include a measure of Si in solution (typically the activity of H_4SiO_4) on one axis, with a function including the activity of, for example, Al^{3+}(log Al^{3+}+ 3pH) or K^+(log K^+ + pH) on the other (Kittrick, 1967, 1973; Churchman, 2000). In practice, there are a number of assumptions made and "experimental and conceptual difficulties" (May et al., 1986) which mark the approach and may limit its applicability to oxides and hydroxides, and only to minerals which are no more complex than kaolin minerals. These assumptions especially detract from its usefulness for assessing and predicting the stability of smectites (Churchman and Jackson, 1976; May et al., 1986; Churchman, 2000). May et al. (1986) found these to include (1) lack of proof of attainment of equilibrium; (2) inability of analytical techniques to measure dissolved species (and to distinguish these from fine colloidal material); (3) use of acidified dissolution media when determining thermodynamic data; (4) erroneous attribution of solution chemistry control to the original solid phase, i.e. when control may be by a secondary phase; and (5) inability to segregate and treat the contributions from exchangeable cations.

Other assumptions are also involved in the approach. They include the reversibility of the dissolutions often used to determine thermodynamic free-energy values. These dissolutions may well be irreversible (Helgeson, 1968). The assumptions do not account for the dynamic nature of the soil solution as moisture levels fluctuate on a seasonal, daily or hourly (often at least) basis due mainly to the fluctuating input of rainfall. The involvement of biological agents and reactive organic matter (e.g. see Section 7.3.10) also leads to difficulties with the purely inorganic approach of equilibrium thermodynamics. Furthermore, the narrowing down of the number of axes to just two, or, at best three, involves assumptions being made about the contributions of other ions to the relationship between solid and solution.

Another assumption that may render the approach unsuitable concerns the sampling of the "true" soil solution. The thermodynamic approach is established on the basis of a high solution: solid ratio when the situation in soils is likely to be the opposite, i.e. solids are mostly bathed in only small volumes of solution, and to different extents in different positions of the reacting solid (e.g., Velde and Meunier, 2008). The solution: solid ratio is different in pores of different sizes and on open faces. The extraction of solutions for analysis may be carried out by a number of methods, each of which has probable drawbacks as representative of an equilibrium solution. These are (1) suction, (2) displacement by a non-aqueous liquid, (3) centrifugation, (4) compaction, (5) saturation pastes, (6) 1:1 or 1:5 water extracts, (7) in lysimeters and (8) suction samplers (Percival, 1985; Churchman, 2000).

Nonetheless, there have been many contributions using thermodynamics – both equilibrium and irreversible types – for the description of the stability of minerals in solutions (e.g. Aja and Rosenberg, 1992a, 1992b, 1996, and Aja, 2018, claim its validity for illites and chlorites; also see Churchman, 2000, p. F-53 for more such contributions). Even so, Churchman (2000) concludes that the use of equilibrium thermodynamics as described for predicting mineral components in soils is fraught with many difficulties, as discussed. Churchman (2000) further concludes that kinetics should be considered alongside or instead of thermodynamics. Indeed, "if any system is left for long enough and if drainage is not seriously impaired, virtually all parent materials, including quartz can be dissolved" (Churchman, 2000, p. F-53), and it is likely that the kaolinite–gibbsite–iron oxide suite of minerals will dominate the products (Weaver, 1989).

7.2.2 Transformation of Minerals

Transformation involves minor modifications of 2:1 phyllosilicate structures where the largest part of the covalently bonded silicate structure is left intact. The initial phases are phyllosilicates from low-grade metamorphic rocks and diagenetically altered sediments, which form sedimentary rocks. For the most part the initiation of crystal transformation occurs with the oxidation of an Fe or Mn divalent ion in the structure that changes the charge balance on the 2:1 structure. Change in charge releases some ionically bonded ions from the interlayer position between the 2:1 layers. Depending upon the extent of oxidation or the amount of oxidisable ions present in the layers of the mineral structure, the charge on the structure is more or less decreased. This can create inhomogeneous minerals where different layers of the mineral have different charges, which can create an apparent mixed layering of mineral types such as mica–smectite or vermiculite–smectite. High-temperature 1:1 phyllosilicates such as serpentine minerals do not form new phases by transformation but by recrystallisation (Caillaud et al., 2006).

2:1 aluminosilicates are most easily altered by so-called transformation. The essence of this process is the alteration of the species which reside in the interlayer region of the minerals. These are mostly cationic to a greater or lesser extent but may also be net uncharged. These changes within the interlayers may be accompanied by changes within the aluminosilicate layers that are subtle but can lead to a change in the (negative) charge on these layers. However, the fundamental structure of the layers remains unchanged. Transformation essentially involves processes occurring within the interlayers of 2:1 aluminosilicate layers in contact with aqueous solutions. The first transformations that occur on weathering involve either or both primary micas or chlorites, which themselves are products of the early alteration of Fe- and Mg-containing primary minerals such as augite, hornblende, biotite and serpentines (Churchman and Lowe, 2012).

In micas, 2:1 aluminosilicate layers with a high negative charge (~0.9–1.0 per formula unit $O_{10}(OH)_2$) are separated and held together by potassium ions K^+, which fit snugly into the networks created by the oxygen atoms associated with silicon atoms in the tetrahedral sheets of adjacent layers. The size of the K^+ ions is ideal for these to occupy sites close to the layers to the exclusion of water and to ensure that the repeat distance between identical spots on adjacent layers is maintained at 10 Å or 1 nm. The potassium ions can be displaced by other cations in excess in the bathing solution, invariably leading to the entry into the interlayers of water associated with the displacing cations. In soils, these are most commonly divalent cations, usually Mg^{++} or Ca^{++}, although Na^+ may also displace K^+ in saline or sodic soils. Displacement of K^+ by divalent cations requires a lower concentration of the displacing cation than displacement by Na^+, but more molecules of water are associated with sodium ions than with either magnesium or calcium ions and their entry into the interlayer can result in more expansion of the clay mineral as a result. The extent of displacement that occurs depends to an extent on the negative charge on the aluminosilicate layers. This can change as a result of other reactions in the soil solution that may occur alongside those of the displacement of potassium ions from the interlayer (e.g., Churchman and Lowe, 2012).

In soils, it is rarely correct to refer to the 2:1 aluminosilicates with a 10 Å spacing for their repeat distance as micas, although this may be the case for soils in cold climates or where weathering is only at the earliest stages, such as underneath recently retreated glaciers. In these soils, weathering may be largely physical in nature, with comminution leading to clay-sized material formed from the physical breakdown of primary rock minerals. Otherwise, it is more common to refer to the fine-grained micaceous minerals with a main spacing in X-ray diffraction of about 10 Å as illite. Illites contain less potassium in their interlayers than primary micas and, although they inevitably contain more interlayer water as a result, they do not show substantial expansion (e.g., Fanning et al., 1989). It is likely that they have a lower layer charge than the primary micas, with the Clay Minerals Society Nomenclature Committee in 1984 suggesting a value of between 0.6 and 0.8 per formula unit (Fanning et al., 1989). However, as these authors point out, layer charges in this particular range (0.6 to 0.9) were considered by another nomenclature committee, that of the Association Internationale Pour L'Etude des Argiles (AIPEA) in 1980, to typify vermiculites. This raises the issue of the particular peculiar properties of soil clays, which is a recurring theme of this book, and also the question of whether there is a necessary link for 2:1 aluminosilicates between interlayer displacements, on the one hand, and changes in layer charge, on the other.

Artificial transformations of biotite micas to expanded 2:1 minerals, especially vermiculite, suggest that it is sufficient to simply displace K^+ ions from micaceous interlayers in order to bring about such a mineralogical change (Fanning et al., 1989). In these experiments, displacement of potassium has been performed through leaching with divalent ions, especially Ba^{++}, precipitation of interlayer K^+ by addition of sodium tetraphenyl boron, addition of molten salts or of organic cations and organic acids (Fanning et al., 1989). Even so, transformations that occur in soils are often accompanied by changes within layers. That of loss of interlayer K^+ is commonly associated with a reduction of layer charge. Mechanisms that are proposed for this change include the incorporation of protons into the layers, exchange of Si for Al in tetrahedral sheets, loss of hydroxyls and the deprotonation of hydroxyl groups, oxidation of Fe^{++} and ejection of Fe, with some of these occurring together (Fanning et al., 1989; Churchman and Lowe, 2012).

It is not surprising that protons play a key role in the release of interlayer K^+. Acidification is often an important effect of plant growth, with protons being released to balance the charge when roots take up an excess of cations over anions (Hinsinger, 2001; Courchesne, 2006; Calvaruso et al., 2009). For example, some tree species take up NH_4^+ in preference to NO_3^-, leading to acidification, especially in the rhizosphere (Calvaruso et al., 2009). In addition, roots, as well as the tips of fungal hyphae, exude organic acids into soils (Calvaruso et al., 2009). Acidification resulting from plant growth may also cause the dissolution of nanocrystalline forms of Al and Fe (Calvaruso et al., 2009) as well as other forms of Al and Si, including those often found in the interlayers of expandable 2:1 minerals (Pai et al., 2004). The reversal of the transformation of 2:1 minerals, in which illites form from expanded smectites, has also been shown, experimentally, to occur as a consequence of the bacterially mediated reduction of Fe(III) in the smectite (Kim et al., 2004).

Trioctahedral micas, including biotite and phlogopite, are altered more readily than dioctahedral micas, including muscovite and most illites. Dioctahedral vermiculites are more common as products of alteration than their trioctahedral counterparts. Even when vermiculitic phases are formed from biotite, these tend to be converted from their trioctahedral to their dioctahedral form as a consequence of a structural rearrangement involving the replacement of some octahedral cations by Al lost from tetrahedral sites (Douglas, 1989; Churchman and Lowe, 2012).

It has long been known that potassium ions can bring about the effective collapse of expanded clays such as vermiculites and smectites, with layer spacings of 14 Å or more, to give spacings that are more suggestive of illites, i.e. ~10 Å, although the collapse may be only partial to give either intermediate spacings or broad bands straddling the range from 10 to 14 Å. The collapse is often aided by heating, even to just ~100°C. The possibility of contraction upon addition of K^+ ions often forms a part of a diagnostic test for clay mineral species (e.g. Whitton and Churchman, 1987).

An apparent 'reversal' of the weathering transformation of micas and illites towards expanded, K-depleted phases has often been found to occur in situations where plants exert a strong influence on soils, generally through their roots. April and Keller (1990) and also Calvaruso et al. (2009) found that non-expanded micas and illites were concentrated in the rhizosphere zone of soils relative to their expanded counterparts, which were more common in the bulk soils. Prehistoric agriculture activity in chernozemic (rich) soils in Germany had apparently led to a distinct increase in illite in A horizons relative to their C horizons (Kleber et al., 2003; Velde and Barré, 2010). He et al. (2008) found that illite was more concentrated in A horizons than in lower horizons of soils from basalt in China. In a review of the literature, Barré et al. (2009) concluded that "the clay mineralogy observed in surface soils is probably largely plant mediated". In particular, the plants perform "nutrient uplift" in order to satisfy their growth requirements. In the instances cited here, this has occurred in order to ensure their adequate supply of potassium, which becomes stored in 2:1 mineral phases that become less expandable as a result. Silica may also be brought to the surface of soils by plants, where it is stored as opal in phytoliths, at the same time as potassium, as observed by Derry et al. (2005) and He et al. (2008). Many instances of silica being uplifted by plants in this way are cited by Churchman and Lowe (2012). Sometimes, the extra silica is incorporated into aluminosilicate clay minerals.

The common agricultural practice of adding fertilisers to soils can also affect soil clay minerals. Thus, in the Morrow experimental plots at the University of Illinois, continuous cropping with corn has led to a marked alteration of illite towards illite-smectite within ~40 years (Velde and Peck, 2002). There was no such change with rotational cropping over the same period. The authors also found that additions of NPK fertiliser after >40 years of continuous cropping with corn restored the clay minerals to their original state. Illite-smectite (I-S) played the role of a relatively quick-acting K buffer for plants in these soils, enabling its release when plants required the nutrient and storing it when there was an excess of K (Velde and Peck, 2002). Pernes-Debuyser et al. (2003) found that illite content increased at the expense of I-S in long-term treatments with potassium from either KCl or manure of soil without plants at Versailles in France. The long-term use of K fertilisers had also increased contents of illite, both as a discrete phase and also within mixed layers and the extent of illitisation reflected the rate of addition of the fertilisers (Simonsson et al., 2009). Illitisation occurred from fertiliser additions except where hydroxyaluminium interlayers were present, presumably blocking K^+ from entering the interlayers (Simonsson et al., 2009).

Hydroxy-interlayed (HI) layers forming chloritic phases, may also be returned to soil clays under suitable conditions. The introduction of the exotic tree species, sequoia, into the parks of French Chateaûx about 150 years ago has resulted in the formation of magnesian chlorite on soils on granite where it does not occur on associated soils under prairie grasses (Barré and Velde, 2011).

What is more, illitic layers in clays are also capable of taking up ammonium ions for their storage and later supply to plants. Although pure NH_4^+-clays occur only rarely in nature (Velde and Barré, 2010), ammonium ions can quite easily displace potassium ions, and, with a similar ionic size, are able to fit favourably in illitic interlayers (e.g. Velde and Barré, 2010). 2:1 aluminosilicates thereby can act as a reservoir for nitrogen, as well as potassium, for supplying plants.

2:1 aluminosilicates, including the range from K-rich illites to more or less K-depleted phases, mostly as mixed-layer minerals, are well suited to agriculture. They provide a reservoir for plants for essential potassium and can also store nitrogen as ammonium ions. It is no accident that productive agricultural soils in the Central United States are primarily composed of illite–smectites among clay minerals (Velde, 2001). Velde and Meunier (2008, p. 142) have stated "the richness of the earth for agricultural mankind resides in 2:1 minerals, illites, smectites and to a lesser extent hydroxy interlayered (HI) minerals".

Thus, there may be a symbiotic relationship between plants and 2:1 aluminosilicates. If not fully depleted of potassium ions and also, in some cases, of ammonium ions, they can supply these essential elements for the growth of plants. In return, plants may serve to replenish potassium – and also silica – from minerals deeper in soil profiles into their surface horizons. In this way, they act to reverse the trend in weathering towards depletion of clay interlayers, particularly of potassium. Even so, as Velde and Barré (2010, p. 217) stated, "The overall trend in weathering is to lose alkalis and silica, no matter how much plants work to overcome this inevitable fact. The soils will become impoverished in the long run".

7.3 EFFECT OF PLANTS ON SOIL CLAY ASSEMBLAGES

The presence of clay minerals in soils is largely driven by the chemistry engendered by plants and other life forms associated with them in the upper portions of an alteration sequence. Usually this zone is identified by the presence of decayed and decaying organic material and hence an enhanced carbon content. The amount of alteration and products of its chemical effect are determined by a combination of climate (water content of the upper zone) and the types of plants present, which often determines the pH of the soil solutions. These chemical parameters are enforced upon the altered material coming from rock types or sediments of various origins. The starting material (rock and sediment minerals) determines the framework of alteration to form soil clays, which is finalised by the chemical constraints imposed by the amount of water present (saturation with dissolved material) and the acidity, thus hydrogen ion content of the soil solutions. Thus, climate and plant regime are the major chemical forces in action in soil systems.

The mineral materials that interact in the plant–soil zone can be of rather different origins. Many are derived from weathering of rocks (rock–water interaction) below the surface horizon. However, many soils are formed on sediment deposits, either water- or airborne. In the case of sediments, the source material (usually dominated by phyllosilicates and oxides) is periodically renewed with the result that the end term of chemical interaction with atmosphere and aqueous solutions is not finalised. Deposits such as loess (airborne phyllosilicate and oxide dominant dust) can be renewed with such frequency that they are not greatly altered during their contact with plants in the soils zone. A case is that of Central China where loess deposits can occur at the rate of 2 mm/year and hence a layer of soil of 20 cm depth is only 10 years old. However, sediment material is usually relatively close to the mineralogical and chemical configuration that is stable in the soil zone, so that there is relatively little change in mineralogy unless there has been a change in climate conditions. Sediment clay minerals tend to be affected by transformation processes more than by dissolution and recrystallisation.

7.3.1 Transformation in Temperate Climates

If we consider the types of mineral change processes produced in temperate climates, transformation of 2:1 phyllosilicates is of great importance, whereas non-phyllosilicates are dissolved and recrystallised in the soil zone. More intense weathering increases the amounts of oxides produced as soil phases and the production of 1:1 minerals. Eventually only oxides and hydroxides are present in zones of intense weathering (i.e. high rainfall). Much of the chemical force is derived and moderated by plants in the soil zone.

7.3.2 Transformation of Pre-Existing Phyllosilicate Minerals of High Temperature Origin

The transformation series: muscovite + chlorite ⟶ vermiculite ⟶ smectite (De Coninck et al., 1987; Wilson, 1999) is classic in most books dealing with surface clay minerals. It is not clear how the reaction occurs, but it is most likely that a change in oxidation state of some of the iron in the chlorites and that in muscovite changes the effective layer charge and in doing so releases the interlayer ions of potassium and magnesium. The sequence is one of decreasing charge and hence more oxidation of the iron ions in the structure. In order to produce the end phase there is no need to dissolve the 2:1 mineral structure, and hence the phyllosilicate content of the soil is maintained to a large extent. The formation of final smectite mineralogy takes place in soils based upon alterites in different biological contexts, forest or prairies at different altitudes and hence average annual temperature ranges. The reaction chlorite ⟶ mixed layered ⟶ smectite (Arocena and Velde, 2009) observed in the laboratory due to mycorrhizal action from plant roots produces mixed-layered minerals (with chlorite, smectite and/or illite) and eventually the smectite minerals observed in soils. This occurs over a period of weeks under various laboratory conditions. The chemical agent is that of oxidation due to hydrogen migration into the structure and oxidation of some of the iron ions. Another similar reaction, biotite ⟶ mixed layered + smectite (Arocena and Velde, 2009), occurs in the same manner producing smectite–mica interlayered minerals after laboratory treatments by growing plants. The reaction chlorite + illite ⟶ mixed-layered phases + smectite + illite (Bain et al., 1993; Righi et al., 1999; Velde and Church, 1999; Egli et al., 2001, 2003) can be seen as a temperature sequence where lower temperatures require more time to effect the transformation reaction (see Figure 7.2). The reactions occur on altered materials and sediments.

In looking closely at the clay mineral transformations where 2:1 phyllosilicates are concerned, it is apparent that chlorite is the mineral most often affected by chemical interaction in the soil zone, as noted by Egli et al. (2001). Normally illite and chlorite are initially found together, and chlorite is replaced by interstratified minerals and smectite. Illite usually decreases in abundance but remains present as an independent phase. The formation of mixed-layered minerals is perhaps not unexpected in that low-temperature chlorites from diagenetic sedimentary rocks, for instance, are rather inhomogeneous from grain to grain such that oxidation will change the layer charge relations variably, which can produce layers of different surface charge that leads to differential interlayer ion occupation and hence interlayering of layer types (Velde and Medhioub, 1988; Velde et al., 1991).

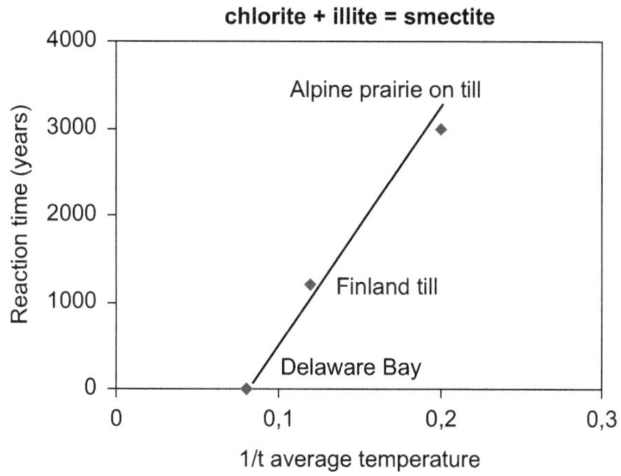

FIGURE 7.2 Time sequence of chlorite loss. Chlorite + illite = smectite observed reaction rates in various sites. (1/t average temperature °C) reaction time (years). Sites on till at Delaware Bay USA, Finland till, and Alpine prairie on till at various sites.

Mixed-layered minerals are typical of soil clay minerals, whereas in the initial stages of weathering (water–rock interactions) they are rare. It is therefore possible that a large portion of mixed-layered and smectite minerals in soils is due to the transformation of trioctahedral chlorites. X-ray diffraction spectra of inhomogeneous trioctahedral minerals quickly lose their reflections which identify their octahedral site occupancy and are therefore identified as being dioctahedral in many cases.

An interesting aspect of these transformation reactions is that they can be reversed when the plant regime changes. Barré et al. (2009) showed that the extraction of potassium from 2:1 soil clay layers is a very common occurrence but also that this element can be replaced under conditions during the cycle of plant growth. Velde and Church (1999) indicate that smaller but significant changes in illite content (potassium layers in the 2:1 clay fraction) can occur in salt marsh sediments over periods of several hundreds of years. Even more rapid is the change in clay mineralogy of smectite layers to form well-defined trioctahedral magnesium chlorite after a 14-year period under a sequoia tree (Velde and Barré, 2010). Adamo et al. (2015) showed potassium loss and recuperation in clays over a single growing season.

These 'mineralogical' relations tend to confirm the view that 2:1 soil clays are of the same structure and highly stable with differences in the interlayer site occupancy: either hydrous cations (vermiculite, smectite), Al hydroxyl ions material (hydroxyl interlayer clays), hydroxyl magnesium ions (chlorite) or anhydrous potassium occupancy and probably also NH_4^+ (illite). The difference is in the charge on the layer which attracts different interlayer ions and the availability of these ions in the soil solutions. There is probably only limited dissolution of materials, but it does occur with loss of the 2:1 mineral structures. The overriding tendency of absorption of silica into the soil solutions is counteracted by plants that bring up silica from water–rock interaction at depth and alterite zones to deposit it as phytolith material (amorphous silica nodules) in the soil zone. This counteracts the normal silica loss to some extent in the temperate regions but is often not sufficient in areas of higher rainfall, such as tropical climate zones.

7.3.3 Formation of Mixed-Layered 2:1 Clays

There are some cases where new 2:1 clay minerals are found in the soil horizon (Viennet et al., 2015). These clays, found in the very fine fraction of the soils, are often three-component interlayered 2:1 species which do not exist in the other parts of the soils. They form by crystallisation of dissolving elements in the soil when the soil solutions become saturated during times of low rainfall input.

An example of soil clay mineral transformation was found by Arocena and Velde (unpublished) in laboratory experiments where plants were inoculated with mycorrhizae in a biotite substrate soil. The biotite is eminently unstable at the surface due to the existence of large amounts of divalent iron ions in the structure. When there is interaction of the biotic assemblage and the biotites, the changes are quite pronounced according to plant type and the presence or not of mycorrhizae. In any event the biotite is changed chemically with the loss of a certain amount of the divalent trioctahedral iron ions when oxidised. As iron is oxidised there is a change in the layer charge on the 2:1 layers and the tendency to retain monovalent potassium ions decreases (see Figure 7.3). The mica is gradually transformed into a mixed-layered smectite-mica structure of various proportions. The amount and type of change is a function of the plant regime operating on the biotites. However, the end product is strongly enriched in mixed-layer 2:1 mineral phases created in the soil zone. One major chemical force is that of oxidation which operates not only on biotites but also on chlorite minerals and to a lesser extent on micas from high-temperature rocks. The diagrams presented show the effects on the micaceous (near 10 Å) phase where the peak widens and shows a second band due to very small grains; Figure 7.3 shows the redistribution of grain sizes during the experiments where the new, potassium poor flakes are much smaller than the initial biotite grains.

Another series of experiments in the same direction is seen in those summarised by Arocena and Velde (2009) where chlorite was the initial mineral subjected to biological chemical forces (mycorrhizae on crop plant roots). Here the same results are seen as the chlorite crystals are partially oxidised releasing interlayer ions from the 2:1 minerals which are Mg and Al hydroxide complexes. In the case of chlorites, the structures tend to remain with hydroxyl interlayered ions giving a near 14 Å spacing, but saturation with potassium produces mixed-layered chlorite-mica phases (Figure 7.4).

7.3.4 Formation of Palygorskite and Sepiolite in Soils

These fibrous minerals are ostensibly 2:1 in proportion, but the ratio is that of Si:Mg, rather than Si:Al as in smectites, illites and vermiculites. They often appear in the parent materials of soils that have developed in arid climates (Singer, 2002; Fraser et al., 2016) and debates rage about whether they may be formed in soils (Singer, 1989, 2002; Owliaie et al., 2006; Bouza et al., 2007; Kadir and Eren, 2008; Churchman and Lowe, 2012; Fraser et al., 2016). They almost invariably

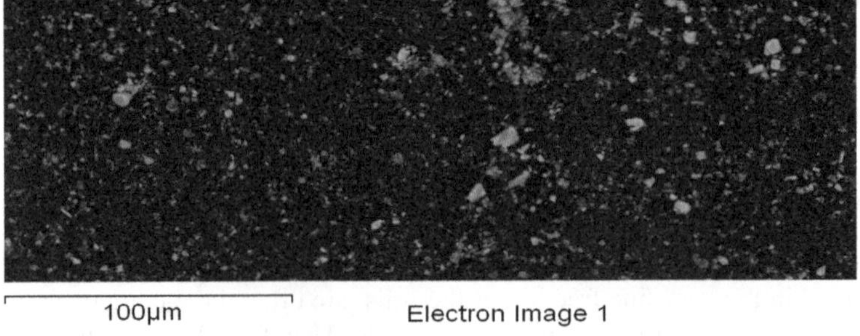

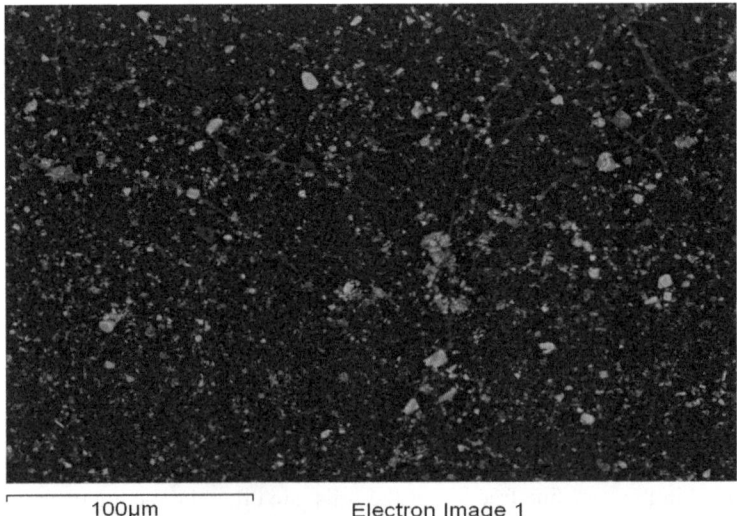

FIGURE 7.3 Biotite grains after the interaction with mycorrhizae. (Arocena and Velde, unpublished.)

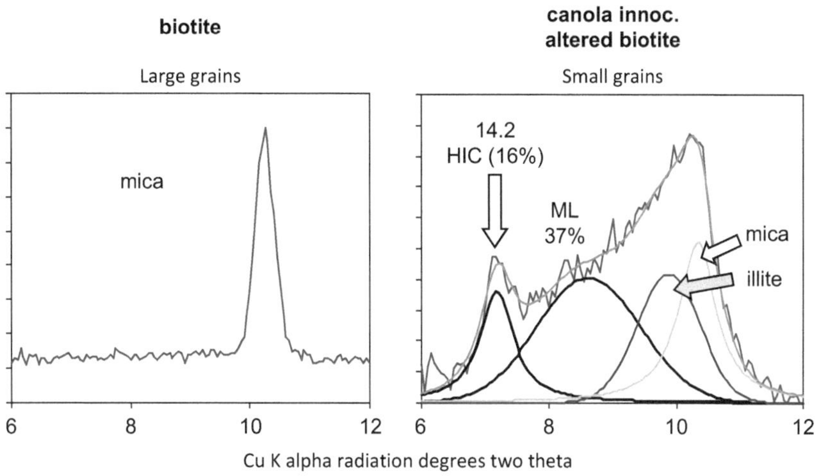

FIGURE 7.4 X-ray diffractograms of biotite alteration, mica. HIC, vermiculite-type interlayering; ML, interlayering between smectite and mica; Mica, biotite.

form in calcareous environments (Singer, 2002). It has been considered (Singer and Norrish, 1974; Singer, 1989, 2002) that palygorskites form in soils when there is (1) a fluctuating saline or alkaline groundwater; (2) strong, continuous evaporation; and (3) a sharp textural transition. Since calcretes (or caliches) provide such transitions in suitable soils, palygorskite is often found in or below these secondary carbonate transitions. Palygorskite has also been found in gypsiferous soils (Khormali and Abtahi, 2003; Owliaie et al., 2006; Churchman and Lowe, 2012). Sepiolite is rarer than palygorskite in soils and its occurrence there can be attributed to inheritance from parent materials rather than pedogenic processes (Fraser et al., 2016). Palygorskite and sepiolite may occur alongside smectites and may disappear in their favour (Paquet and Millot, 1972; Khormali and Abtahi, 2003; Owliaie et al., 2006; Churchman and Lowe, 2012). They are commonly found in subsoil horizons but rarely in topsoils. However, where plants are rare or absent because of poor conditions for their growth, such as a high pH and/or high salinity, both palygorsite and sepiolite can occur in surface soils (Fraser et al., 2016). Both have been shown to give way to smectites upon an increase in rainfall, while smectite led to palygorskite upon evaporation or an increase in salinity (Hillier and Pharande, 2008).

7.3.5 Formation and Transformation of 2:1 to 1:1 Mixed-Layer Clays

Interstratifications of both kaolinite and smectite and also halloysite and smectite each occur in some soils. The origin of each of them appears to be different.

7.3.5.1 Interstratifications of Kaolinite and Smectite and Their Evolution

Interstratifications of kaolinite and smectite, commonly known as kaolin-smectites, generally occur as intermediates in the transformation of smectites to kaolinite. These have been observed in soils across drainage sequences or in catena where smectites are found at the lowest level or foot of the sequences or catena, and kaolinites at the summit or top of these sequences or catena. Kaolin-smectites may then occur in intermediate positions between these two extremes in the sequence or catena (Herbillon et al., 1981; Bühmann and Grubb, 1991; Delvaux and Herbillon, 1995; Vingiani et al., 2004; Churchman and Lowe, 2012). They may also occur at intermediate depths or horizons, generally within deep profiles, in which smectites or more smectitic forms of kaolin-smectites occur at depth, with kaolinite or more kaolinitic kaolin-smectites occurring in the upper soil horizons (Churchman et al., 1994; Churchman and Lowe, 2012; Vingiani et al., 2004).

Two main hypotheses have been put forward to explain the evolution from smectites to kaolin-smectites, and from these to kaolinites (Churchman and Lowe, 2012). These may be regarded as a subtraction transformation or an addition crystallisation (Churchman and Lowe, 2012).

Ryan and Huertas (2009), who studied a chronosequence including this type of mineral, proposed that subtraction could occur by layer-by-layer transformation of the smectite crystal by means of a localised dissolution of a 2:1 layer, with a loss of Fe and Mg from the octahedral sheet and of Al from the tetrahedral sheet. These authors suggested that addition might occur by a deposition of hydroxy-Al species from solution onto an octahedral sheet to give a tetrahedral sheet conjoined to the octahedral sheet. Some authors (Wada and Kakuto, 1983, 1989) have favoured a subtraction mechanism, while others (Altschuler et al., 1963; Poncelot and Brindley, 1967) have preferred the addition alternative. Support for the subtraction alternative comes from an observation in a mineral deposit of opal having formed during the alteration of smectite to a kaolin mineral (halloysite) that occurred therein (Watanabe et al., 1992). The opal was presumed to have been lost as the 2:1 phase transformed to a 1:1 phase. A study by Vigniani et al. (2004) of the smectite being transformed showed that it comprised a more ferric type and a more aluminous type. As evolution took place to kaolinite, the ferrous type dissolved relative to the aluminous type, thence a subtraction mechanism prevailed.

7.3.5.2 Halloysite-Smectites

Interstratifications of halloysites with smectites, while forming under the same conditions as those giving rise to kaolin-smectites with kaolinite (e.g., Watanabe et al., 1992) often also occur under different conditions. Halloysite-smectites arise from volcanic ash under some restricted drainage, but under drainage that is not as restricted as that giving rise to smectites per se (Delvaux and Herbillon, 1995). They form part of a drainage sequence on volcanic ash (Churchman, 2006), as follows.

Smectite	Halloysite-smectite	Halloysite, 1:1 Allophane	2:1 Allophane	Gibbsite
Si:Al: 2	1-2	1	0.5	0

7.3.5.3 Interstratification of Kaolins with Other 2:1 Minerals

Kaolin minerals can also participate in interstratifications with other 2:1 minerals besides smectite. These include vermiculite and micas (Jaynes et al., 1989; Singh and Gilkes, 1992; Melo et al., 2001 and Kanket et al., 2005). A consequence of these interstratifications may be a rise in the potassium contents of the minerals over those of kaolinite. Some kaolinites in soils in Queensland, Australia, were shown by HRTEM to include smectite layers terminating the stacks of kaolinite layers, leading to higher cation exchange capacities (16–34 $cmol_{(+)}kg^{-1}$) than for kaolinite, although the effect on X-ray diffraction was undetectable (Ma and Eggleton, 1999). There may also be interstratifications of kaolinite with more than one other crystal, including those of illite and smectite (Hong et al., 2015).

7.3.6 Crystallisation of 1:1 Clays in Soils

Most often, kaolinites in soils have formed out of soil solutions by crystallisation *in situ*, i.e. neogenesis. Kaolinites in deposits have similarly formed by crystallisation out of solutions, but soil kaolinites are almost invariably much smaller and often also less well-ordered than those formed in deposits (see Figure 7.5). This is because soil-forming environments are heterogeneous in composition, containing Al, Fe and/or Mn and other elements in solution and their oxides, oxyhydroxides and hydroxides, as well as organic species in the solid or colloidal phase.

Results in Figure 7.6 of a study of clays formed in infill out of the saprolite from rocks (in this case, volcanic tuff) in Hong Kong show the effect of the inorganic impurities alone. There is a clear contrast between (1) the large kaolinite plates formed in the white infill, the colour of which indicates it is devoid of iron and manganese oxides (Figure 7.6A); and (2) the small plates formed in the

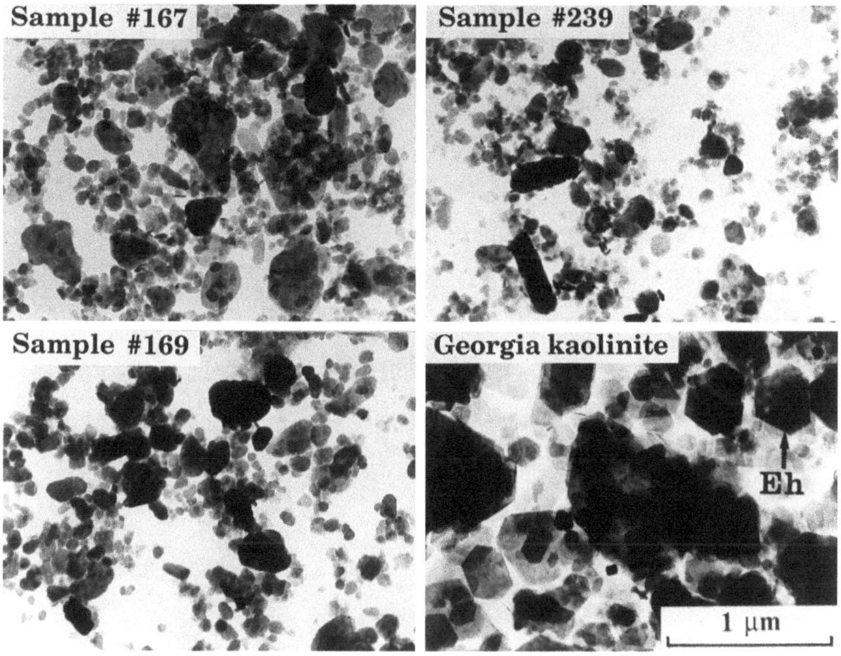

FIGURE 7.5 Transmission electron micrographs showing comparisons between some soil kaolinites from Western Australia and a reference kaolinite (from Georgia). Eh indicates a euhedral hexagonal shape for particles in the reference sample. (Micrographs reproduced from Singh, B., and R.J. Gilkes, 1992, *Journal of Soil Science* 43: 645–667. Thanks to Dr. Balwant Singh. With permission.)

reddish-brown infill, the colour of which indicates that it contains substantial iron oxides (Figure 7.6B). This effect of crystal growth being constrained by impurities in the solution in which growth occurs is similarly noted for halloysite formed in infill in granite saprolite also in Hong Kong (see Chapter 8, Figure 8.6). Once again, while the figures show only the effects of inorganic impurities such as iron, the effect of organic matter whether with (more likely) or without inorganic impurities is also likely to constrain crystal growth (see Figure 7.5). As with halloysite (Churchman et al., 2016), the constraining effects are likely to arise both from the inclusion of impurities, especially Fe, in the crystal structure and also the constraint on growth imposed by impurities, including Fe and also organic matter, in the crystal-forming environment.

Kaolinites are formed by neogenesis most commonly in highly weathered soils; they therefore occur in soils formed under warm humid climates. However, kaolinite is virtually ubiquitous in soils (Churchman and Lowe, 2012). Kaolinite – and also halloysite – in soils may not have formed *in situ* but was carried in by wind (and deposited by rain, alongside other small particles; Rex et al., 1969; Churchman et al., 1976; Collyer et al., 1984) or water. In this case the concentrations of these 1:1 minerals would be anomalously high in the topsoil horizons.

Halloysite can form from a variety of parent materials (Churchman and Lowe, 2012). It is most commonly formed in volcanic parent materials. Continuous moisture is required for halloysite to form (Churchman et al., 2016). Kaolinite occurs instead on drier sites (Churchman and Lowe, 2012). Halloysite is often found at depth in weathering profiles on a variety of parent materials and in a variety of climates, including tropical, temperate and Mediterranean or xeric (Churchman and Lowe, 2012). Some early workers proposed that halloysite was transformed from allophane, but recent work has suggested otherwise. Instead, allophane could give way to halloysite by dissolution and recrystallisation, consistent with their quite different crystal structures (Churchman and Lowe, 2012). On volcanic ash, drainage sequences, which often reflect rainfall sequences, have shown that allophane occurs with strong leaching, while halloysite occurs on less well-leached sites (Parfitt et al., 1983, 1984;

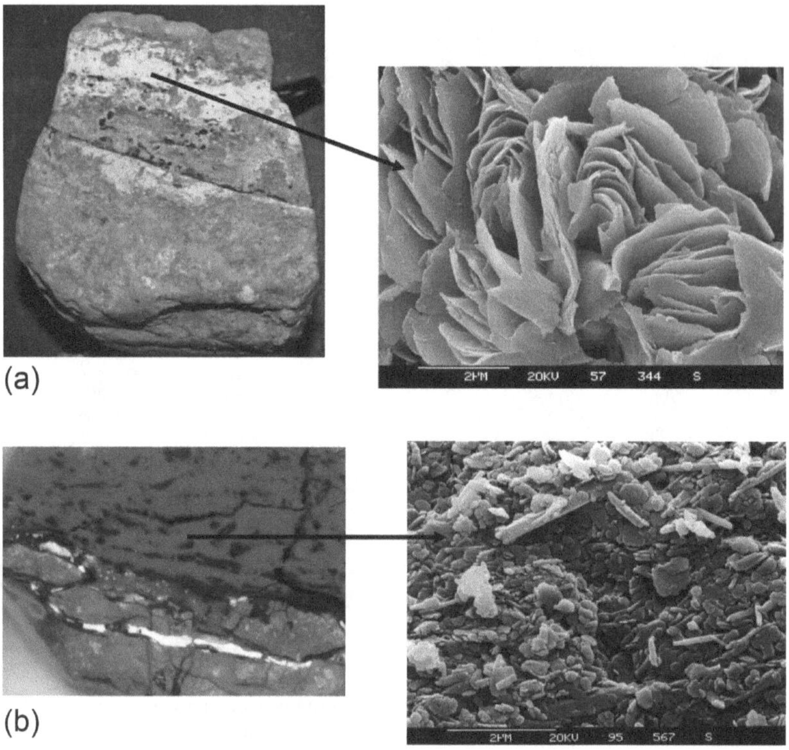

FIGURE 7.6 (See colour insert.) (A) Scanning electron micrograph (right) of kaolinite clay formed in the white infill in volcanic tuff (left), showing large plates. (With permission from Geotechnical Engineering Office, Civil Engineering and Development Department, Hong Kong.) (B) Scanning electron micrograph (right) of kaolinite clay formed in the reddish-brown infill in volcanic tuff (left), showing small plates (and some halloysite tubes). (With permission from Geotechnical Engineering Office, Civil Engineering and Development Department, Hong Kong.)

Parfitt, 1990, 2009). Halloysite appears to require a low pH at formation in order for interlayer water to be taken up and held within the structure (Churchman et al., 2016). At pH <~8, the octahedral sheet is positively charged and the tetrahedral sheet negatively charged, with opposing charges providing the force to attract and hold H_2O molecules in the interlayer region (Churchman et al., 2016).

7.3.7 Formation of Oxyhydroxide Al- and Fe-Dominated Soil Clay Assemblages

Iron oxides and oxyhydroxides (collectively 'iron oxides') are almost ubiquitous in soils. They are the products of the oxidation of ferrous iron in primary minerals and are effectively insoluble at the pHs of most soils (Churchman and Lowe, 2012). Among the iron oxides, goethite is almost universally found in soils, but is most common in cool and temperate climates and where organic matter is abundant. Hematite, by contrast, is common in soils in the tropics and subtropics that are aerated, and also in soils in arid and semi-arid climates as well as in Mediterranean or xeric climates. Goethite is favoured at low pHs; hematite at high pHs. Al substitution is common in both goethite and hematite, but goethite can accommodate more Al in its structure than hematite.

Poorly crystallised or nanocrystalline ferrihydrite is also widespread in soils and particularly abundant in soils from tephra and in illuvial or placic B horizons of podzols. It is formed by the rapid oxidation of Fe (II) and from solutions where both inorganic impurities and organic matter hinder the crystallisation of other Fe oxides. Ferrihydrite and hematite are seldom found together. Lepidocrocite forms when Fe(II) accumulates and there is a slow rate of oxidation. It therefore forms in soils which are seasonally anaerobic. Maghemite commonly occurs when other Fe oxides are heated in the presence of organic matter, hence in bush and forest fires. Magnetite may also give rise to maghemite on burning, but magnetite is usually present as a primary mineral, with its pedogenic formation rarely reported (Churchman and Lowe, 2012). Acid sulphate soils may give rise to schwertmannite, and green rusts can occur in highly hydromorphic soils.

Generally, iron oxides can be subject to quite rapid reduction and oxidation as redox conditions change. This process is governed by the activity of a wide variety of microorganisms (Robert and Berthelin, 1986; Schwertmann and Taylor, 1989; Robert and Chenu, 1992; Fortin and Langley, 2005). Iron oxides are distinguished by their close association with other soil components and particularly aluminosilicate minerals (Churchman and Lowe, 2012). Iron can also be closely associated with organic matter.

Among aluminium oxides, oxyhydroxides and hydroxides, the hydroxide gibbsite occurs most widely in soils. It forms when Si is absent or in short supply. Thus, it occurs under strong leaching. Typically, gibbsite occurs in topsoils at high altitudes, with kaolinite sometimes occurring in the corresponding subsoils (e.g., Norfleet et al., 1993, in temperate mountain soils in South Carolina; Herbillon et al., 1981, in the humid tropics; Huang et al., 2002, in some Oxisols; Kleber et al., 2007, in strongly weathered andesite in Costa Rica). In the topsoils in the latter soils, gibbsite occurs alongside halloysite, kaolinite, quartz, cristobalite and iron oxides. It is considered that a plant-based biological resilication has taken place to restore Si-rich phases to the surface alongside gibbsite in these soils (Kleber et al., 2007). Gibbsite may also be inherited from saprolites (Huang et al., 2002; Simas at al., 2006), in which case its content decreases towards the surface of soil profiles. Gibbsite occurs in volcaniclastic materials on weathering (Takahashi et al., 1993) alongside aluminosilicate minerals including imogolite, allophane and/or halloysite in a xeric climate, likely because gibbsite forms in the wet season with strong leaching, but the more Si-rich minerals form in the dry season.

Al oxyhydroxides and oxides form under especially strong weathering. Thus, the oxyhydroxide boehmite has been found in the surfaces of lateritic profiles (e.g. Churchman and Gilkes, 1989) and also in tropical soils (Huang et al., 2002). The Al oxide corundum occurs along with maghemite in some lateritic profiles, and it is likely that corundum formed from gibbsite and/or boehmite by dehydration in bush fires (Anand and Gilkes, 1987).

7.3.8 Formation of Other Oxides

Manganese oxides are widespread in soils but usually occur in small amounts. Reduced Mn (Mn(II)) is very soluble and mobile in soils, leading to Mn oxides becoming coated on rocks, ped surfaces and soil particles, as well as forming concretions, segregations, pans and nodules, often along with Fe oxides (McKenzie, 1989; Dixon and White, 2002; Churchman and Lowe, 2012). Common Mn oxide forms include birnessite, lithiophorite and todorokite.

The relatively easy leaching of Si from topsoils means that Si can be reprecipitated deeper in profiles as overgrowths on quartz and carbonate minerals and as quartz per se (Drees et al., 1989; Churchman and Lowe, 2012). Si cements occur as hardpans and duripans (Chadwick et al., 1987). In the extreme they form massive silcretes in acidic environments (Thiry and Simon-Coinçon, 1996). Secondary silica sometimes occurs as opal, either opal-CT or opal-A. The latter is biogenic and includes plant opals (phytoliths). Much of the secondary silica in soils derives from plant phytoliths (Farmer et al., 2005; Churchman and Lowe, 2012). Cristobalite has also been seen to derive from silica by leaching (Elsass et al., 2000).

Most forms of titanium oxides in soils are inherited from primary rocks. However, anatase and pseudorutile may form by neogenesis (Milnes and Fitzpatrick, 1989; Fitzpatrick and Chittleborough, 2002; Churchman and Lowe, 2012). Pseudorutilile mainly forms as an alteration product of ilmenite. Rutile and anatase can form, in turn, by the breakdown of pseudorutile (Grey and Reid, 1975; Fitzpatrick and Chittleborough, 2002). Some titanium can become incorporated in other minerals, particularly the Fe oxides, goethite and hematite.

Zircon, as the most common Zr-containing mineral, is usually inherited in soils from parent materials (Fitzpatrick and Chittleborough, 2002).

7.3.9 Formation of Other Compounds in Soils

Phosphates are quite rare in soils. They derive from the weathering of the forms of apatite, fluorapatite and hydroxyapatite, or from reactions between soil minerals and phosphate fertilisers (Lindsay et al., 1989; Harris, 2002; Churchman and Lowe, 2012). They are generally constituted as apatite where the Ca cation is replaced by other cations: Al in variscite; Al and Fe in barrandite; Fe(II) in vivianite; Fe in strengite; Ca and Al in crandallite; Pb, Al and others in plumbogummite; Ba and Al in gorceixite; and K and Fe in leucophosphite. An Al-hydroxy form, wavellite, occurs in soils from phosphoritic marine sediments (Harris, 2002).

Sulphide minerals in soils are the product of the bacterial reduction of sulphate in seawater (Doner and Lynn, 1989; Fanning et al., 2002; Churchman and Lowe, 2012). Pyrite is the most common of these minerals. The sulphides, in turn, can be oxidised to sulphates on exposure to air and drying. The most common of these are the hydroxy Fe sulphates jarosite and schwertmannite. Native gypsum (as distinct from added agricultural gypsum) occurs in soils by inheritance from gypsiferous parent materials. Barite, of unknown origin, occurs in some soils (Fanning et al., 2002).

Pyrophyllite and talc are rare in soils, where they tend to break down (Zelazny at al., 2002; Churchman and Lowe, 2012). Talc weathers to nontronite or to Fe oxides.

Zeolites may be present in small amounts (Boettinger and Ming, 2002). Apart from their occurrence by inheritance from parent rocks, some have formed pedogenically (Ming and Mumpton, 1989; Boettinger and Ming, 2002; Claridge and Campbell, 2008; Churchman and Lowe, 2012). Most often analcime, but also chabazite, mordenite and phillipsite, have been found in salt-affected alkaline soils containing sodium carbonate. In Antarctica, Claridge and Campbell (2008) found stilbite as a weathering product of dolorite; phillipsite, weathered from tephra; and chabazite, from thin saline intergranular films.

Soluble minerals found in soils include calcium carbonate, mainly as calcite, but also as aragonite and Mg calcite (Doner and Grossl, 2002; Fraser et al., 2016), and also magnesite and dolomite. Highly soluble halites, sulphates and some carbonates occur in saline soils and particularly in soils in arid environments such as Antarctica (Churchman and Lowe, 2012). The extremely arid Atacama Desert has given rise to nitrate, borate, chromate and perchlorite salts (Erickson, 1983; Doner and Grossl, 2002)

7.3.10 Biology and Its Effect on Clays and Clay Associations in Soils

Biology drives the direction of alteration in soils. As already discussed herein, the growth of plants has often been seen to bring about the loss of potassium from the interlayers of micas to give more or less expandable 2:1 minerals, including vermiculites and smectites, by the process of 'transformation' in the solid state (Churchman and Lowe, 2012).

In the first place, biota provides a fundamental reactant for weathering through the production of carbon dioxide. CO_2 then dissolves in water to give hydrogen ions:

$$CO_2 + H_2O = H^+ + HCO3^- \tag{7.1}$$

It is the action of the hydrogen ions in water that affects the breakdown of the feldspars, with subsequent precipitation of kaolinite. The process can be seen as dissolution followed by recrystallisation or neoformation.

$$CaAl_2Si_2O_8 + 2H^+ + H_2O = Al_2Si_2O_5(OH)_4 + Ca^{2+} \tag{7.2a}$$

$$2NaAlSi_3O_8 + 2H^+ + 9H_2O = Al_2Si_2O_5(OH)_4 + 2Na^+ + 4H_4SiO_4 \tag{7.2b}$$

Although not strictly in soils, crystalline minerals have also been observed as newly formed phases in sediments via microbial-mediated pathways. The Fe-rich smectite nontronite was found embedded in polysaccharides associated with microbial cells in deep-sea sediments, and its synthesis could be achieved by analogy in groundwater containing Si-bearing Fe hydroxides that had been seeded with polysaccharides (Ueshima and Tazaki, 2001). In another approach, incubation of freshwater sediments containing kaolinite and feldspar among more recalcitrant minerals led, over a period of months, to their dissolution and the recrystallisation (neoformation) of halloysite (Tazaki, 2005). Given sufficient time, kaolinite has also been synthesised through the mediation of bacteria from solutions containing Si and Al together with complexing organic ligands (Fiore et al., 2011). Biota can perform the synthesis of all major types of clay minerals, whether by transformation or neoformation (Konhauser and Riding, 2012). Their role may be to concentrate elemental mineral components to catalyse the precipitation of minerals from supersaturated solutions, hence passive. However, bacteria may otherwise mediate the formation of minerals to play a physiological or structural role, when mineral formation occurs despite the contacting bulk solutions being thermodynamically unfavourable for their formation and stability. The bacteria play an active role in this case.

Organic matter and minerals evolve together in soils, leading often to organo-mineral associations. Even so, minerals are not simply adsorbents for organic carbon (Kleber, 2010). They "may play an ecological function in relation to microbiota ... as rate controlling constituents of micro-bioreactors" (Kleber, 2010, p. 1767). In other words, "they have a role in the respiratory functions of some ... microbial fauna" (Andrade et al., 2018, p. 97). In exchange, bacteria – and maybe also fungi (Wilson et al., 2008) – act to reduce (e.g., Hong et al., 2016; Andrade et al., 2018), oxidise (e.g., Minyard et al., 2011) and synthesise (e.g. Tazaki, 2005, 2006, 2013) many clay minerals, including by nucleation on the microorganisms (e.g. Minyard et al., 2011; Tazaki, 2013). For example, Fe(III) in illite has been seen to be reduced to help form illite-smectite (Hong et al., 2016); in kaolinite to help form kaolinite-smectite, illite-smectite and then (Fe-rich) illite (Andrade et al., 2018); and in nontronite to help form illite (Jaisi et al., 2011). Oxidation of Fe(II) in illite has been observed to help form smectite and, ultimately, kaolinite (Zhao et al., 2017). Synthesis of halloysite has occurred biologically (Tazaki, 2005), as has that of kaolinite (Fiore et al., 2011), Fe/Mn oxides, imogolite and allophane (Tazaki, 2006, 2013) and nontronite (Ueshima and Tazaki, 2001; Tazaki, 2006). Close association between microorganisms and clay minerals in these interactions is often achieved through biofilms and microbial mats from extracellular polymeric substances secreted by microbial cells (Tazaki, 2006, 2013).

7.3.10.1 Soil Structure

Soil structure concerns the presence of microaggregates, often deriving from organo-mineral associations and the association of these microaggregates into larger aggregate units (macroaggregates), giving pores between and sometimes also within the aggregates. As wetting and drying cycles progress there is generally a tendency for silicate clay-rich soils to form larger aggregates with larger interaggregate porosites (Li and Velde, 2003). The macroaggregate-based porosity can be useful to plants, forming pathways for root growth; or destructive when too few pores are present forming large, empty pores (soils cracks) where root growth is not possible. Soil structure (the presence and distribution of aggregates) is an often-studied aspect of soil research. A study of the nature of soil science (Churchman, 2010) found that aggregates, along with horizons and characteristic soil colloids, comprised the unique objects of study for soil science as a discipline. Aggregates and soil structure are further discussed in Chapter 10.

REFERENCES

Adamo, O., P. Barré, C. Vincenza, D. Vincenzo, and B. Velde. 2015. Short term clay mineral release and re-capture of potassium in a *Zea mays* field experiment. *Geoderma* 264: 54–60.

Ahn, J.-H., and D.R. Peacor. 1987. Kaolinization of biotite: TEM data and implications for an alteration mechanism. *American Mineralogist* 72: 353–356.

Aja, S. 2018. The thermodynamic stability of clay minerals: A retrospective, p. 32. In: Program and Abstracts, 55th Annual Meeting, The Clay Minerals Society, University of Illinois at Urbana-Champaign, June 11–14, 2018. The Clay Minerals Society, Chantilly, Virginia.

Aja, S.U., and P.E. Rosenberg. 1992a. The thermodynamic status of compositionally-variable clay minerals: a discussion. *Clays and Clay Minerals* 40: 292–299.

Aja, S.U., and P.E. Rosenberg. 1992b. The thermodynamic status of compositionally-complex clay minerals: Discussion of clay mineral thermometry – A critical perspective. *Clays and Clay Minerals* 44: 560–568.

Aja, S.U., and P.E. Rosenberg. 1996. The thermodynamic status of compositionally-complex clay minerals: A discussion. *Clays and Clay Minerals* 44: 560–568.

Altschuler, Z.S., E.J. Dwornik, and H. Kramer. 1963. Transformation of montmorillonite to kaolinite during weathering. *Science* 141: 148–152.

Anand, R.R., and R.J. Gilkes. 1987. The association of maghemite and corundum in Darling Range laterites, Western Australia. *Soil Research* 25: 303–311.

Andrade, G.R.P., J. Cuadros, C.S.M. Partiti, R. Cohen, and P. Vidal-Torrado. 2018. Sequential mineral transformation from kaolinite to Fe-illite in Brazilian mangrove soils. *Geoderma* 309: 84–99.

April, R., and D. Keller. 1990. Mineralogy of the rhizosphere in forest soils of the eastern United States. *Biogeochemistry* 9: 1–18.

Arocena, J.M., and B. Velde. 2009. Transformation of chlorites by primary biological agents: A synthesis of X-ray diffraction studies. *Geomicrobiology Journal* 26: 382–388.

Bain, D.C., A. Mellor, M.S.E. Robertson-Rintoul, and S.T. Buckland. 1993. Variations in weathering processes and rates with time in a chronosequence of soils from Glen Freshie, Scotland. *Geoderma* 57: 275–293.

Banfield, J.F., and R.A. Eggleton. 1988. Transmission electron microscope study of biotite weathering. *Clays and Clay Minerals* 36: 47–60.

Barré, P., G. Berger, and B. Velde. 2009. How element translocation by plants may stabilize illitic clays in the surface of temperate soils. *Geoderma* 151: 22–30.

Barré, P., and B. Velde. 2011. Clays developed under *Sequoia Gigantia* and prairie soils: 150 years of soil-plant interaction in the parks of French châteaux. *Clays and Clay Minerals* 58: 803–812.

Boettinger, J.L., and D.W. Ming. 2002. Zeolites, p. 585–610. In: J.B. Dixon, and D.G. Schulze (eds.), *Soil Mineralogy With Environmental Applications*. Soil Science Society of America, Madison, Wisconsin.

Bouza, P.J., M. Simón, J. Aguilar, H. Del Valle, and M. Rostagno. 2007. Fibrous-clay mineral formation and soil evolution in Aridosols of northeastern Patagonia, Argentina. *Geoderma* 139: 38–50.

Bűhmann, C., and P.L.C. Grubb. 1991. A kaolin-smectite interstratification sequence from a red and black complex. *Clay Minerals* 26: 343–358.

Caillaud, J., D. Proust, and D. Righi. 2006. Weathering sequences of rock-forming minerals in serpentinite: Influences of microsystems on clay mineralogy. *Clays and Clay Minerals* 54: 87–100.

Calvaruso, C., L. Mareschal, M.-P. Turpault, and E. Leclerc. 2009. Rapid clay weathering in the rhizosphere of Norway spruce and oak in an acid forest ecosystem. *Soil Science Society of America Journal* 73: 331–338.

Carroll, D. 1970. *Rock Weathering. Monographs in Geoscience*. Plenum Press, New York.

Chadwick, O.A., D.M. Hendricks, and W.D. Nettleton. 1987. Silica in duric soils: I. A depositional model. *Soil Science Society of America Journal* 51: 975–982.

Chamley, H. 1989. *Clay Sedimentology*. Springer-Verlag, Berlin.

Churchman, G.J. 2000. The alteration and formation of soil minerals by weathering, p. F3–F76. In: M.E. Sumner (ed.), *Handbook of Soil Science*. CRC Press, Boca Raton, Florida.

Churchman, G.J. 2006. Soil phases: the inorganic solid phase, p. 23–44. In: G. Certini, and R. Scalenghe (eds.), *Soils: Basic Concepts and Future Challenges*. Cambridge University Press, New York.

Churchman, G.J. 2010. The philosophical status of soil science. *Geoderma* 157: 214–221.

Churchman, G.J., R.N. Clayton, K. Sridhar, and M.L. Jackson. 1976. Oxygen isotopic composition of aerosol-sized quartz in shales. *Journal of Geophysical Research* 81: 381–386.

Churchman, G.J., and R.J. Gilkes. 1989. Recognition of intermediates in the possible transformation of halloysite to kaolinite in weathering profiles. *Clay Minerals* 24: 579–590.

Churchman, G.J., and M.L. Jackson. 1976. Reaction of montmorillonite with acid aqueous solutions: Solute activity control by a secondary phase. *Geochimica et Cosmochimica Acta* 40: 1251–1259.

Churchman, G.J., and D.J. Lowe. 2012. Alteration, formation and occurrence of minerals in soils, p. 20.1–20.72. In: P.M. Huang, Y. Li, and M.E. Sumner (eds.), *Handbook of Soil Sciences: Properties and Processes*, 2nd edn. CRC Press/Taylor & Francis Group, Boca Raton, Florida.

Churchman, G.J., P. Pasbakhsh, D.J. Lowe, and B.K.G. Theng. 2016. Unique but diverse: Some observations on the formation, structure and morphology of halloysite. *Clay Minerals* 51: 395–416.

Churchman, G.J., P.G. Slade, P.G. Self, and L.J. Janik. 1994. Nature of interstratified kaolin-smectites in some Australian soils. *Soil Research* 32: 805–822.
Claridge, G.G.C., and I.B. Campbell. 2008. Zeolites in Antarctic soils: Examples from Coombs hills and marble point. *Geoderma* 144: 66–72.
Collyer, F.X., B.G. Barnes, G.J. Churchman, T.S. Clarkson, and J.T. Steiner. 1984. A trans-Tasman dust transport event. *Weather and Climate* 4: 42–46.
Courchesne, F. 2006. Factors of soil formation: Biota. As exemplified by case studies on the direct imprint of trees on trace metal concentrations, p. 165–179. In: G. Certini, and R. Scalenghe (eds.), *Soils: Basic Concepts and Future Challenges.* Cambridge University Press, New York.
De Coninck, F., W. Jensen, and C. De Kimpe. 1987. Evolution of silicates and especially phylosilicates during podsolization, p. 147–162. In: D. Righi, and A. Chauvel (eds.), *Podzols and Podzolization.* Association Français pour L' Etude du Sol, INRA, Versailles.
Delvaux, B., and A.J. Herbillon. 1995. Pathways of mixed-layer kaolin-smectite formation in soils, p. 457–461. In: G.J. Churchman, R.W. Fitzpatrick, and R.A. Eggleton (eds.), *Clays: Controlling the Environment.* Proceedings of the 10th International Clay Conference, Adelaide, Australia (1993). CSIRO Publishing, Melbourne.
Derry, L.A., A.C. Kurtz, K. Ziegler, and O.A. Chadwick. 2005. Biological control of terrestrial silica cycling and export fluxes to watersheds. *Nature* 433: 728–731.
Dixon, J.B., and G.N. White. 2002. Manganese oxides, p. 367–388. In: J.B. Dixon, and D.G. Schulze (eds.), *Soil Mineralogy with Environmental Applications.* Soil Science Society of America, Madison, Wisconsin.
Doner, H.E., and P.R. Grossl. 2002. Carbonates and evaporates, p. 199–228. In: J.B. Dixon, and D.G. Schulze (eds.), *Soil Mineralogy with Environmental Applications.* Soil Science Society of America, Madison, Wisconsin.
Doner, H.E., and W.C. Lynn. 1989. Carbonate, halide, sulfate, and sulfide minerals, p. 279–330. In: J.B. Dixon, and S.B. Weed (eds.), *Minerals in Soil Environments*, 2nd edn. Soil Science Society of America, Madison, Wisconsin.
Douglas, L.A. 1989.Vermiculites, p. 635–674. In: J.B. Dixon, and S.B. Weed (eds.), *Minerals in Soil Environments*, 2nd edn. Soil Science Society of America, Madison, Wisconsin.
Drees, L.R., L.P. Wilding, N.E. Smeck, and A.L. Senkayi. 1989. Silica in soils: Quartz and disordered silica polymorphs, p. 913–974. In: J.B. Dixon, and S.B. Weed (eds.), *Minerals in Soil Environments*, 2nd edn. Soil Science Society of America, Madison, Wisconsin.
Egli, M., A. Mirabella, and P. Fitze. 2001. Weathering and evolution of soils formed from granitic, glacial deposits: Results from chronosequences of Swiss alpine environments. *Catena* 45: 19–47.
Egli, M., A. Mirabella, A. Sartori, and P. Fitze. 2003. Weathering rates as a function of climate: Results from a climosequence of Val Genova (Trentino, Italian Alps). *Geoderma* 111: 99–121.
Elsass, F., D. Dubroeucq, and M. Thiry. 2000. Diagenesis of silica minerals from clay minerals in volcanic soils of Mexico. *Clay Minerals* 35: 477–489.
Erickson, G.E. 1983. The Chilean nitrate deposit. *American Science* 71: 366–374.
Fanning, D.S., V.Z. Keramidas, and M.A. El-Desoky. 1989. Micas, p. 551–634. In: J.B. Dixon, and S.B. Weed (eds.), *Minerals in Soil Environments*, 2nd edition. Soil Science Society of America, Madison, Wisconsin.
Fanning, D.S., M.C. Rabenhorst, S.N. Burch, K.R. Islam, and S.A. Tangren. 2002. Sulfides and sulfates, p. 229–260. In: J.B. Dixon, and D.G. Schulze (eds.), *Soil Mineralogy with Environmental Applications.* Soil Science Society of America, Madison, Wisconsin.
Farmer, V.C., E. Delbos, and J.D. Miller. 2005. The role of phytolith formation and dissolution in controlling concentrations of silica in soil solutions and streams. *Geoderma* 127: 71–79.
Fiore, S., S. Dumontet, F.J. Huertas, and V. Pasquale. 2011. Bacteria-induced crystallization of kaolinite. *Applied Clay Science* 53: 566–571.
Fitzpatrick, R.W., and D.J. Chittleborough. 2002. Titanium and zirconium minerals, p. 667–690. In: J.B. Dixon, and D.G. Schulze (eds.), *Soil Mineralogy with Environmental Applications.* Soil Science Society of America. Madison, Wisconsin.
Fortin, D., and S. Langley. 2005. Formation and occurrence of biogenic iron-rich minerals. *Earth-Science Reviews* 72: 1–19.
Fraser, M.B., G.J., Churchman, D.J. Chittleborough, and P. Rengasamy. 2016. Effect of plant growth on the occurrence and stability of palygorskite, sepiolite and saponite in salt-affected soils on limestone in South Australia. *Applied Clay Science* 124–125: 183–196.
Grey, I.E., and A.F. Reid. 1975. The structure of pseudorutile and its role in the natural alteration of ilmenite. *American Mineralogist* 60: 898–906.

Harris, W.G. 2002. Phosphate minerals, p. 637–665. In: J.B. Dixon, and D.G. Schulze (eds.), *Soil Mineralogy with Environmental Applications*. Soil Science Society of America, Madison, Wisconsin.

He, Y., D.C. Li, B. Velde, Y.F. Yang, C.M. Huang, Z.T. Gong, and G.L. Zhang. 2008. Clay minerals in a soil chronosequence derived from basalt on Hainan Island China. *Geoderma* 148: 206–212.

Helgeson, H.C. 1968. Evaluation of irreversible reactions in geochemical processes involving minerals and aqueous solutions - I. Thermodynamic relations. *Geochimica et Cosmochimica Acta* 32: 853–877.

Herbillon, A.J., R. Frankart, and L. Vielvoye. 1981. An occurrence of interstratified kaolinite-smectite minerals in a red-black soil toposequence. *Clay Minerals* 16: 195–201.

Hillier, S., and A.L. Pharande. 2008. Contemporary pedogenic formation of palygorskite in irrigation-induced saline-sodic, shrink-swell soils of Maharashta, India. *Clays and Clay Minerals* 56: 531–548.

Hinsinger, P. 2001. Bioavailability of soil inorganic P in the rhizosphere as affected by root-induced chemical changes: A review. *Plant and Soil* 237: 173–195.

Hong, H., F. Cheng, K. Yin, G.J. Churchman, and C. Wang. 2015. Three-component mixed-layer illite/smectite/kaolinite (I/S/K) minerals in hydromorphic soils, south China. *American Mineralogist* 100: 1883–1891.

Hong, H., Q. Fang, L. Cheng, C. Wang, and G.J. Churchman. 2016. Microorganism-induced weathering of clay minerals in a hydromorphic soil. *Geochimica et Cosmochimica Acta* 184: 272–288.

Huang , P.M., M.K. Wang, N. Kämpf, and D.G. Schulze. 2002. Aluminum hydroxides, p. 261–290. In: J.B. Dixon, and D.G. Schulze (eds.), *Soil Mineralogy with Environmental Applications*. Soil Science Society of America, Madison, Wisconsin.

Jaisi, D.P., D.D. Eberl, H. Dong, and J. Kim. 2011. The formation of illite from nontronite by mesophoilic and thermophilic bacterial reaction. *Clays and Clay Minerals* 55: 21–33.

Jaynes, W.F., J.M. Bigham, N.E. Smeck, and M.J. Shipitalo. 1989. Interstratified 1:1-2:1 mineral formation in a polygenetic soil from southern Ohio. *Soil Science Society of America Journal* 53: 1888–1894.

Kadir, S., and M. Eren. 2008. The occurrence and genesis of clay minerals associated with quaternary caliches in the Mersin area, southern Turkey. *Clays and Clay Minerals* 56: 244–258.

Kanket, W., A. Suddhiprakarn, I. Kheoruenromne, and R.J. Gilkes. 2005. Chemical and crystallographic properties of kaolin from Ultisols in Thailand. *Clays and Clay Minerals* 53: 478–489.

Khormali, F., and A. Abtahi. 2003. Origin and distribution of clay minerals in calcareous arid and semi-arid soils of Fars Province, southern Iran. *Clay Minerals* 38: 511–527.

Kim, J., H.L. Dong, J. Seabaugh, S.W. Newell, and D.D. Eberl. 2004. Role of microbes in the smectite-to-illite reaction. *Science* 303: 830–832.

Kittrick, J.A. 1967. Gibbsite-kaolinite equilibria. *Soil Science Society of America Proceedings* 31: 314–316.

Kittrick, J.A. 1973. Mica-derived vermiculites as unstable intermediates. *Clays and Clay Minerals* 21: 479–488.

Kleber, M. 2010. Minerals and carbon stabilization: towards a new perspective of mineral-organic interactions in soils, p. 1767–1769. In: R. Gilkes, and N. Prakonkep (eds.), Proceedings of the 19th World Congress of Soil Science, Brisbane, Australia, 1–6 August 2010. Curran Associates, Red Hook, New York.

Kleber, M., J. Röner, C. Chenu, B. Glaser, H. Knicker, and R. Jahn. 2003. Prehistoric alteration of soil properties in a central German chernozemic soil: In search of pedologic indicators for prehistoric activity. *Soil Science* 168: 292–306.

Kleber, M., L. Schwendenmann, E. Veldkamp, J. Röβner, and R. Jahn. 2007. Halloysite versus gibbsite: silicon cycling as a pedogenetic process in two lowland rain forest soils of La Selva, Coast Rica. *Geoderma* 138: 1–11.

Konhauser, K., and R. Riding. 2012. Bacterial biomineralization, p. 105–130. In: A.H. Knoll, D.E. Canfield, and K.O. Konhauser (eds.), *Fundamentals of Geobiology*. Wiley-Blackwell, Chichester, United Kingdom.

Li, D., and B. Velde. 2003. Aggregation in some clay-rich agricultural soils as seen in 2D image analysis. *Geoderma* 118: 191–120.

Lindsay, W.L., P.L.G. Vlek, and S.H. Chien. 1989. Phosphate minerals, p. 1089–1130. In: J.B. Dixon, and S.B. Weed (eds.), *Minerals in Soil Environments*, 2nd edn. Soil Science Society of America, Madison, Wisconsin.

Ma, C., and R.A. Eggleton. 1999. Surface layer types of kaolinite: A high-resolution transmission electron microscope study. *Clays and Clay Minerals* 47: 181–191.

May, H.M., D.G. Klnniburgh, P.A. Helmke, and M.L. Jackson. 1986. Aqueous dissolution, solubilities and thermodynamic stabilities of common aluminosilicate clay-minerals: Kaolinite and smectites. *Geochimica et Cosmochimica Acta* 50: 1667–1677.

McKenzie, R.M. 1989. Manganese oxides and hydroxides, p. 439–465. In: J.B. Dixon, and S.B. Weed (eds.), *Minerals in Soil Environments*, 2nd edn. Soil Science Society of America, Madison, Wisconsin.

Melo, V.F., B. Singh, C.E.G.R. Schaefer, R.F. Novais, and M.P.F. Fontes. 2001. Chemical and mineralogical properties of kaolinite-rich Brazilian soils. *Soil Science Society of America Journal* 65: 1324–1333.

Meunier, A., and B. Velde. 1976. Mineral reactions at grain contacts in early stages of granite weathering. *Clay Minerals* 11: 235–240.

Milnes, A.R., and R.W. Fitzpatrick. 1989. Titanium and zirconium minerals, p. 1131–1205. In: J.B. Dixon, and S.B. Weed (eds.), *Minerals in Soil Environments*, 2nd edn. Soil Science Society of America, Madison, Wisconsin.

Ming, D.W., and F.A. Mumpton. 1989. Zeolites in soils, p. 873–911. In: J.B. Dixon, and S.B. Weed (eds.), *Minerals in Soil Environments*, 2nd edn. Soil Science Society of America, Madison, Wisconsin.

Minyard, M.L., M.A. Bruns, C.E. Martínez, L.J. Liermann, H.L. Buss, and S.L. Brantley. 2011. Halloysite nanotubes and bacteria at the saprolite-bedrock interface, Rio Icacos watershed, Puerto Rico. *Soil Science Society of America Journal* 75: 348–356.

Norfleet, M.L., A.D. Karathanasis, and B.R. Smith. 1993. Soil solution composition relative to mineral distribution in Blue Ridge Mountain soils. *Soil Science Society of America Journal* 57: 1375–1380.

Owliaie, H.R., A. Abtahi, and R.J. Heck. 2006. Pedogenesis and clay mineralogical investigation of soils formed on gypsiferous and calcareous materials, on a transect, southwestern Iran. *Geoderma* 134: 62–81.

Pai, C.W., M.K. Wang, H.B. King, C.Y. Chiu, and J.-L. Hwong. 2004. Hydroxy-interlayered mineral of forest soils. *Geoderma* 123: 245–255.

Paquet, H., and G. Millot. 1972. Geochemical evolution of clay minerals in the weathered products in soils of Mediterranean climate, p. 199–206. In: J.M. Serratosa (ed.), *Proceedings of the International Clay Conference*, Madrid. Division de Ciencias C.S.I.C., Madrid.

Parfitt, R.L. 1990. Allophane in New Zealand – A review. *Soil Research* 28: 343–360.

Parfitt, R.L. 2009. Allophane and imogolite: Role in soil biogeochemical processes. *Clay Minerals* 44: 135–155.

Parfitt, R.L., M. Russell, and G.E. Orbell. 1983. Weathering sequence of soils from volcanic ash involving allophane and halloysite, New Zealand. *Geoderma* 29: 41–57.

Parfitt, R.L., M. Saigusa, and J.D. Cowie. 1984. Allophane and halloysite formation in a volcanic ash bed under differing moisture conditions. *Soil Science* 138: 360–364.

Paton, T.R., G.S. Humphreys, and P.B. Mitchell. 1995. *Soils: A New Global View.* UCL Press, London.

Pauling, L. 1929. The principles determining the structure of complex ionic crystals. *Journal of the American Chemical Society* 51: 1010–1026.

Percival, H.J. 1985. Soil solutions, minerals, and equilibria. New Zealand Soil Bureau Scientific Report 69. 21pp.

Pernes-Debuyser, A., M. Pernes, B. Velde, and D. Tessier. 2003. Soil mineralogy evolution in the INRA 42 plots experiment (Versailles, France). *Clays and Clay Minerals* 51: 577–584.

Poncelot, G.M., and G.W. Brindley. 1967. Experimental formation of kaolinite from montmorillonite at low temperatures. *American Mineralogist* 52: 1161–1173.

Rex, R.W., J.K. Syers, M.L. Jackson, and R.N. Clayton. 1969. Eolian origin of quartz in soils of Hawaiian Islands and in Pacific pelagic sediments. *Science* 163: 273–274.

Righi, D., K. Huber, and C. Keller. 1999. Clay formation and podzol development from postglacial moraines in Switzerland. *Clay Minerals* 34: 319–332.

Robert, M., and J. Berthelin. 1986. Role of biological and biochemical factors in soil mineral weathering, p. 453–495. In: P.M. Huang, and M. Schnitzer (eds.), *Interactions of Soil Minerals with Natural Organics and Microbes.* Soil Science Society of America Special Publication No. 17. Soil Science Society of America, Madison, Wisconsin.

Robert, M., and C. Chenu. 1992. Interactions between soil minerals and microorganisms, p. 307–404. In: G. Stotsky, and J.-M. Bollag (eds.), *Soil Biochemistry.* Marcel Dekker, New York.

Ryan, P.C., and F.J. Huertas. 2009. The temporal evolution of pedogenic Fe-smectite to Fe-kaolin via interstratified kaolin-smectite in a moist tropical soil chronosequence. *Geoderma* 151: 1–15.

Schwertmann, U., and R.M. Taylor. 1989. Iron oxides, p. 379–438. In: J.B. Dixon, and S.B. Weed (eds.), *Minerals in Soil Environments*, 2nd edn. Soil Science Society of America, Madison, Wisconsin.

Simas, F.N.B., C.E.G.R. Schaefer, V.F. Melo, M.B.B. Guerra, M. Saunders, and R.J. Gilkes. 2006. Clay-sized minerals in permafrost-affected soils (Cryosols) from King George Island, Antarctica. *Clays and Clay Minerals* 54: 721–736.

Simonsson, M., S. Hillier, and I. Öborn. 2009. Changes in clay minerals and potassium fixation capacity as a result of release and fixation of potassium in long-term field experiments. *Geoderma* 151: 109–120.
Singer, A. 1989. Palygorskite and sepiolite group minerals, p. 829–872. In: J.B. Dixon, and S.B. Weed (eds.), *Minerals in Soil Environments*, 2nd edn. Soil Science Society of America, Madison, Wisconsin.
Singer, A. 2002. Palygorskite and sepiolite, p. 555–584. In: J.B. Dixon, and D.G. Schulze (eds.), *Soil Mineralogy with Environmental Applications*. Soil Science Society of America, Madison, Wisconsin.
Singer, A., and K. Norrish. 1974. Pedogenic palygorskite occurrences in Australia. *American Mineralogist* 59: 508–517.
Singh, B., and R.J. Gilkes. 1991. A potassium-rich beidellite from a lateritic pallid zone in Western Australia. *Clay Minerals* 26: 233–244.
Singh, B., and R.J. Gilkes. 1992. Properties of soil kaolinites from south-western Australia. *Journal of Soil Science* 43: 645–667.
Takahashi, T., R. Dahlgren, and P. van Susteren. 1993. Clay mineralogy and chemistry of soils formed in volcanic materials in the xeric moisture regime of northern California. *Geoderma* 59: 131–150.
Tazaki, K. 2005. Microbial formation of a halloysite-like mineral. *Clays and Clay Minerals* 53: 224–233.
Tazaki, K. 2006. Clays, microorganisms and bioremediation, p. 477–497. In: F. Bergaya, B.K.G. Theng, and G. Lagaly (eds.), *Handbook of Clay Science*. Developments in Clay Science, vol. 1. Elsevier, Amsterdam.
Tazaki, K. 2013. Clays, microorganisms and bioremediation, p. 613–653. In: F. Bergaya, and G. Lagaly (eds.), *Handbook of Clay Science*, 2nd edn. Developments in Clay Science, vol. 5A. Elsevier, Amsterdam.
Thiry, M., and R. Simon-Coinçon. 1996. Tertiary paleoweatherings and silcretes in the southern Paris Basin. *Catena* 26: 1–26.
Ueshima, M., and K. Tazaki. 2001. Possible role of microbial polysaccharides in nontronite formation. *Clays and Clay Minerals* 49: 292–299.
Velde, B. 2001. Clay minerals in the agricultural surface soils in the Central United States. *Clay Minerals* 36: 277–294.
Velde, B., and P. Barré. 2010. *Soils, Plants and Clay Minerals: Mineral and Biologic Interactions*. Springer-Verlag, Berlin.
Velde, B., and T. Church. 1999. Rapid clay transformations in Delaware salt marshes. *Applied Geochemistry* 14: 559–568.
Velde, B., N. El Moutaouakkil, and A. Iijima. 1991. Compositional homogeneity in low-temperature chlorites. *Contributions to Mineralogy and Petrology* 107: 21–26.
Velde, B., and M. Medhioub. 1988. Approach to chemical equilibrium in diagenetic chlorites. *Contributions to Mineralogy and Petrology* 98: 122–127.
Velde, B., and A. Meunier. 2008. *The Origin of Clay Minerals in Soils and Weathered Rocks*. Springer-Verlag, Berlin.
Velde, B., and T. Peck. 2002. Clay mineral changes in the Morrow experimental plots, University of Illinois. *Clays and Clay Minerals* 50: 364–370.
Viennet, J.-C., F. Hubert, E. Ferrage, E. Tertre, A. Legout, and M.-P. Turpault. 2015. Investigation of clay mineralogy in a temperate acidic soil of a forest using X-ray diffraction profile modeling: Beyond the HIS and HIV description. *Geoderma* 241–242: 75–86.
Vingiani, S., O. Righi, S. Petit, and F. Terribile. 2004. Mixed-layer kaolinite-smectite minerals in a red-black soil sequence from basalt in Sardinia (Italy). *Clays and Clay Minerals* 52: 473–483.
Wada, K., and Y. Kakuto. 1983. Intergradient vermiculite-kaolin mineral in a Korean Ultisol. *Clays and Clay Minerals* 31: 183–190.
Wada, K., and Y. Kakuto. 1989. "Chloritized" vermiculite in a Korean Ultisol studied by ultramicrotomy and transmission electron microscopy. *Clays and Clay Minerals* 37: 263–268.
Watanabe, T., Y. Sawada, J.D. Russell, W.J. McHardy, and M.J. Wilson. 1992. The conversion of montmorillonite to interstratified halloysite-smectite by weathering in the Omi acid clay deposit, Japan. *Clay Minerals* 27: 159–173.
Weaver, C.E. 1989. *Clays, Muds and Shales*. Developments in Sedimentology 44. Elsevier, Amsterdam.
Whitton, J.S., and G.J. Churchman. 1987. Standard methods for mineral analysis of soil survey samples for characterisation and classification in N.Z. Soil Bureau. N.Z. Soil Bureau Scientific Report 79.
Wilson, M.J. 1999. The origin and formation of clay minerals in soils: Past, present and future perspectives. *Clay Minerals* 34: 7–25.
Wilson, M.J., G. Certini, C.D. Campbell, I.C. Anderson, and S. Hillier. 2008. Does the preferential microbial colonisation of ferromagnesian minerals affect mineral weathering in soil? *Naturwissenschaften* 95: 851–858.

Zelazny, L.W., P.J. Thomas, and C.L. Lawrence. 2002. Pyrophyllite-talc minerals, p. 415–430. In: J.B. Dixon, and D.G. Schulze (eds.), *Soil Mineralogy with Environmental Applications*. Soil Science Society of America, Madison, Wisconsin.

Zhao, L., H. Dong, R.E. Edelmann, Q. Zeng, and A. Agrawal. 2017. Coupling of Fe(II) oxidation in illite with nitrate reduction and its role in clay mineral transformation. *Geochimica et Cosmochimica Acta* 200: 353–366.

8 Nature and Origin of Surface Soil Clays

We glibly talk of nature's laws
but do things have a natural cause?
Black earth turned into yellow crocus
is undiluted hocus-pocus.

Piet Hein (1905–1996)

The properties and occurrence of secondary minerals found most often in soils are presented in Table 8.1 and described next.

8.1 ILLITES (2:1)

Illites in soils mainly originate from micas in the parent rocks. They are the products of partial transformation of these micas, which have lost some, but not all, of their interlayer K ions. If only a small proportion of the potassium ions are lost, they can be referred to as clay-sized micas. Strictly speaking, all other illites comprise interstratifications of mica and vermiculite. Since they are abundant in mica schists, which derive from shales and slates, and since these rock types comprise ~40%–60% of all sedimentary rocks (by volume) and ~80% of all crustal weathering products (Fairbridge and Bourgeois, 1978), illites are very common in soils. As illite-vermiculites (a more encompassing term), they have indeed been found to be widespread in surface soils (Robert et al., 1991; Velde, 2001). Further transformation, probably involving oxidation and possibly also ejection of iron, can lead to illite-smectites and smectites per se, most likely beidellitic (Churchman, 1980; Churchman and Lowe, 2012).

Illites may also form by neogenesis. Quite often they appear in close proximity to feldspars and their formation is regarded as a 'sericitisation' of the feldspars, with the product sometimes denoted as 'sericite'. In Australia, illite from a soil (from Willalooka) has been found to comprise very small, quite uniform particles, similar in appearance and size to an illite (from Muloorina) found to have formed in a seasonally dry lake. It is thought that both formed by neogenesis (Norrish and Pickering, 1983; Churchman and Lowe, 2012).

8.2 VERMICULITES (2:1)

Vermiculites in soils form predominantly by the transformation of micas and also of chlorite. Vermiculite in surface soils is nearly always dioctahedral. However, biotite (Fe-rich), phlogopite (Mg-rich) and chlorite generally give rise to trioctahedral vermiculites in the alterite zone of profiles, but these are altered to the dioctahedral form in the surface soil (Wilson, 2004; Churchman and Lowe, 2012). In soils with pHs between 4.6 and 5.8, vermiculites commonly attract hydroxy-Al ions into their interlayers to form a non-expandable species given a variety of names including soil or pedogenic chlorite, aluminous vermiculite, hydroxy-interlayered vermiculite (HIV), chloritised vermiculite, or a 2:1–2:2 intergrade (Barnhisel and Bertsch, 1989; Churchman and Lowe, 2012) (see also Section 8.4). There have been only a few reports of vermiculites forming by neogenesis (Churchman and Lowe, 2012).

TABLE 8.1
Secondary Minerals Formed in Soil Alteration Sequences

a. Products of Oxidation

Group	Mineral Type	Chemical Formula (Typical)	SSA (m^2g^{-1})	CEC (cmol (+)kg^{-1})	Distinguishing Features	Soils of Common Occurrence
Fe oxyhydroxides, oxides	Goethite	αFeOOH	14–77 (syn) 70–250 (soils)	pH-variable PZC 7-9	Yellow to yellow-brown colour	Widespread – most common iron 'oxide'
	Lepidocrocite	γFeOOH	Unknown, probably high	pH-variable PZC 5–7	Orange colour	Noncalcareous soils that are seasonally anaerobic
	Hematite	αFe_2O_3	35–45 (syn) 70–250 (soils)	pH-variable PZC 7–9	Bright red colour	Soils of warmer climates
	Maghemite (also titanomagnetite –related)	γFe_2O_3	pH-variable	Unknown, probably high	Ferrimagnetic	Tropical and subtropical soils, from fires, possible from bacteria
	Ferrihydrite	$Fe_5HO_8 \cdot 4H_2O$	200–500	pH-variable PZC[c] ~8	Nanocrystalline: spherical nanoparticles, pinkish- to yellowish-red colour	Widespread, where Fe(II) is oxidised rapidly
Mn oxide	Birnessite (also todorokite, hollandite, lithiophorite, pyrolusite)	$Na_{0.7}Ca_{0.3}Mn_7O_{14} \cdot 2.8H_2O$ (birnessite)	Unknown, probably high	pH-variable	(Blue) black colour	Widespread but usually minor component; in concretions, segregations, pans and nodules

b. Products of Transformations in the Solid State

Group	Mineral Type	Chemical Formula (Typical)	SSA (m^2g^{-1})	CEC (cmol (+)kg^{-1})	Distinguishing Features, Related Phases	Soils of Common Occurrence
2:1 Si:Al	Illite	$K_{0.6}(Ca,Na)_{0.1}Si_{3.4}Al_2Fe^{III}_{0.4}Mg_{0.2}O_{10}(OH)_2$	55–195	10-40	Blocky particles	Widespread, especially in weakly weathered soils
	Illite-vermiculite	Variable, depending on proportions of I, V	Unknown	Unknown	Often as regular, mica–vermiculite interstratification	Eluvial horizons of podzols

(Continued)

TABLE 8.1 (CONTINUED)
Secondary Minerals Formed in Soil Alteration Sequences

	Illite-smectite	Variable, depending on proportions of I, S	Unknown	Unknown	Either regular, mica-beidellite, or random, possibly with single layers	Regular: eluvial horizons of podzols Random: very widespread, often in agricultural soils
	Vermiculite (dioctahedral)	$K_{0.2}Ca_{0.1}Si_{3.2}Al_{0.8}(Al_{1.6}Fe_{0.2}Mg_{0.2})Al_{1.5}[OH]_4O_{10}(OH)_2$	Unknown	pH variable		Leached, mildly acid soils
	Vermiculite (trioctahedral)	$M^{II}_x(Mg,Fe)_3(Al_xSi_{4-x})O_{10}(OH)_2{\cdot}4H_2O$	50–150	100–210		Early stages of weathering, especially below soil zone
	Smectite-beidellite (also montmorillonite)	$M^{II}_{0.25}Si_{3.5}Al_{2.5}O_{10}(OH)_2$	~800 (total) 15–160 (interlayer)	45–160	Swelling	Acid-leached upper horizons, especially eluvial horizons of podzols. Often occur as ferribeidellites is soils, but heterogeneous
2:1–1:1	Kaolin-smectite	Variable, depending on proportions of K, S	Unknown	30–70	Either kaolinite–smectites, or halloysite–smectites; may also have kaolin–illites, kaolin–vermiculites,kaolin–illite–smectite	Intermediates in transitions between 2:1 and 1:1 minerals; often in moderately poorly drained soils
2:1–2:2 Si:Al (Mg, Fe)	Chlorite-vermiculite	Variable, depending on proportions of C, V	Unknown	Unknown		At very early stages of weathering
	Chlorite-smectite *(or chlorite swelling chlorite)*	Variable, depending on proportions of C, S	Unknown	Unknown		At very early stages of weathering
2:2 Si:Al (Mg, Fe)	Pedogenic chlorite *(or HIV, 2:1-2:2 intergrade, chloritised vermiculite)**	Variable, depends on whether layers are vermiculite or smectite and interlayered species	Unknown	Unknown		Intermediate pHs, ~4.6–5.8, with wetting and drying, low organic matter

(Continued)

TABLE 8.1 (CONTINUED)
Secondary Minerals Formed in Soil Alteration Sequences

c. Products of Neogenesis (including Precipitation and Evaporation)

Group	Mineral Type	Chemical Formula (Typical)	SSA (m^2g^{-1})	CEC (cmol(+) kg^{-1})	Distinguishing Features	Soils of Common Occurrence
2:1 Si:Al	Smectite-montmorillonite, (also nontronite, saponite, beidellite)	$M^{II}_{0.25}Si_4Al_{1.5}Mg_{0.5}O_{10}(OH)_2$	~800 (total) 15–160 (interlayer)	45–160	Swelling	Where drainage restricted and pH high
2:1–1:1 Si:Al	Halloysite-smectite	Variable, depending on proportions of H, S	Unknown	30–70		On volcanic materials with some restrictions to drainage (less than for smectites)
1:1 Si:Al	Kaolinite	$Al_2Si_2O_5(OH)_4$	6–40	0–8	Small euhedral particles, associated with Fe oxides	Widespread, especially in well-weathered soils
	Halloysite	$Al_2Si_2O_5(OH)_4{\cdot}2H_2O$	20–60	5–10	Generally small tubular, also spheroidal and lamellar	Where humid, especially from tephra
1:2–1:1 Si:Al	Imogolite	$SiO_2.Al_2O_3{\cdot}2H_2O$	1500	pH variable	Tubular	Mainly from pumice; some in podzol B horizons
	Allophane	Similar to imogolite but Si:Al 1:1–1:2	700–1500		Made up of many small spheres	From tephra, podzol B horizons, strongly leached soils
2:1 Si:Mg	Palygorskite	$Si_8Mg_5O_{20}(OH)_2(OH_2)_4{\cdot}4H_2O$	140–190	3–30	Fibrous	In dry, usually calcareous soils, at texture break

(Continued)

TABLE 8.1 (CONTINUED)
Secondary Minerals Formed in Soil Alteration Sequences

	Sepiolite	$Si_{12}Mg_8O_{30}(OH)_2(OH_2)_4 \cdot 8H_2O$	260–330	20–45	Fibrous	In dry, usually calcareous soils
Aluminium oxides	Gibbsite	$Al(OH)_3$	Unknown	pH variable	Hexagonal crystals	Where Si low, strong leaching
	Boehmite	$\alpha AlOOH$	Unknown, probably high	pH variable		Strongly weathered, ferricrete (laterite)
Manganese oxides	Birnessite, todokorite, lithiophorite, etc.	$Na_{0.7}Ca_{0.3}Mn_7O_{14} \cdot 2 \cdot 8H_2O$ (birnessite)	Unknown, probably high	pH variable	Blue/black colour	Widespread, usually minor
Sulphide	Pyrite	FeS_2	Unknown	Unknown		Coastal regions, and from unoxidised sediments
Sulphate	Gypsum,	$CaSO_4$	Unknown	Unknown		In desert soils
	jarosite	$KFe_3(OH)_6(SO_4)_2$	Unknown	Unknown		In acid sulphate soils, acid seeps
Phosphate	Plumbogummite	$PbAl_3(PO_4)_2(OH)_5 \cdot H_2O$	Unknown	Unknown		Rare, from breakdown of rock phosphates
Titanium oxide	Anatase, rutile	TiO_2	Unknown	Unknown		Widespread, in trace amounts
Chloride	Halite	$NaCl$	Unknown	Unknown		Seasonally dry saline sols
Zeolite	Analcime	$NaAlSi_2O_6 \cdot H_2O$	Unknown	Unknown		High pH saline soils

Notes: Most data from non-soil sources. SSA, specific surface area; CEC, cation exchange capacity; PZC, point of zero charge.
* May also be based on smectite layers, hence, e.g. HIS, chloritised smectite.

8.3 SMECTITES (2:1)

The conditions for the formation of smectites in soils by neogenesis are well established. They involve the occurrence in the originating solutions of high concentrations of Si and also basic cations such as Mg and a relatively high pH. There must also be a source of Al and, quite often in soils, some Fe also contributes to the composition of the neoformed smectite (Borchardt, 1989). These conditions are generally achieved where there is little leaching and impeded drainage. Bentonite deposits also form in bodies of water such as lakes and lagoons or embayments which have been supplied with the same basic ingredients by volcanic materials, either airborne or fluvial (Churchman and Lowe, 2012). Calcareous and also volcanic rocks often provide suitable environments for the formation of smectites, but they can originate by neogenesis from many rock types (Churchman, 2000; Churchman and Lowe, 2012). In particular soils, often named in soil classification schemes by a *vert-* prefix (e.g. Vertisols, Vertosols), neogenetic smectites dominate both the clay mineral composition and the soil behaviour, due to their high capacity for both cations and water. Hence these soils are subject to volume expansion, arising from their relatively high charge and extensive surfaces.

Many other soils contain some smectite, which can also arise by transformation and also often by inheritance from the soil-forming rocks (Wilson, 1999). Smectites are a common component of sedimentary rocks, either from earlier, eroded soils or as a result of marine diagenesis (Borchardt, 1989). Most smectites found in soils are dioctahedral. While beidellitic smectites are more common as a result of the transformation of micas and chlorites, they may also form by neogenesis (Borchardt, 1989). Montmorillonite can also form by the transformation of biotite. Nontronite and saponite occur rarely in soils, even if saponite may appear as an early weathering product of serpentinites; it gives way to a dioctahedral smectite within the soil profile (Churchman and Lowe, 2012). Smectites formed in soils, although heterogeneous in composition, tend to have more tetrahedral Al and more octahedral Fe than the montmorillonite usually found in deposits of bentonite (Wilson, 1987; Weaver, 1989). There is a tendency towards a ferribeidellite composition for smectites in soils (Churchman and Lowe, 2012).

8.4 MIXED-LAYERED 2:1 MINERALS

It has been well demonstrated in experiments in pots that plants can bring about the loss of interlayer potassium from micas, particularly biotite (Mortland et al., 1956; Hinsinger et al., 1992, 1993; Hinsinger and Jaillard, 1993) and mychorrhizal fungi are also effective in this regard (Arocena et al., 2012) (see also Chapter 7, Section 7.3.3 and Figure 7.4). These experiments have tended to show that the products of the displacement of potassium are rarely simple phases such as vermiculite. Instead, they are most often mixed-layer phases such as illite-vermiculite, and close examinations have revealed that the products of weathering of micas in soils very often comprise mixed layers.

In rare cases, the mixed-layered products of the alteration of micas – and of chlorites – in soil formation are regular in type, so that the product layer with altered interlayers occurs as alternate layers alongside the unaltered primary mineral. This is the case for weathering that has taken place by leaching in cool climates, so that the changes occur slowly. Regularly interstratified derivatives of micas, including mica-vermiculite and mica-smectite have been identified in upland Croatia, in Scotland and in Norway (Churchman, 2000) and also in South Island, New Zealand (Churchman, 1980, 2000). They tend to be based on mica, rather than illite, because the (nominally) unaltered component is also little altered from its primary mineral precursor. Regular interstratification comes about because replacement of the potassium ions in one interlayer by a hydrated cation leads to a strengthening of the bonds between the K^+ ions and the tetrahedral sheets of the two adjacent interlayers. These bonds are thought to strengthen because the orientation of the structural hydroxyl groups in the octahedral sheets shift towards the opened interlayer region (Norrish, 1973; Churchman and Lowe, 2012). A tendency to form regular interstratifications of chlorite and vermiculite at early stages of weathering of chlorites in cold climates has been analogously explained by the removal of interlayer oxyhydroxides of Mg and Fe leading to a layer shift of –a/3 occurring in the two adjacent layers that stabilise their remaining incorporated interlayer oxyhydroxides against easy removal (Banfield and Murakami, 1998; Churchman and Lowe, 2012)

The X-ray diffraction (XRD) patterns for regularly interstratified minerals are typified by a high-spacing [001] peak for the combined spacings of the constituent layers together with a [002] peak at half this spacing. Figure 8.1 shows patterns from Churchman (1978) for clay fractions of the several A horizons of a soil which is dominantly mica–smectite (spacing of 24–25 Å or 2.4–2.5 nm, expandable to 27–28 Å or 2.7 nm with glycerol) in the three upper A horizons and is dominantly mica–vermiculite (non-expandable spacing of 24 Å or 2.4 nm) in the lowest A horizon.

Typically, the leaching process that leads to the loss of potassium ions to give expanded layers, as occurs in the formation of these regularly interstratified phases, takes place at a low pH. At high altitudes where precipitation is high but the temperature is low, microbial breakdown of organic matter is slowed or halted (e.g. Theng et al., 1986) and pH can be low enough to bring about the alteration of the 2:1 layer into a smectite by eluviation, which concentrates Al in the structure, hence a beidellite. The formation of a smectite layer from first a micaceous layer, then a vermiculitic layer,

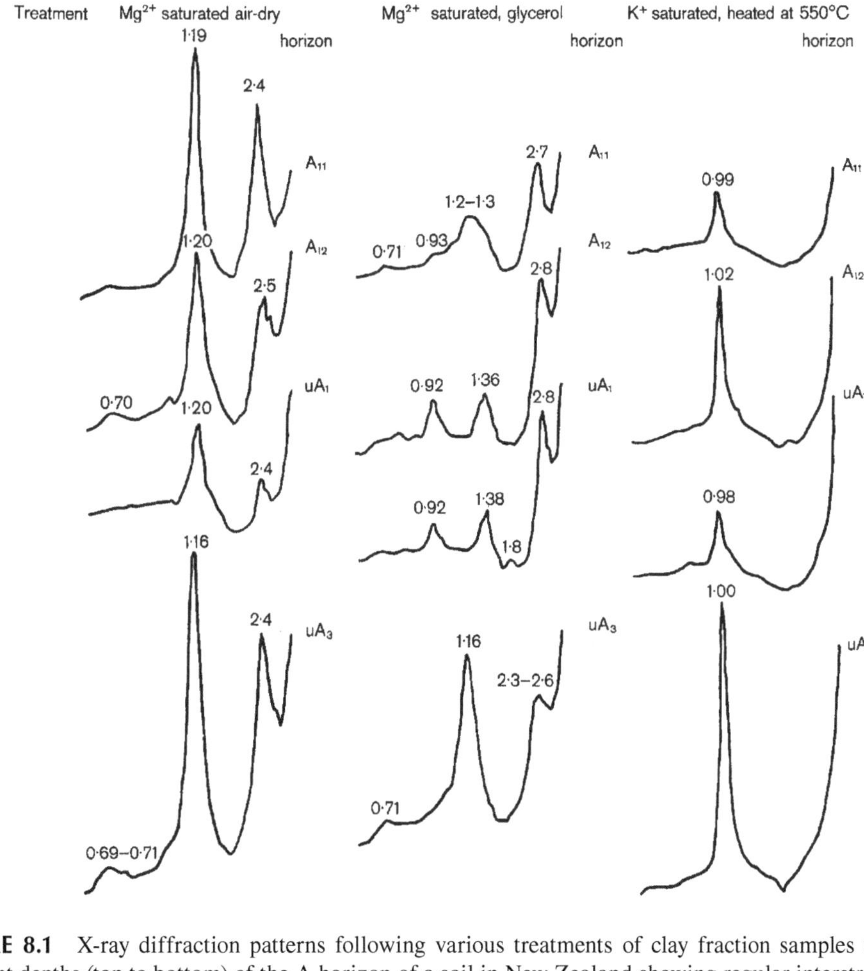

FIGURE 8.1 X-ray diffraction patterns following various treatments of clay fraction samples from four different depths (top to bottom) of the A horizon of a soil in New Zealand showing regular interstratification with mica–smectite dominating the top three horizons and mica–vermiculite, the lowest horizon. Peaks are indicated in nanometres (1 nm = 10 Å). (From Churchman, G.J., 1978, *New Zealand Journal of Science* 21: 467–480.)

involves a decrease in layer charge, with smectite layers having charges in the range of 0.2–0.6 per formula unit.

Regularly interstratified phases, which are essentially 1:1 combinations of their constituent layer types, enable easy quantification of changes that have taken place over monosequences in which the influence of one environmental factor is thought to dominate other factors. Churchman (2000) plotted the proportion of micaceous layers which are depleted to give expanded layers, in relation to mean annual precipitation, over a range of soils on parent rocks containing micas. There is some variation – from greywacke through semi-schist to schist and granite – in effects between rock types. Generally, the effect of rock type was found to be minor compared to that of precipitation. There is also some effect due to variations in type of vegetation, with more expanded layers tending to form under trees than under tussock grassland, the type of vegetation on most sites. Two chronosequences studied under a high precipitation, one under tussock grassland and one under trees, also show that a certain time is needed (500–1000 years under an extremely high precipitation of 11,000 mm y^{-1}) for micaceous layers to become depleted to the 'equilibrium' extent consistent with that expected from the apparently overriding influence of the level of precipitation at a given site.

Much more commonly, mixed-layer products of the transformation of micas and chlorites are quite random. They may not even come about by the replacement of all of the K^+ from complete layers, although this can occur. As well, it is possible for replacement of K^+ to occur at the edges of several layers simultaneously, giving a number of frayed, partially opened edges, while the cores of the layers remain clamped together by interlayer K^+ (e.g. Fanning et al., 1989). If pH is relatively low (4.5–5.5 range) and organic matter contents not too high, Al becomes mobilised as hydroxy cations. These are strongly attracted to the opened interlayers and, especially under alternate wetting and drying conditions that occur often in soils, whether seasonally or on different days, have a strong tendency to hold layers together at a spacing of about 14 Å. The phase formed this way, even if it forms only part of a crystal, is a pedogenic chlorite, or a hydroxy-interlayer (HI) mineral such as hydroxyl-interlayer vermiculite (HIV) or hydroxyl-interlayer smectite (HIS). In reality, 2:1 minerals in surface soils can 'toggle' between types, e.g. HI to illite, smectite to illite, HI to smectite or smectite to HI, depending in large part upon changes in the soil solution (Righi and Velde, unpublished).

HI minerals can also alter to 1:1 minerals. High-resolution transmission electron micrographs (HRTEM) in Figure 8.2 illustrate a range of mineralogical changes occurring within crystals of clay minerals in a hydromorphic soil in China (from Han et al., 2014). There is alteration of illite (10 Å lattice fringes) into HIV (12 Å) (Figure 8.2a), HIV into kaolinite (7 Å) (Figure 8.2b) and also illite directly into kaolinite without obvious intermediates (Figure 8.2c). In Figure 8.2a, illite incorporates hydroxy-Al cations to give HIV, here represented by 12 Å (rather than 14 Å) fringes probably as a result of heating in preparation for HRTEM. In Figure 8.2b, an HIV layer probably splits into two kaolinite layers, following the addition mechanism described in Section 8.5. The apparently direct alteration from illite to kaolinite is more curious. It may imply that a possible HIV intermediate has been rapidly degraded to kaolinite so that no intermediate stage (i.e. HIV) is shown or that dissolution/precipitation has occurred followed by a topotactic growth of kaolinite. The latter would be more credible if there were gaps in the images, as were noted for the alterations of biotite to kaolinite (Dong et al., 1998) and vermiculite to kaolinite (Aspiandiar and Eggleton, 2002), but they are not seen here. The observation of the close proximity of the layers for kaolinite and HIV in Figure 8.2c suggests that direct transformation of 2:1 minerals to 1:1 minerals is possible, illustrating the wide range of possible pathways for the alterations of soil clays. The overall effect on XRD patterns (Han et al., 2014) is of the occurrence of peaks from end-member illite, HIV and kaolinite minerals and also those from the interstratifications of I and HIV and probably also HIV and K. These latter, and any from I–K interstratifications, are not clear against overlaps of the end-member peaks.

Even a lack of formation of Al-hydroxy interlayers can suggest the acidification of soils. In forest soils in subtropical Taiwan (Lin et al., 2002; Pai et al., 2004) as well as in the cooler Polish mountains (Skiba, 2007), there was no interlayering of 2:1 Si:Al aluminosilicates in the surface

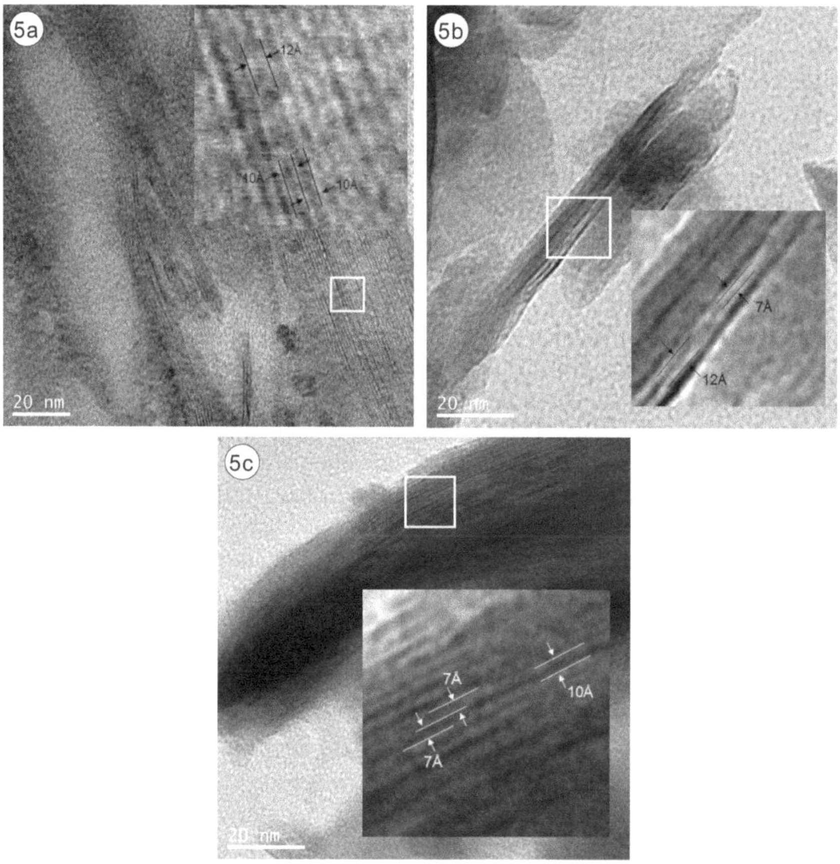

FIGURE 8.2 HRTEM images of clay minerals in a hydromorphic soil in China, indicating alterations occurring within crystals. (a) Lattice fringes of I-HIV, the inset is a locally magnified zone showing two illite layers altering into a HIV layer. (b) Lattice fringes of HIV-K. The inset is a locally magnified zone showing lateral transition of one HIV layer into kaolinite. (c) Lattice fringes of I-K. The inset is a locally magnified zone showing two layers of kaolinite terminating at a single layer of altered illite. (From Han, W., et al., 2014, *Clay Minerals* 49: 379–390. Thanks to Professor Hanlie Hong. With permission from The Mineralogical Society of Great Britain and Ireland.)

O, A and E horizons, but interlayering, which was shown to involve both Al and Fe, occurred in deeper horizons where pH rose high enough (>~4.4). Either or both a low pH or chelation of Al and Fe by abundant organic material could have precluded the formation of interlayers in the expandable aluminosilicates near the surface, but not in the lower horizons with higher pHs and lower contents of organic carbon. In a later study, Pai et al. (2007) found that acidification of the surface of other forest soils in Taiwan led to a decrease in the charges on the expanded layers in these soils, which occurred either as discrete vermiculite or in mixed layers with illite. In the North Island of New Zealand, Jongkind and Buurman (2006) found that weathering under kauri (*Agathis australis*) trees, which produce especially low pHs (4.0 $\pm$ 0.2) in the underlying soil, depleted the hydroxyl interlayers of vermiculites, and, through charge reduction, led to the conversion of vermiculite layers to those of smectite. Churchman (1980) had earlier found that vermiculite was converted to a smectite (beidellite) in soils under native beech (*Nothofagus* sp.) forest in the South Island of New Zealand. Several studies (including Mirabella and Egli, 2003) have identified trees as very effective agents of acid leaching in comparison with other plants, e.g. grasses.

The parent material can also influence the nature of the mixed-layered minerals formed. Andrade et al. (2019) found that soils formed on shales in a subtropical climate in southern Brazil varied in clay mineralogy according to whether the parent shales were Al-rich or Mg-rich, albeit that mixed-layered illitic minerals prevailed in each type of soil. Al-rich shales gave rise to kaolinite-illite phases, with increasing kaolinisation occurring towards the surface. By contrast, Mg-rich shales led to dominantly illite-smectite interstratifications, with a progressive dominance of smectitic layers occurring toward the surface. A subtle change in the composition of the parent material has led to a significant change in the nature of the secondary product.

Pronounced and prolonged acid leaching of soils often results in podzolisation. In this process, cations (of Al, Fe, Mg and Si) are mobilised and translocated as either organic or inorganic complexes, or both, from acidified surface horizons to deeper horizons at higher pH, where some metals (Al and Fe in particular) are deposited in high concentrations as oxides, oxyhydroxides, hydroxides and nanocrystalline aluminosilicates (especially allophane and imogolite) (e.g. Russell, 1973; Farmer et al., 1980; Churchman, 2000; Lundström et al., 2000; Churchman and Lowe, 2012).

In the upper horizons of podzolised soils, acid leaching typically depletes potassium, magnesium, aluminium and iron preferentially from the interlayers and often also from the layers of micas and chlorites. A smectite, as a discrete phase, but quite often also in mixed layers with illite, often results from the eluviation of the upper horizons of podzolised soils. The resulting smectite and other expanded 2:1 aluminosilicates can be heterogeneous in composition (Righi et al., 1999; Gillot et al., 2001; Mirabella and Egli, 2003; Egli et al., 2004; Churchman and Lowe, 2012). According to Gillot et al. (2001), the heterogeneity was the effect of different precursors leading to different types of smectites upon transformation: ferruginous beidellite from biotite and aluminous vermiculite from dioctahedral mica. In time, high-charge trioctahedral smectite dissolves and a progressive decrease in the charge of the dioctahedral expanded phases occurs. Righi et al. (1999) found the nature of the 2:1 aluminosilicate minerals resulting from podzolisation in the eluvial A and E horizons differed from those in their B horizons. The former contained a mixed-layered mica-smectite and, given time, also discrete smectite from the transformation of dioctahedral mica. By contrast, the B (illuvial) horizons contained mixed-layered mica–vermiculite and smectite from the transformation of trioctahedral minerals, notably biotite and chlorite. The nature of the acids involved as weathering agents in the two different parts of the soil profile explained the different products and processes. Organic acids, occurring only in the upper, eluvial horizons, are more aggressive than the carbonic acid in the lower horizons (Righi et al., 1999). The organic acids are responsible both for transformation of dioctahedral micas and also for dissolution of the products of weathering of the trioctahedral minerals. The products of their dissolution are translocated to the deeper Bh and Bs horizons where they accumulate as 'amorphous' (i.e. probably nanocrystalline) Fe and Al oxides; these latter are recrystallised as gibbsite in these particular soils.

A study conducted throughout Asia of minerals formed in soils from parent materials containing micas (Nakao et al., 2009) has found that the extent to which micaceous layers are 'opened up' by loss of interlayer K^+ relates to the extent of leaching – related to intensity of precipitation – that occurs, on the one hand, and the length of the dry season, on the other. The extent of opening up (or vermiculitisation) of layers was measured by the capacity of the product phase to incorporate (radioactive) Cs^+ into altered layers, known as the radiocesium interception potential (RIP). Where leaching was intense throughout the year, the RIP was high, reflecting a high rate of loss of K^+ from micaceous interlayers. Where leaching was intense for part of the year (the 'wet season'), but there was also a lengthy dry season, RIP was nonetheless low, because although much K^+ was removed from interlayers, this was largely replaced by hydroxyl interlayers, which resisted replacement by Cs ions. In another case, a lower intensity of leaching had brought about little replacement of interlayer K^+ and hence a low RIP value.

So-called primary chlorites, with interlayers rich in Mg and/or Fe, in the main, are often altered in an analogous way to micas, with the interlayer species being stripped to be replaced by hydroxy-Al interlayers or hydrated divalent cations (Velde and Barré, 2010; Churchman and Lowe, 2012). A trioctahedral mineral, Mg-Fe chlorite, typically gives way to a dioctahedral mineral, whether another 2:2 aluminosilicate, Al-chlorite or, quite often, a smectite (Herbillon and Makumbi, 1975;

Carnicelli et al., 1997; Velde and Barré, 2010) or a vermiculite (Loveland and Bullock, 1975; Murakami et al., 1996). The pH is an important factor affecting the particular product formed and, where strongly acid, as in a podzolised soil, chlorite can dissolve, typically leaving a residue of iron oxides and oxyhydroxides (e.g. Bain, 1977; Ross et al., 1982; Righi et al., 1993; Carnicelli et al., 1997).

Unlike for regular mixed-layer phases (Figure 8.4), XRD peaks for diagnostic basal plane peaks from their random counterparts tend to be poorly resolved. Furthermore, there are many different combinations of the basic layer types possible, both qualitatively and quantitatively. Illites and smectites, illites and either vermiculites, or, more probably, hydroxyl-interlayered phases of the latter and smectites may be combined in any combination between 1:0 and 0:1 in each pair. As well, multiphase combinations are possible. Each of these combinations, as well as the end-member minerals, micas and illites, with 10 Å spacings, at one extreme, and smectites, with either 14–15 Å in water or 17–18 Å spacings (in glycerol or ethylene glycol), at the other, give rise to diagnostic basal peaks with spacings over a very limited range. It is often the case that the soil clay of interest comprises a combination of unaltered mica or only partially altered illite as well as layers or crystals of these that have been altered to different extents, giving a set of mixed-layer combinations of (generally) illite and vermiculite, smectite and/or hydroxyl-interlayered phases, in addition to phases showing complete alteration within interlayers, or a vermiculite, as well as some that are altered within layers, hence smectite. The resulting melange of overlapping peaks for spacings between 10 and 14–18 is difficult to resolve by eye, let alone to enable quantitative estimates to be made of the individual contributing phases.

Recently, however, computer software (see Annex) has been written to enable decomposition of the complex XRD traces obtained from many soil clays representing mixtures of 2:1 aluminosilicates in various stages of transformation from the original micas (Lanson, 1997). This approach should also be applicable to the products of transformation of primary chlorites. The principles upon which peak decomposition can be employed for this purpose are summarised by Velde and Barré (2010) as follows:

The background radiation, which can vary with angle, is first subtracted.

1. A small peak widening at the base of a sharp peak is most likely due to the presence of another mineral peak (rather than to asymmetry due to physical effects on X-ray diffraction).
2. Use the minimum number of bands that can possibly simulate the complex spectrum.
3. Use only bands that could represent a known mineral.
4. Only peak widths that range from 0.2 to 2° 2 theta at half height are realistic; wider peaks are unrealistic.

Application of the peak decomposition approach has revealed that plants, in the course of extracting the essential element potassium from soils, can effect a change in the relative proportion of different 2:1 minerals in soils. This is shown over 83 years of continuous corn growth on the long-term unfertilised plots at the University of Illinois (Figure 4.4 in Velde and Barré, 2010), where the balance of the 2:1 minerals in the soil has shifted from 30% illite phases (well crystallised illite [WCI] + poorly crystallised Illite [PCI]) in 1913 to only 4% WCI and no PCI in 1996, while there have been increases in illite–smectite and smectite–illite phases over the same time period.

8.5 KAOLIN-SMECTITES INTERSTRATIFIED (1:1–2:1)

Kaolin-smectites, which are interstratifications between 1:1 Si:Al kaolin minerals, including both kaolinite and halloysite, on the one hand, and 2:1 Si:Al smectites, on the other, have been found worldwide in soils, including in North America, Central America, Africa, India, China, Japan, Australia and Europe (Churchman, 2000; Churchman and Lowe, 2012). Their occurrence has probably been often overlooked because a common effect of the interstratification on XRD patterns is to broaden and weaken the peaks from the dominant phase, whether the kaolin or smectite component. A major effect of their lack of identification is that their cation exchange capacity (CEC) and the expandability of the clay mineral may be underestimated, if they are misidentified as poor, or fine particle, kaolins (or as halloysite, in the case of a halloysite-smectite, Sakharov and Drits, 1973; Parfitt and Churchman, 1988). Their CECs are often much greater than expected from their XRD patterns, which give peaks near 7 Å (Figure 8.3). They could also be misidentified as poor, or fine particle, smectites. To our knowledge, only random interstratifications of kaolins and smectites (Figure 8.3) have been found.

Kaolin-smectites have been found in intermediate positions in several toposequences spanning from kaolinitic soils on summits and steep slopes (with good drainage) to smectitic soils in hollows where drainage is poor (Churchman and Lowe, 2012). These are often labelled 'red-black' sequences on account of their typical range of colours, and kaolin-smectite mixed-layer phases occur in soils between the red and black extremes. Kaolin-smectite interstratifications have also been found at intermediate depths within a number of profiles in which the clay mineralogy has evolved from more smectitic at depths to more kaolinitic towards the surface (Churchman et al., 1994). Kaolinites in a number of soils in Australia even showed a tendency to occur in stacks where the end-members of the stacks were revealed, on close examination with HRTEM, as single smectite layers, imparting elevated CECs to the overall combination (Ma and Eggleton, 1999).

Early commentaries on the mechanism of formation of kaolin–smectites (Churchman, 2000) suggested two different types of mechanisms, characterised by Churchman and Lowe (2012) as addition or subtraction (see also Chapter 7, Section 7.3.5.1). According to the former, smectite interlayers incorporate hydroxyl-Al species, which then form the basis of octahedral sheets for kaolin layers within the kaolin-smectite interstratified phase. This particular mechanism would require acid conditions and also inversion of some of the tetrahedral sheets in smectite layers to effect the synthesis of a new kaolin layer (e.g. Ryan and Huertas, 2009). The alternative, subtraction mechanism involves the dissolution or, more likely, partial dissolution (Dudek et al., 2006) of

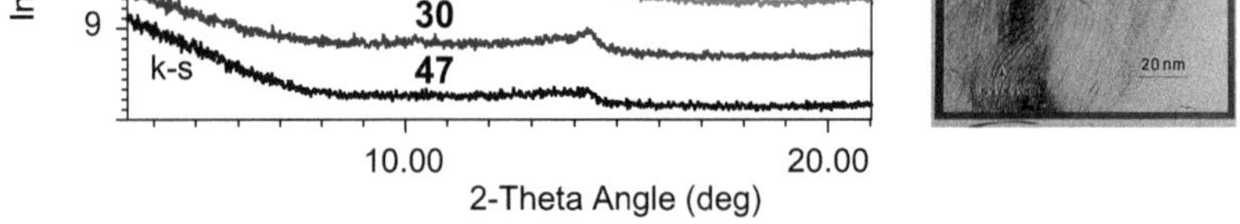

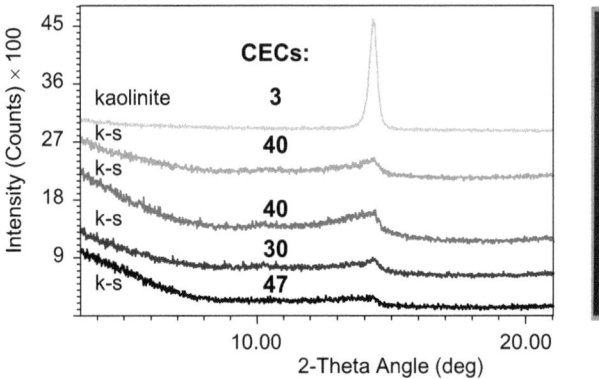

FIGURE 8.3 (Left) XRD patterns with CoKα radiation of oriented samples of a number of kaolin–smectites from different Australian soils compared with that of a standard non-soil kaolinite (top pattern). CECs are given in $cmol_{(+)}kg^{-1}$. (Right) TEM of a portion of a clay fraction from a soil with an interstratified kaolin–smectite (K-S) having 60% K-S with K:S 40:60 plus 40% discrete K (and a CEC of 70 $cmol_{(+)}kg^{-1}$, with an XRD patterns showing broad peaks near 18, 9 and 7.2 Å). Kaolinite layers are denser than smectite layers so the contrast in densities within crystals shows they are mixed together in varying sequences, e.g. KSSKSSK (lower left). Elsewhere in this sample the sequence was KSSKKSK, so the interstratification is random. Crystals of pure K are also evident. (See Churchman et al., 1994, for more details and examples). (From Churchman, G.J. et al.,1994, *Soil Research* 32: 805–822. With permission from CSIRO Publishing).

some tetrahedral sheets in smectites, leading to the alteration of some smectite layers to those of kaolins. There is considerable evidence favouring the more straightforward subtraction or desilication mechanism over the addition or 'alumination' suggestion. This includes the observation that opal formation accompanied the conversion of smectite (S) to kaolin–smectite (K-S) (Watanabe et al., 1992), and the selective loss of the more ferruginous smectite layers, with accompanying formation of Fe oxyhydroxides as a result of the S to K-S conversion (Ryan and Huertas, 2009). It may be that the process follows slightly different paths under different environmental conditions, such as those of climate and pH, and Ryan and Huertas (2009) put forward some evidence for part at least of the newly formed kaolin layers having a halloysitic character.

Other types of interstratified phases involving 1:1 and 2:1 aluminosilicate layers in mixed-layering have been described. They include kaolinite interstratified with vermiculite (Jaynes et al., 1989) and also with illite and HIV (Figure 8.2). In some soils in Australia (Singh and Gilkes, 1992a), Brazil (Melo et al., 2001; Andrade et al., 2019) and Thailand (Kanket et al., 2005), elevated potassium contents along with high CECs suggest that kaolins there are interstratified with mica/illite layers as well as expanded 2:1 layers. However, in view of the ease of transformation between different 2:1 micaceous phases, their expanded products (vermiculite and smectite) and interlayered phases, both primary chlorites and hydroxyl-interlayered vermiculites and smectites (see earlier), it is not surprising that there is a very wide variety in soils of minerals involving interstratifications of 1:1 and 2:1 aluminosilicates together and also of different 2:1 minerals with one another, such as chlorite–mica (Li et al., 2003; Velde and Meunier, 2008). Furthermore, calculations, profile fitting and peak decomposition of XRD patterns (Cradwick and Wilson, 1978; Sakharov et al., 1999; Hubert et al., 2012; Dumon et al., 2014) have indicated that interstratifications may involve more than two components in one sample. This is confirmed visually in Figure 8.4 by the occurrence side by side in a single crystal of layers of illite, smectite and kaolin, as shown by lattice fringes in HRTEM for 10, 17 and 7 Å in a sample from a hydromorphic soil in China (Hong et al., 2015).

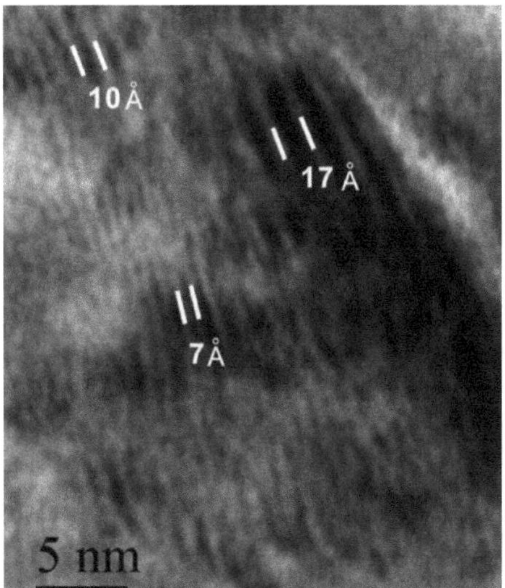

FIGURE 8.4 HRTEM showing interstratified 10, 17, and 7 Å layers in clay crystallites of a glycolated sample of a red earth hydromorphic soil (with Oxisol characteristics) from China. (From Hong, H., et al., 2015, *American Mineralogist* 100: 1883–1891. With thanks to Professor Hong Hanlie. With permission from the Mineralogical Society of America.)

8.6 KAOLINITE (1:1)

Kaolinite is the most ubiquitous secondary phyllosilicate in soils (White and Dixon, 2002; Churchman and Lowe, 2012). Kaolinites in soils tend to be highly disordered, giving many poorly resolved peaks in their XRD patterns (Hughes and Brown, 1979; Churchman and Lowe, 2012). The Hinckley Index, which is commonly used to assess the degree of crystallinity of kaolinites, generally cannot be used with soil kaolinites because of the poor resolution of the relevant peaks in XRD. Instead, Hughes and Brown (1979) devised an alternative, empirical index (which has been named after them) that is used to measure the relative degrees of crystallinity among kaolinites – and also halloysites – in soils.

Kaolinites in soils tend to be both smaller in particle size and also more irregular in shape than kaolinites from geological deposits such as the type kaolinites from either the state of Georgia in the United States or Cornwall in the United Kingdom. Figure 7.5 (Chapter 7) illustrates these differences in TEMs of some soil kaolinites in comparison with a type kaolinite from the commercial geological deposits of Georgia. As with halloysites (Section 8.7 and Figure 8.6), this tendency is likely to be boosted by high contents of impurities at the time of formation of the kaolinites. The Fe_2O_3 contents were 5%–15% in soils from New Zealand with >80% kaolinite in their clay fractions (Churchman, 2010, Table 1b), and there were notable amounts of goethite and hematite identified in soils from Australia that were similarly concentrated in kaolinite in their clay fractions (Churchman, 2010, Table 1a).

Kaolinites are usually the product of strong weathering and are therefore commonly formed in warm, humid climates and as a result of quite strong leaching. Kaolinite is the weathering product of many different types of minerals (Dixon, 1989). It can be found in minor proportions in quite young soils, although it may occur there because of its prior formation under an earlier warmer climate, as in Scotland (Wilson et al., 1984). It may also occur through the transport of dust over hemispheric distances (see Chapter 3, Section 3.3.4.3). Dixon (1989) proposed that it required much more than 10,000 years to produce kaolinite as a major component within a soil. Kaolinite is abundant in laterites and bauxites, as examples of products of highly advanced weathering, and in soils that typify a strongly weathered origin, e.g. Ultisols and Oxisols within the Soil Taxonomy system of soil classification (Allen and Hajek, 1989; Dixon, 1989).

When kaolinites were isolated from soils in a number of studies of kaolinite-rich soils from Western Australia, Indonesia and Thailand (two studies) through the use of oxidation (to destroy organic matter) and reduction (to destroy oxides), and their CECs and specific surface areas (SSAs) were determined, these two indices were almost always greater than those of commercial geological kaolinites from both Georgia and Cornwall. Figure 8.5 shows the mismatch between the data for soil kaolinites and those for commercial, geological kaolinites. It also shows that there was a wide spread in values for the soil kaolinites. It was expected that CEC would reflect SSA, with smaller particles having more surfaces for the development of (mostly) variable charge, and it appears from the figure that this is broadly the case in each set of soils.

Kaolinite has also formed, or at least been preserved, as a result of the uplift of Si by plants in the surfaces of soils undergoing strong leaching and hence desilication. This has apparently occurred in forest soils in Brazil for which the underlying saprolite was composed of gibbsite and hence was Si-poor (Furian et al., 2002). Within Costa Rica, halloysite is maintained in topsoils, while subsoils are gibbsitic (Kleber et al., 2007).

8.7 HALLOYSITE (1:1)

While commonly formed from volcanic materials, halloysite has been identified as a weathering product of a wide range of rock types, including granite, gneiss, dolerite, nephelinite, schist, greywacke, greenstone, granodiorite, gabbro, shale and amphibolite (Churchman and Lowe, 2012). It is often noticed under electron microscopy because of an unusual morphology, which is commonly

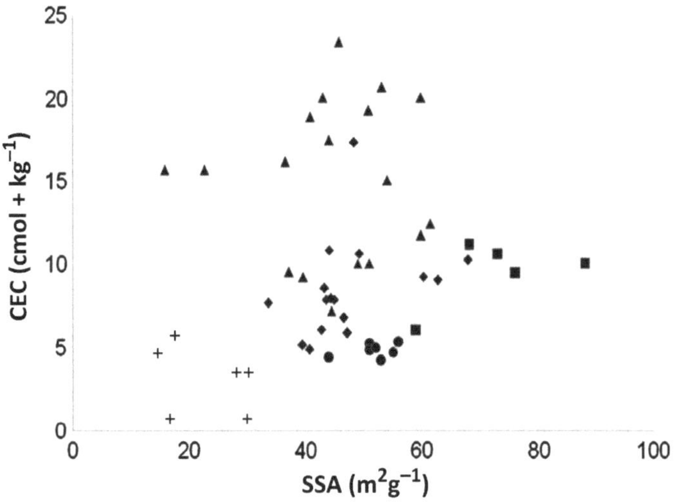

FIGURE 8.5 Cation exchange capacity (CEC) versus specific surface areas (by N_2 – BET) of predominantly kaolinitic clay fractions extracted after deferration and oxidation of soils from Western Australia (circles; data from Singh, B., and R.J. Gilkes, 1992, *Journal of Soil Science* 43: 645–667), Indonesia (squares; data from Hart, R.D., et al., 2002, *Clays and Clay Minerals* 50: 198–207) and Thailand (diamonds; data from Hart, R.D., et al., 2003, *Clay Minerals* 38: 71–94; and also triangles; data from Kanket, W., et al., 2005, *Clays and Clay Minerals* 53: 478–489). Data also plotted as plus signs for standard, non-soil kaolinites from Georgia and Cornwall (data from data from Hart, R.D., et al., 2002, *Clays and Clay Minerals* 50: 198–207). (From Churchman, G.J., 2010, *Physics and Chemistry of the Earth, Parts A/B/C* 35: 927–940. With permission from Elsevier.)

tubular but may also be spheroidal (Cravero and Churchman, 2016). It can also be lamellar, but often with corrugations and is sometimes seen to be prismatic (Hillier et al., 2016). It may even occur in a platy form, with crystals stacked in a booklike arrangement, more commonly associated with kaolinite (Cunningham et al., 2016). Particles with these various shapes also range in both length and width or diameter (e.g. Joussein et al., 2005; Churchman, 2015). With its crystal shape being highly variable, hopes arising from early applications of electron microscopy that it might provide a definition of this kaolin-group 1:1 Si:Al aluminosilicate were dashed in favour of a definition that is consistent with its original identification by Berthier (1826) as a mineral that differed from kaolinite due to its higher water content (Churchman and Carr, 1975; Churchman, 2015). Halloysites are "those minerals with a kaolin layer structure which either contain interlayer water in their natural state or for which there is unequivocal evidence for their formation by dehydration from kaolin minerals containing interlayer water" (Churchman and Carr, 1975; Churchman, 2015). Halloysites containing their full complement of interlayer water (2 moles per kaolin $Si_2Al_2O_5(OH)_4$ unit cell) have a repeat distance of 10 Å along their c-axis and are known as 'halloysite-10 Å'. Many studies have shown how it is the incorporation of interlayer water in halloysites that leads to a tendency to form their characteristic shapes, given that the loss of interlayer water is irreversible, giving rise to 'halloysite-7 Å', with a repeat distance near 7 Å, as in kaolinite (Churchman and Lowe, 2012; Churchman, 2015; Guggenheim, 2015). The effect of prior intercalation by water on ease of uptake of other polar compounds, such as their rapid uptake of formamide (Churchman et al., 1984), has also enabled easy tests for the distinction of halloysite-7 Å from kaolinite.

While less intense leaching of volcanic materials than is required for allophane normally gives rise to halloysite, there are situations where this does not occur. In particular, soils formed in quite recent basaltic volcanic materials in South Australia comprised Al-rich allophane in the tops of their profiles, where water throughput was about 280 mm/year, hence similar to the requirements for Al-rich allophane according to Parfitt et al. (1984) and Lowe (1995). However, they comprise

1:1 Si:Al allophane rather than halloysite in the lower parts of their profiles where seasonal drying in a xeric or Mediterranean climate and impediments to drainage has led to a slower throughput of water (Churchman and Lowe, 2012). These soils are formed on limestone and occur at a high pH, whereas halloysites generally occur at low pHs. Their formation may need to occur in acid pHs in order for the layers to acquire a sufficiently high charge to attract polar water to the interlayers to form halloysite. Bailey (1990) has suggested that their charge arises from substitution of Al for Si in tetrahedral layers, but Newman et al. (1994) have shown that halloysites do not necessarily have more tetrahedral Al than kaolinites. Alternative explanations come from the possibility that halloysites contain more Fe(II) than kaolinites while still wet, or that the Al-octahedral sheet has acquired a significant positive charge from its formation at a pH that is well below that of its point of zero charge (~8.5, according to Abdullayev and Lvov, 2013) while the tetrahedral sheet remains negatively charged. Then layers with a positively charged sheet opposite a negatively charged sheet can attract the polar molecule H–OH into the interlayer region (see Churchman et al., 2016, for the detailed evidence for this argument for the origin of interlayer water in halloysites). This mechanism seems most likely because that involving Fe(II) hardly explains the interlayer water in a halloysite ('patch clay') with an Fe(total) content as low as 0.11% (Norrish, 1995). Furthermore, a mechanism involving Fe(II) would require interlayer cations in the same way as Bailey's (1990) mechanism involving Al (IV), and Bordello et al. (2008) have shown that the quasi-elastic scattering data for a halloysite are consistent with there being no interlayer cations in this mineral. Notably also, smectite occurs in some of the deepest layers in the soils from basaltic volcanic materials in South Australia, indicating that if leaching and drainage are sufficiently low, build-up of Si concentration can lead to the neogenesis of 2:1 Si:Al smectites (Churchman and Lowe, 2012).

From a genetic point of view, easy availability of water at the time of their formation seems to be a central requirement for the formation of halloysite rather than the anhydrous 1:1 Si:Al aluminosilicate mineral kaolinite (Churchman and Lowe, 2012). It is for this reason that there is a general trend in many weathering profiles and soils for halloysite to occur and dominate at depth, where moisture is retained, while kaolinite is more common and more abundant towards the surfaces, where drying may occur at least seasonally (Churchman and Lowe, 2012). It also explains why, in the alteration of granite and volcanic tuff under the same climate within the limited area of Hong Kong, halloysite was formed in saprolites where poorer drainage precluded drying, as indicated by the absence of either manganese or iron oxides (Churchman et al., 2010). By contrast, while drying led to the appearance of these oxides with their distinctive black and red colours, it also gave rise concurrently to the occurrence of kaolinite in similarly weathered samples (Churchman et al., 2010; Churchman and Lowe, 2012). Many occurrences of halloysites are complex. They may occur alongside kaolinites, although the two minerals could originate at different stages of the weathering process (e.g. Churchman and Gilkes, 1989). They may also occur in different morphologies within the same sample (e.g. Saigusa et al., 1978; Churchman and Theng, 1982). Different forms of halloysite may change into other forms with time. Papoulis et al. (2004) has proposed that halloysite formed first as very small spheres which later coalesced into tubes and these later became platy in shape. Cunningham et al. (2016) found halloysites occurring in plates with a booklike form that is more often associated with kaolinite. Many studies have found that the content of impurity elements, particularly Fe, but also perhaps Ti, strongly affects both the shape and size of particles (Joussein et al., 2005; Churchman and Lowe, 2012; Churchman, 2015; Cunningham et al., 2016), so these changes would likely involve dissolution and recrystallisation.

When halloysites were observed in veins, denoting infill, in saprolites in granite in Hong Kong, there was an apparent relationship between the lengths of tubular halloysite and impurities within the veins alongside halloysite (and sometimes also, some kaolinite) (Churchman et al., 2010). This is shown in Figure 8.6. Where veins were white, indicating the absence of the coloured oxides of Fe and Mn (Figure 8.6a), energy-dispersive X-ray (EDX) analyses confirmed that neither of these were present (Figure 8.6b). The tubular particles were very long in this case (Figure 8.6a). By the same token, where they were dark greyish brown in colour (Figure 8.6c), there was substantial

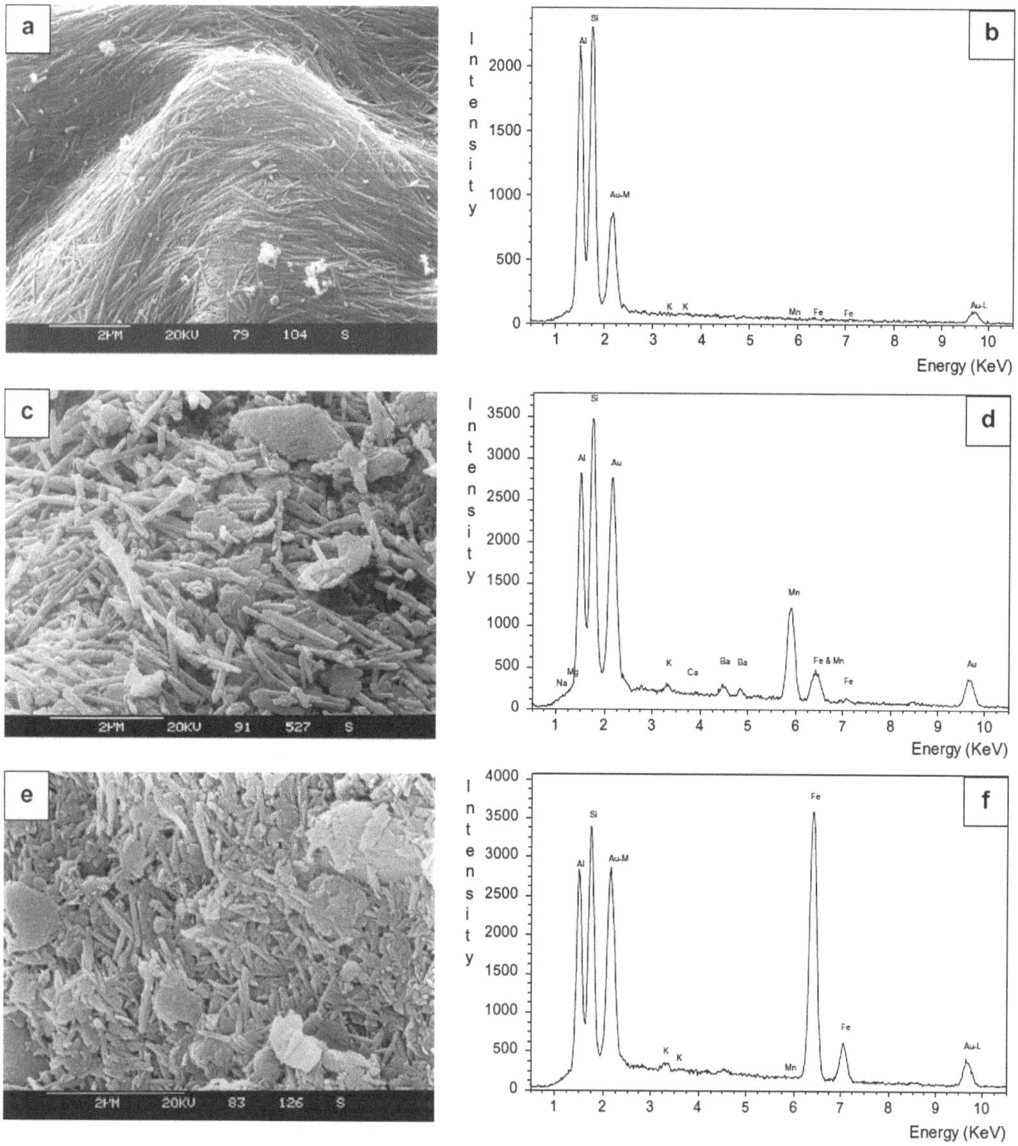

FIGURE 8.6 Scanning electron micrographs at similar magnifications of halloysites, with associated minerals within infill veins in saprolites in weathered granite at different locations in Hong Kong (a, c, e), with EDX analyses of the areas alongside (b, d, f). Panel (a) is from a white, mottled pink vein (Munsell colour 7.5YR 8/4), (c) is from a very dark greyish-brown vein (Munsell colour 2.5Y3/2) and (e) is from a reddish-yellow vein (Munsell colour 5YR 6/8). (From Churchman, G.J., 2010, *Physics and Chemistry of the Earth, Parts A/B/C* 35: 927–940. With permission from Elsevier. Permission for reuse also kindly given by the Geotechnical Engineering Office, Civil Engineering and Development Department, Hong Kong.)

Mn (and some Fe) present according to the EDX analyses (Figure 8.6d). In this case, tubular particles were much smaller (Figure 8.6b). They were smaller again in a sample of a red vein (Figure 8.6e) and in this case Fe content was very high by EDX analyses (Figure 8.6f). The correspondence between contents of Mn and/or Fe, clearly as oxides by their colour, means that the presence of these oxides in the veins, in which halloysite and sometimes also kaolinites have formed, took place in an impure environment represented by Figure 8.6c and e. In contrast halloysite was formed in a vein free of oxide impurities in Figure 8.6a. This illustrates the general principle that minerals (not just halloysite) formed in an impure soil environment will also be small. This is borne out by TEMs of soil kaolinites (Figure 7.5 herein) and a plot of CEC versus SSA for kaolinites (Figure 8.5). Impurities in the environment of formation constrain crystal growth for all minerals. These impurities will also include organic matter, as well as Fe (and Mn) in the soil zone, leading to generally small particles in halloysites formed there (Figure 8.6). The soil halloysite shown in Figure 8.7 has a high Fe content, but most of the Fe is dithionite-extractable (see caption for Figure 8.7), hence as oxides, hydroxides and oxyhydroxides external to the halloysite crystals. These have greatly restrained the growth of the crystals. Halloysites in soils are commonly associated with high

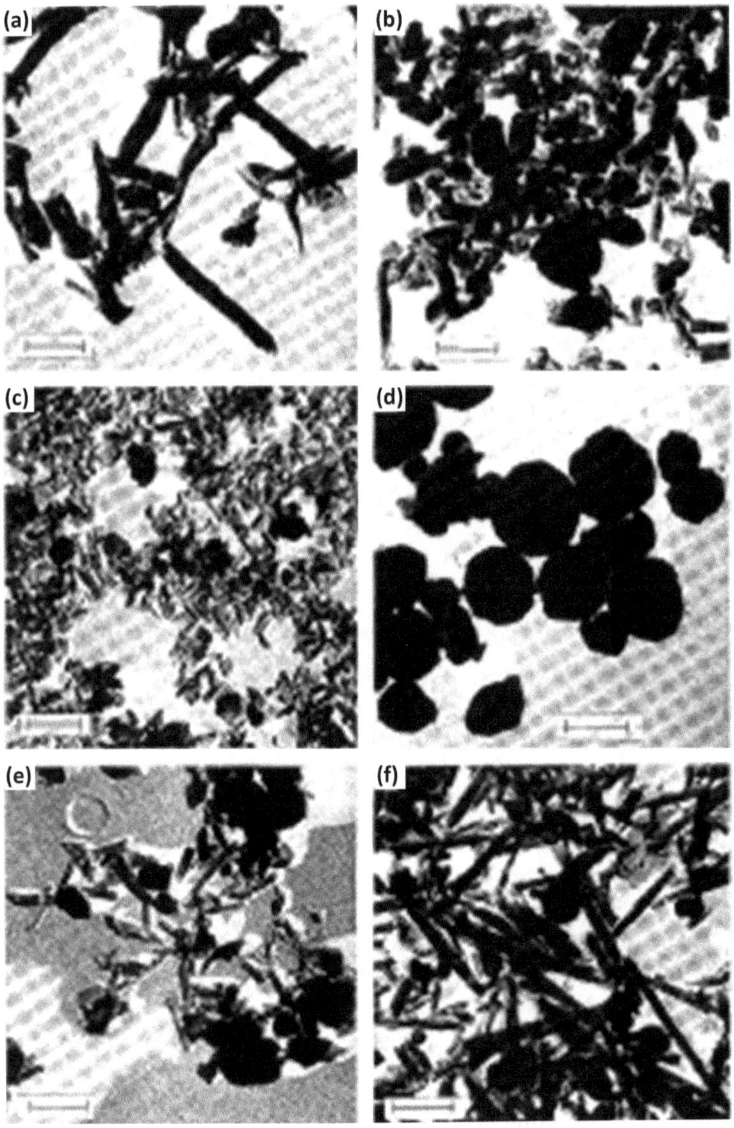

FIGURE 8.7 Transmission electron micrographs of halloysites from different sources (five non-soil and one soil) in New Zealand. The soil halloysite is shown in (c) (centre left). The Fe contents (%), before and after ([]) reduction with dithionite of the halloysites are (left to right and top to bottom): (a) 0.5 [0.4], (b) 0.7 [0.6], (c) 5.6 [1.0], (d) 2.8 [2.6], (e) 2.3 [2.2], (f) 1.5 [0.6]. The scale bar is 0.5 μm. (From Churchman, G.J., and B.K.G. Theng, 1984, *Clay Minerals* 19: 161–175. With permission from the Mineralogical Society of Great Britain & Ireland.)

contents of iron oxides, as illustrated by Fe_2O_3 contents of 5%–7% in soils from New Zealand with >60% halloysite (Churchman, 2010, Table 2).

Papoulis et al.'s (2004) sequence also involved the ultimate conversion of the platy halloysites, considered to be unstable, into kaolinite plates, and then perhaps to the booklike forms often typical of well-crystallised kaolinites. Albeit in weathered tephra and not strictly in soils, halloysite has also been found in booklike forms where it appears to have originated in a sequence from spheroids, tubes and plates (Cunningham et al., 2016). Hillier et al. (2016), studying 21 different halloysites, concluded that the continued growth of tubular halloysites leads to a prismatic form. Several studies with electron microscopy (Churchman and Lowe, 2012; Churchman, 2015) have observed an apparent conversion of platy kaolinite particles into tubular particles, considered to be halloysite (e.g. Robertson and Eggleton, 1991; Singh and Gilkes, 1992b). While this change is contrary to the greater thermodynamic stability of kaolinite to halloysite, it also implies that uptake of water into kaolin interlayers occurs to effect this change (Robertson and Eggleton, 1991). With considerable evidence pointing to subtle but defining differences between the aluminosilicate layer structure of halloysites from that of kaolinites (Churchman, 2015; Churchman et al., 2016), the apparent change from kaolinite to halloysite would seem to involve a dissolution/recrystallisation sequence (Churchman and Lowe, 2012).

Within Costa Rica, halloysite occurs in topsoils, while subsoils are gibbsitic (Kleber et al., 2007), implying that desilication is prevented by the uplift of Si by plants, as for kaolinites in forest soils in Brazil (Furian et al., 2002; also see Section 8.6 herein).

8.8 ALLOPHANE (1:2 TO 1:1) AND IMOGOLITE (1:2)

Allophane is variable in composition and also in its degree of order. It has been defined as "a group of clay minerals with short-range order which contain silica, alumina and water in chemical combination" (Parfitt, 1990, p. 344) and is best described as 'nanocrystalline'. Allophanes vary in composition between Si:Al 1:2 and 1:1. High-resolution transmission electron micrographs in Figure 8.8 show that allophane and imogolite each display distinctive particle shapes.

Allophane was once referred to as 'amorphous'. However, it has shown some regularity of morphology where it appears as small aggregated spheres in HRTEM (Figure 8.8, left) and some instances of it can also show a few broad peaks in XRD.

Imogolite is rarer than allophane in soils. It is considered that imogolite and allophane, particularly in its 1:2 combination, have the same basic structure. Imogolite, which is tubular in morphology

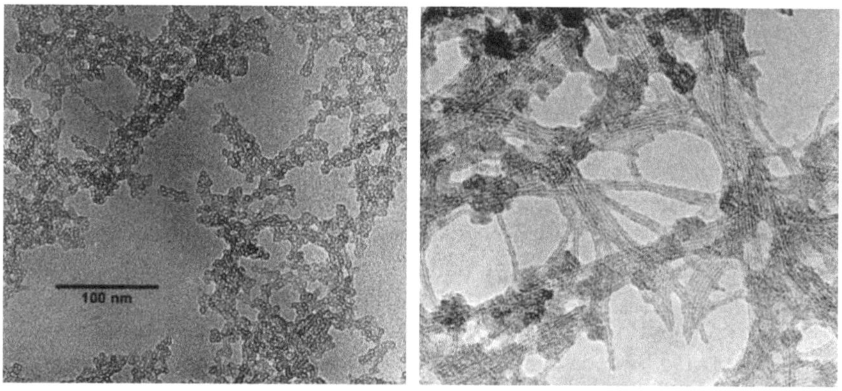

FIGURE 8.8 (Left) Electron micrograph of an Al-rich soil allophane; external diameter of the spherules is 4–5 nm. (Right) Electron micrograph of imogolite (external diameter of the tubes is ~2.1 nm and internal diameter is ~0.7 nm). (From Parfitt, R.L., 2009, *Clay Minerals* 44: 135–155. With permission from the Mineralogical Society of Great Britain & Ireland.)

(Figure 8.8, right), has a structure with concentric rings of alumina, which forms the outer ring, and silica, which forms the inner ring. 1:2 allophanes, which have the most common form, "appear to be made up of fragments having the imogolite atomic structure over a short range" (Parfitt and Henmi, 1980; also see Lowe, 1995, Figure 2). Albeit that imogolite shows several more peaks in XRD than allophanes, they have been described together as having a crystallinity that is "intermediate between Al_{13} polymers and well-crystallised clay minerals" (Harsh et al., 2002, p. 291).

Strong leaching of the readily soluble glass from volcanic parent materials and especially volcanic ash leads to ideal conditions for the formation of allophane with an Si:Al ratio of 1:2, i.e. Al-rich or 'proto-imogolite' allophane (Parfitt et al., 1983, 1984). Although the ultimate arbiter of the nature of the secondary product formed by neogenesis is probably the composition of the solution resulting from the breakdown of rock minerals, the extent of leaching, reflecting the efficiency of drainage in a weathering environment, often controls the nature of the products that recrystallise out of these solutions. Under most natural conditions, where pH is not extremely low (<~3) or extremely high (>~9), Si is more soluble than Al, so is leached more easily. Hence strong leaching with no impediments to drainage favours the formation of Al-rich minerals. In the extreme, gibbsite is formed from a variety of parent materials and at both early and late stages of weathering (Churchman and Lowe, 2012). If volcanic ash is subjected to less intense leaching (less than ~250 mm/year, according to Parfitt et al., 1984), then the clay mineral halloysite, with an Si:Al ratio of 1:1, forms instead of allophane.

For a long time, the idea that Al-rich allophane was formed and persisted in soils because of favourable chemical conditions, i.e., low Si and high Al, in originating solutions, was not the favoured view of its origin in soils from volcanic ash. Instead, it was thought (Fieldes, 1955) that allophane was the fast-forming amorphous precursor to crystalline halloysite. However, it has often been found that halloysite can form very quickly and also that allophane can persist in very old weathered tephra sequences (Churchman and Lowe, 2012). As well, many depth sequences on volcanic parent materials have shown allophane in surface horizons but a tendency towards halloysite formation at depth. This is considered to occur as a result of a build-up of Si that is transported to depths where drainage becomes impeded. The effect has been seen in New Zealand, Japan and the Western United States, among other volcanic zones (Churchman and Lowe, 2012). In some of these zones, namely Ecuador, the Western United States and Rwanda, allophane has been found to occur at higher altitudes, with stronger leaching, and halloysite at lower altitudes. Both may occur together at intermediate altitudes. Allophane has been found to persist in tephra in New Zealand that represents warm, wet periods in vertical sequences, while halloysite has been found to occupy tephra within the same sequences that correlate with cold, dry periods (Stevens and Vucetich, 1985; Lowe, 1986; Churchman and Lowe, 2012). In these and many situations, the possibility exists that allophane may have been formed first and halloysite later, either through a build-up of Si from the surface, through the development of impedance to drainage, or through drying arising from climate change. Across the island of Java, in Indonesia, there is a mineralogical change from allophane to halloysite that correlates with a change in (present day) climate from shorter to longer annual dry periods proceeding from west to east of the island (Utama et al., 2019). Across the Hawaiian island chain, at sites with the same precipitation and temperature, there is a change in mineralogy from non-crystalline minerals on the newest basaltic surfaces in the eastern part of the chain to halloysite in the western-most island (Kauai) (Torn et al., 1997). The most recent, hence also thinnest, basalt (on the island of Hawaii) gives rise to mainly depleted primary minerals and mineral-organic complexes, while basaltic deposits of up to 1.4 million years old have given rise to clay mineralogies that predominantly comprise allophane and ferrihydrite (Mikutta et al., 2009). It is on Kauai that the basalt, with an age of 4.1 million years, and which constitutes the thickest deposit among the island chain (>3 m deep, according to Torn et al., 1997), has given rise to a predominantly halloysitic clay mineralogy. Superficially, it appears that allophane, once formed, might have given rise to halloysite by a solid-phase transformation with the passage of time. However, the sequence from east to west along the island chain is a depth sequence, and it is likely that there is a build-up of Si from impeded drainage as weathering proceeds to greater depths.

Structural considerations also argue against the solid-state transformation of allophane to halloysite. Allophane and halloysite are quite different structurally, with (Al-rich) allophane comprising isolated silica tetrahedra inside a coiled spherule of gibbsite composition, while halloysite comprises linked silica tetrahedra generally curved around alumina octahedra (Churchman and Lowe, 2012). Hence changes from allophane to halloysite must have occurred by a dissolution and recrystallisation. Furthermore, there may be other products formed from the weathering of volcanic ash besides allophane and halloysite. Competition for Al that precludes allophane formation can come from expanded micaceous layers of vermiculite and smectite, which incorporate Al into HI interlayers under favourable pH conditions (Ndayiragije and Delvaux, 2003; Kleber et al., 2007). In soils with high OM contents, Al may also become incorporated into Al-humus complexes rather than being available for the formation of allophane (Dahlgren et al., 1997; Johnson-Maynard et al., 1997).

The other situation in soils where allophane commonly occurs is in the illuvial, B_h and B_s lower horizons of podzolised soils. Podzolisation arises from the intense leaching of soils with acidic solutions. There is contention about whether the process is primarily organic or primarily inorganic, but, in any case, aluminium and also iron are transported from surface horizons to lower positions in soil profiles (Churchman and Lowe, 2012). Allophane is formed in the lower illuvial horizons as a result of deposition of the Al, transported as either or both an organic complex or an inorganic sol ('proto-imogolite'; Farmer et al., 1980, 1983, 1985). Unlike that of the formation of allophane from strong leaching of silicon out of volcanic ash, leaving an Al-rich residual solution from which allophane is recrystallised, the formation of allophane during podzolisation involves that of the selective leaching of aluminium and its later deposition.

8.9 PALYGORSKITE AND SEPIOLITE (2:1 SI:MG)

Palygorskite and sepiolite are two ('hormite') minerals that are fibrous hydrated magnesium silicates containing a continuous two-dimensional tetrahedral sheet of composition Si_2O_5, but differ from the other layer silicates because they lack continuous octahedral sheets. The structures of palygorskite and sepiolite are alike. The apical oxygens alternately point up and down such that they sandwich discontinuous strips of octahedrally coordinated cations (overwhelmingly Mg) in the same way as in the typical 2:1 (Si:Al) structure (Wilson, 2013). They can be regarded as consisting of narrow strips or ribbons of 2:1 (Si:Mg) layers that are linked stepwise at the corners (Brigatti et al., 2006). Palygorskite contains Al, and sometimes also Fe (III) in varying low proportions, while sepiolite has little or no substitution of Mg. Further details of their structures can be sought elsewhere (e.g. Brigatti et al. 2006, 2013; Wilson, 2013).

Sepiolite and palygorskite are both formed in evaporitic areas of the alterite zone in arid, including desert, areas. They are commonly associated with carbonates and these minerals often occur in sedimentary rocks found in this type of environment. They often form in these rocks within lacustrine, perimarine and marine basins (Galán and Pozo, 2015). They probably occur most commonly in soils as a result of their inheritance from these sedimentary rocks (Singer, 2002). Nonetheless, palygorskite, in particular, has been found to form within soils in many countries, including Israel, Syria, Egypt, Morocco, Iraq, Iran, Turkey, Spain, Portugal, the United States, Mexico, Australia, South Africa and Argentina (Churchman and Lowe, 2012). The main conditions for the formation of palygorskite are (1) a fluctuating saline or alkaline groundwater, (2) strong and continuous evaporation and (3) a sharp textural transition (Singer and Norrish, 1974; Singer, 1989, 2002; Churchman and Lowe, 2012). The requirement of a strong textural transition is often provided by an underlying calcrete horizon (Verrecchia and Le Costumer, 1996; Singer, 2002; Kadir and Eren, 2008). This ensures the maintenance of a saturated solution to enable the precipitation of the palygorskite, even in an arid climate.

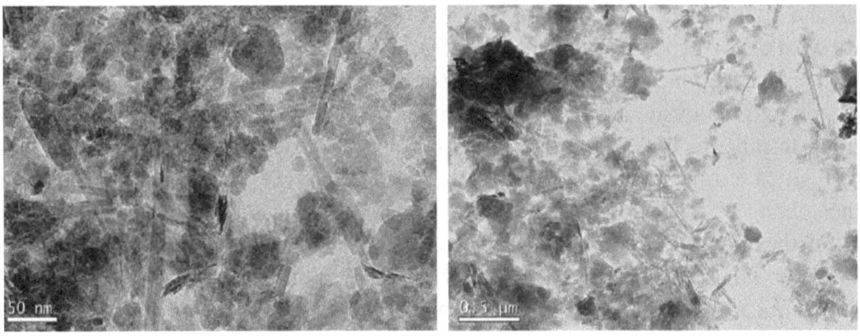

FIGURE 8.9 Fibrous rods for palygorskite and/or sepiolite, together with smectite, illite and kaolinite, in transmission electron micrographs (at different scales) of <2 μm fractions of topsoils of highly alkaline soils with no or little plant growth. (From Fraser, M.B., et al., 2016, *Applied Clay Science* 124–125: 183–196. With permission from Elsevier.)

The minerals are not stable when in a plant-active chemical context in the soil zone of alteration sequences (see Velde, 1985) due to acidity developed by plant and bacterial action. Under slightly acidic conditions they tend to be transformed into other clays such as kaolinite or smectite (Khadema and Arocena, 2008) in the soil zones. In fact, the presence of palygorskite or sepiolite at the surface inhibits plant growth to a certain extent (Neaman and Singer, 2004). Palygorskite and sepiolite may be found in surface deposits due to surface erosion, but they will not be stable when in contact with actively growing plants. Hence, we do not consider them as soil zone clays, although they can be present due to surface transportation by wind or rainwash. Nonetheless, they may appear in surface or near-surface soil in which plant growth is suppressed by a high pH and a low porosity from the action of associated swelling smectitic clays (Figure 8.9) (Fraser et al., 2016).

Gypsiferous soils can also give rise to the neogenesis of palygorskite (Khormali and Abtahi, 2003; Owliaie et al., 2006). The palygorskite and sepiolite are found at depth in alterite or sedimentary materials along with other evaporite phases such as carbonates. However, in arid climates they appear to be unstable in the plant–soil zone (Bigham et al., 1980).

Sepiolite is much rarer than palygorskite in soils, although Bouza et al. (2007) reported that sepiolite formed after palygorskite in Patagonia in Argentina. By contrast, sepiolite has often been synthesised in the laboratory (Churchman, 2000), but syntheses of palygorskite appear to be at least more difficult, if not impossible (Singer, 2002; Churchman and Lowe, 2012). Mg and Si are essential for both sepiolite and palygorskite and, while sepiolite is largely composed of these cations to the exclusion of others, palygorskite also requires some Al and Fe (Singer, 1989). In nature, Mg is most probably supplied by either dolomite or Mg-calcite within calcareous or gypsiferous zones, and Al and Fe may be available for the formation of palygorskite but their presence may preclude the formation of sepiolite in most situations. It is noteworthy that smectites and also palygorskite and/or sepiolite, either inherited or neogenetic, often occur together in soils. Both are favoured by similar chemical requirements of high concentrations of Mg and Si and a high pH. However, there tends to be an inverse relationship between the occurrence of smectites, on the one hand, and the hormite minerals, on the other (Churchman and Lowe, 2012). Paquet and Millot (1972) suggested that the latter gave way to smectites when the mean annual precipitation (MAP) exceeded only ~300 mm, while Khormali and Abtahi (2003) suggested that plants produced this change through the effect of their transpiration on soil-available water. By contrast, palygorskite was observed to form in just 40–50 years when rain-fed smectitic soils in India that were waterlogged as a result of sodicity were irrigated and hence became more readily drained (Hillier and Pharande, 2008). Palygorskite and sepiolite become unstable, giving rise to kaolinite, when exposed to the (acidic) rhizospheres of agricultural crops (Khadema and Arocena, 2008).

8.10 OXIDES AND RARER MINERALS

Iron oxides and oxyhydroxides occur in almost all soils (Allen and Hajek, 1989). The particular type of iron oxide occurring in soils depends upon the conditions under which they formed. Goethite is found in soils almost everywhere but is most abundant in cool and temperate climates and where organic matter contents are high. Hematite is most common in aerated soils in warm climates, including the tropics and subtropics and in arid and semi-arid areas; and also in areas with a Mediterranean (xeric) climate of hot, dry summers and cool, wet winters. Goethite tends to occur at low pHs, hematite at high pHs and there are soils which have (yellow) goethite in upper horizons with (red) hematite in lower horizons (Churchman and Lowe, 2012). Al substitution occurs in all Fe oxides but more in goethite than in hematite.

Ferrihydrite is also common in soils. It is most prevalent in soils from volcanic tephra and in the subsoils (illuvial horizons) of podzolised soils. It can occur alongside goethite and lepidocrocite, but rarely alongside hematite. This is surprising because ferrihydrite can be converted to hematite in the laboratory, thus suggesting that its conversion to hematite in soils is very rapid.

Lepidocrocite is formed in noncalcareous soils that are seasonally anaerobic. Its formation requires an accumulation of Fe^{2+} and a slow rate of oxidation. It often occurs alongside goethite. Maghemite is found in tropical soils and mostly occurs in the surface of these soils. It is thought to be a product of bush and forest fires, through the oxidation of primary magnetite.

Other iron oxides include schwertmannite, found in acid sulphate soils, and 'green rust', which was identified in a highly hydromorphic soil (Churchman and Lowe, 2012). Iron oxides are often formed and also dissolved by microbial action.

Manganese oxides are widespread but often in trace amounts. Mn^{2+} is very soluble and is mobile, so that Mn oxides often occur as coatings on rocks and soil particles as well as in concretions, segregations, pans and nodules, often along with Fe oxides (Churchman and Lowe, 2012). Their appearance in these (oxidised) forms indicates prior drying. Among the Mn oxides, birnessite, lithiophorite and todorokite are most common in soils.

Gibbsite is the most prevalent of aluminium oxides and hydroxides in soils. This hydroxide of Al occurs when Si is in short supply in soils. It therefore occurs where there is strong leaching. Its appearance alongside other secondary minerals may indicate polygenesis from several sources or at least a change having occurred in the leaching regime over time (Churchman and Lowe, 2012). The Al oxyhydroxide, boehmite, has been found at the top of laterite profiles and in tropical soils. Other Al oxyhydroxides, bayerite, norstrandite and doyleite, are rarer in soils. The Al oxide corundum has been found alongside maghemite in ferallitic duricrusts in Western Australia, and it has been proposed (Anand and Gilkes, 1987) that it was formed by the dehydration of gibbsite in bush fires.

Pyrophyllite, talc and zeolites are rare in soils, and when they occur it is usually by inheritance from parent materials. However, some zeolites, particularly analcime, may form in salty alkaline soils containing carbonates (Churchman and Lowe, 2012).

Since Si is often leached from soils, it may reprecipitate lower in profiles to form overgrowths and silica cements, in hardpans, duripans or fragipans (Churchman and Lowe, 2012). In the extreme, it forms massive silcretes. Sometimes, silica reappears deep in profiles as an opal. Plant opals also occur as phytoliths within plants, resulting from the uptake of Si into plants.

In soils, titanium oxide can form by neogenesis as anatase and rutile or brookite, and the Fe-Ti minerals ilmenite, pseudorutile, titanomaghemite and others. It can also be inherited from parent materials. Pseudorutile, which occurs widely in soils, mainly arises by the alteration of ilmenite (Fitzpatrick and Chittleborough, 2002). Some Ti^{4+} can be incorporated in other minerals, notably goethite.

Zircon, the most common Zr-containing mineral, may be formed by neogenesis, but is most reliably reported as an inherited phase.

The sulphide pyrite can occur in soils by inheritance from anaerobic sediments. It can form in acid sulphate soils, which often contain the sulphate jarosite and also the related natrojarosite.

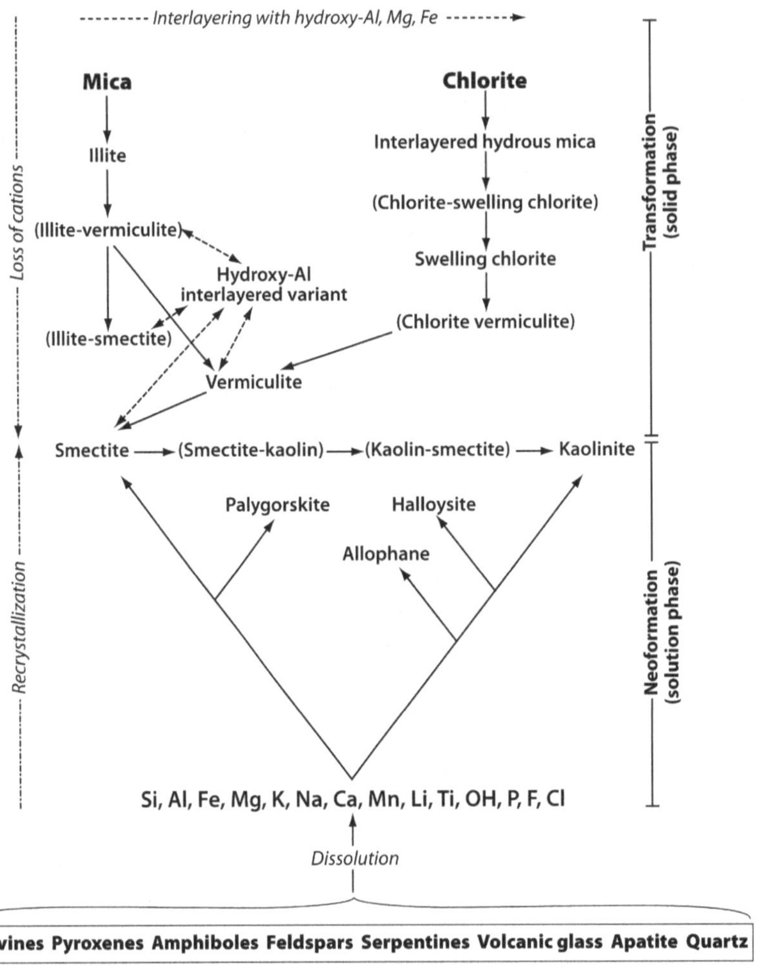

FIGURE 8.10 Main pathways for phyllosilicate formation in soils via either transformation (upper part of figure) or neoformation (lower part) (after Churchman, 1978; Wilson, 2004). (From Churchman, G.J., and D.J. Lowe, 2012, Alteration, formation and occurrence of minerals in soils, p. 20.1–20.72, in P.M. Huang et al., eds., *Handbook of Soil Sciences: Properties and Processes*, 2nd edn., CRC Press, Boca Raton, Florida. With permission from CRC Press/Taylor & Francis.)

Phosphates are rare and arise from the breakdown of rock phosphate. They include plumogummite, containing Pb, variscite (Al), strengite (Fe), crandallite (Ca) and gorcexite (Ba).

Soluble minerals include carbonates dissolved out of rocks and reprecipitated, and also simple salts, like halite and gypsum, found in extremely arid environments like Antarctica and the Atacama Desert in Chile, where nitrate, borate, chromate and perchlorate have been found (Churchman and Lowe, 2012).

The main routes for the formation of phyllosilicates in soil sequences, i.e. those of transformation and neogenesis, are summarised in Figure 8.10 (Churchman and Lowe, 2012)).

REFERENCES

Abdullayev, E., and Y. Lvov. 2013. Halloysite clay nanotubes as a ceramic "skeleton" for functional biopolymer composites with sustained drug release. *Journal of Materials Chemistry B* 1: 2804–2903.

Allen, B.L., and B.F. Hajek. 1989. Mineral occurrence in soil environments, p. 199–278. In: J.B. Dixon, and S.B. Weed (eds.), *Minerals in Soil Environments*, 2nd edn. Soil Science Society of America, Madison, Wisconsin.

Anand, R.R., and R.J. Gilkes. 1987. The association of maghemite and corundum in Darling Range laterites, Western Australia. *Soil Research* 25: 303–311.

Andrade, G.R.P., A.C. de Azevedo, J.K. Lepchak, and T.C. Assis. 2019. Weathering of Permian sedimentary rocks and soil clay minerals transformations under subtropical climate, southern Brazil (Paraná State). *Geoderma* 336: 31–48.

Arocena, J.M., B. Velde, and S.J. Robertson. 2012. Weathering of biotite in the presence of arbuscular mycorrhizae in selected agricultural crops. *Applied Clay Science* 64: 12–17.

Aspandiar, M.F., and R.A. Eggleton. 2002\. Weathering of chlorite I: Reactions and products in microsystems controlled by primary minerals. *Clays and Clay Minerals* 50: 685–698.

Bailey, S.W. 1990. Halloysite – A critical assessment. *Science Géologiques* 86: 89–98.

Bain, D.C. 1977. The weathering of ferruginous chlorite in a podzol from Argyllshire, Scotland. *Geoderma* 17: 193–208.

Banfield, J.F., and T. Murakami. 1998. Atomic-resolution transmission electron microscope evidence for the mechanism by which chlorite weathers to 1:1 semi-regular chlorite-vermiculite. *American Mineralogist* 83: 348–357.

Barnhisel, R.I., and P.M. Bertsch. 1989. Chlorites and hydroxy-interlayered vermiculite and smectite, p. 729–788. In: J.B. Dixon, and S.B. Weed (eds.), *Minerals in Soil Environments*, 2nd edn. Soil Science Society of America, Madison, Wisconsin.

Berthier, P. 1826. Analyse de l'halloysite. *Annales de Chemie et de Physique* 32: 332–335.

Bigham, J.M., W.F. Jaynes, and B.L. Allen. 1980. Pedogenic degradation of sepiolite and palygorskite on the Texas High Plains. *Soil Science Society of America Journal* 44: 159–167.

Borchardt, G. 1989. Smectites, p. 675–727. In: J.B. Dixon, and S.B. Weed (eds.), *Minerals in Soil Environments*, 2nd edn. Soil Science Society of America, Madison, Wisconsin.

Bordallo, H.N., L.P. Aldridge, G.J. Churchman, W.P. Gates, M.T.F. Telling, K. Kiefer, P. Fouquet, T. Seydel, and S.A.J. Kimber. 2008. Quasi-elastic neutron scattering studies on clay interlayer-space highlighting the effect of the cation in confined water dynamics. *The Journal of Physical Chemistry C* 112: 13982–13991.

Bouza, P.J., M. Simón, J. Aguilar, H. Del Valle, and M. Rostagno. 2007. Fibrous-clay mineral formation and soil evolution in Aridisols of northeastern Patagonia, Argentina. *Geoderma* 139: 38–50.

Brigatti, M.F., E. Galán, and B.K.G. Theng. 2006. Structures and mineralogy of clay minerals, p. 19–86. In: F. Bergaya, B.K.G. Theng, and G. Lagaly (eds.), *Handbook of Clay Science*, 1st edn. Elsevier, Amsterdam.

Brigatti, M.F., E. Gálan, and B.K.G. Theng. 2013. Structure and mineralogy of clay minerals, p. 21–81. In: F. Bergaya, and G. Lagaly (eds.), *Handbook of Clay Science: A. Fundamentals*, 2nd edn. Elsevier, Amsterdam.

Carnicelli, S., A. Mirabella, G. Cecchini, and G. Sanesi. 1997. Weathering of chlorite to a low-charge expandable mineral in a Spodosol on the Apennine Mountains, Italy. *Clays and Clay Minerals* 45: 28–41.

Churchman, G.J. 1978. Studies on a climosequence of soils in tussock grasslands. 21. Mineralogy. *New Zealand Journal of Science* 21: 467–480.

Churchman, G.J. 1980. Clay minerals formed from micas and chlorites in some New Zealand soils. *Clay Minerals* 15: 59–76.

Churchman, G.J. 2000. The alteration and formation of soil minerals by weathering, p. F3–F76. In: M.E. Sumner (ed.), *Handbook of Soil Science*. CRC Press, Boca Raton, Florida.

Churchman, G.J. 2010. Is the geological concept of clay minerals appropriate for soil science? A literature-based and philosophical analysis. *Physics and Chemistry of the Earth, Parts A/B/C* 35: 927–940.

Churchman, G.J. 2015. The identification and nomenclature of halloysite (a historical perspective), p. 51–67. In: P. Pasbakhsh, and G.J. Churchman (eds.), *Natural Mineral Nanotubes*. Apple Academic Press, Oakville, Canada.

Churchman, G.J., and R.M. Carr. 1975. The definition and nomenclature of halloysites. *Clays and Clay Minerals* 23: 382–388.

Churchman, G.J., and R.J. Gilkes. 1989. Recognition of intermediates in the possible transformation of halloysite to kaolinite in weathering profiles. *Clay Minerals* 24: 579–590.

Churchman, G.J., and D.J. Lowe. 2012. Alteration, formation and occurrence of minerals in soils, p. 20.1–20.72. In: P.M. Huang, Y. Li, and M.E. Sumner (eds.), *Handbook of Soil Sciences: Properties and Processes*, 2nd edn. CRC Press, Boca Raton, Florida.

Churchman, G.J., P. Pasbakhsh, D.J. Lowe, and B.K.G. Theng. 2016. Unique but diverse: Some observations on the formation, structure and morphology of halloysite. *Clay Minerals* 51: 395–416.

Churchman, G.J., I.R. Pontifex, and S.G. McClure. 2010. Factors affecting the formation and characteristics of halloysites or kaolinites in granitic and tuffaceous saprolites in Hong Kong. *Clays and Clay Minerals* 58: 220–237.

Churchman, G.J., P.G. Slade, P.G. Self, and L.J. Janik. 1994. Nature of interstratified kaolin-smectites in some Australian soils. *Soil Research* 32: 805–822.

Churchman, G.J., and B.K.G. Theng. 1984. Interactions of halloysites with amides: Mineralogical factors affecting complex formation. *Clay Minerals* 19: 161–175.

Churchman, G.J., J.S. Whitton, G.G.C. Claridge, and B.K.G. Theng. 1984. Intercalation method using formamide for differentiating halloysite from kaolinite. *Clays and Clay Minerals* 32: 241–248.

Cradwick, P.D., and M.J. Wilson. 1978. Calculated x-ray diffraction curves for the interpretation of a three-component interstratified system. *Clay Minerals* 13: 53–65.

Cravero, F., and G.J. Churchman. 2016. The origin of spheroidal halloysites: A review of the literature. *Clay Minerals* 51: 417–427.

Cunningham, M.J., D.J. Lowe, J.B. Wyatt, V.G. Moon, and G.J. Churchman. 2016. Discovery of halloysite books in altered silicic Quaternary tephras, northern New Zealand. *Clay Minerals* 51: 351–372.

Dahlgren, R.A., J.P. Dragoo, and F.C. Ugolini. 1997. Weathering of Mt. St. Helens tephra under a cryic-udic climatic regime. *Soil Science Society of America Journal* 61: 1519–1525.

Dixon, J.B. 1989. Kaolin and serpentine group minerals, p. 467–525. In: J.B. Dixon, and S.B. Weed (eds.), *Minerals in Soil Environments*, 2nd edn. Soil Science Society of America, Madison.

Dong, H.L., D.R. Peacor, and S.F. Murphy. 1998. TEM study of progressive alteration of igneous biotite to kaolinite throughout a weathered profile. *Geochimica et Cosmochimica Acta* 62: 1881–1887.

Dudek, T., J. Cuadros, and S. Fiore. 2006. Interstratified kaolin-smectite: Nature of the layers and mechanism of smectite kaolinization. *American Mineralogist* 91: 159–170.

Dumon, M., A.R. Tolossa, B. Capon, C. Detavernier, and E. Van Ranst. 2014. Quantitative clay mineralogy of a vertic planosol in southwestern Ethiopia: Impact on soil formation hypotheses. *Geoderma* 214–215: 184–196.

Egli, M., A. Mirabella, A. Mancabelli, and G. Sartori. 2004. Weathering of soils in alpine areas as influenced by climate and parent material. *Clays and Clay Minerals* 52: 287–303.

Fairbridge, R.W., and J. Bourgeois. 1978. *The Encylopedia of Sedimentology*. Dowden, Hutchinson and Ross, Stroudsberg.

Fanning, D.S., V.Z. Keramidas, and M.A. El-Desoky. 1989. Micas, p. 551–634. In: J.B. Dixon, and S.B. Weed (eds), *Minerals in Soil Environments*, 2nd edn. Soil Science Society of America, Madison, Wisconsin.

Farmer, V.C., W.J. McHardy, L. Robertson, A. Walker, and M.J. Wilson. 1985. Micromorphology and submicroscopy of allophane and imogolite in a podzol Bs horizon: Evidence for translocaton and origin. *Journal of Soil Science* 36: 87–95.

Farmer, V.C., J.D. Russell, and M.L. Berrow. 1980. Imogolite and proto-imogolite allophane in spodic horizons: Evidence for a mobile aluminium silicate complex in podzol formation. *Journal of Soil Science* 31: 673–684.

Farmer, V.C., J.D. Russell, and B.F.L. Smith. 1983. Extraction of inorganic forms of translocated Al, Fe and Si from a podzol Bs horizon. *Journal of Soil Science* 34: 571–576.

Fieldes, M. 1955. Clay mineralogy of New Zealand soils. Part II: Allophane and related mineral colloids. *New Zealand Journal of Science and Technology* 37: 336–350.

Fitzpatrick, R.W., and D.J. Chittleborough. 2002. Titanium and zirconium minerals, p. 667–690. In: J.B. Dixon, and D.G. Schulze (eds.), *Soil Mineralogy with Environmental Applications*. Soil Science Society of America, Madison, Wisconsin.

Fraser, M.B., G.J. Churchman, D.J. Chittleborough, and P. Rengasamy. 2016. Effect of plant growth on the occurrence and stability of palygorskite, sepiolite and saponite in salt-affected soils on limestone in South Australia. *Applied Clay Science* 124–125: 183–196.

Furian, S., L. Barbiéro, R. Boulet, P. Curmi, M. Grimaldi, and C. Grimaldi. 2002. Distribution and dynamics of gibbsite and kaolinite in an oxisol of Sierra do Mar, southeastern Brazil. *Geoderma* 106: 83–100.

Galán, E., and M.A. Pozo 2015. The mineralogy, geology and main occurrences of sepiolite and palygorskite clays, p.117–130. In: P. Pasbakhsh, and G.J. Churchman (eds.), *Natural Mineral Nanotubes: Properties and Applications*. Apple Academic Press, Oakville, Canada.

Gillot, F., D. Righi., and M.L. Räisänen. 2001. Layer-charge evaluation of expandable clays from a chronosequence of podzols in Finland using an alkylammonium method. *Clay Minerals* 36: 571–584.

Guggenheim, S. 2015. Phyllosilicates used as nanotube substrates in engineered materials: Structures, chemistries and textures, p. 3–48. In: P. Pasbakhsh, and G.J. Churchman (eds.), *Natural Mineral Nanotubes*. Apple Academic Press, Oakville, Canada.

Han, W., H.L. Hong, K. Yin, G.J. Churchman, Z.H. Li, and T. Chen. 2014. Pedogenic alteration of illite in subtropical China. *Clay Minerals* 49: 379–390.

Harsh, J.B., J. Chorover, and E. Nizeyimana. 2002. Allophane and imogolite, p. 291–322. In: J.B. Dixon, and D.G. Schulze (eds.), *Soil Mineralogy with Environmental Applications*. Soil Science Society of America, Madison, Wisconsin.

Hart, R.D., R.J. Gilkes, S. Siradz, and B. Singh. 2002. The nature of soil kaolins from Indonesia and Western Australia. *Clays and Clay Minerals* 50: 198–207.

Hart, R.D., W. Wiriyakitnateekul, and R.J. Gilkes. 2003. Properties of soil kaolins from Thailand. *Clay Minerals* 38: 71–94.

Herbillon, A.J., and M.N. Makumbi. 1975. Weathering of chlorite in a soil derived from a chlorito-schist under humid tropical conditions. *Geoderma* 13: 89–104.

Hillier, S., R. Brydson, E. Delbos, T. Fraser, N. Gray, H. Pendlowski, I. Phillips, J. Robertson, and I. Wilson. 2016. Correlations among the mineralogical and physical properties of halloysite nanotubes (HNTs). *Clay Minerals* 51: 325–350.

Hillier, S., and A.L. Pharande. 2008. Contemporary pedogenic formation of palygorskite in irrigation-induced saline-sodic, shrink-swell soils of Maharashta, India. *Clays and Clay Minerals* 56: 531–548.

Hinsinger, P., F. Elsass, B. Jaillard, and M. Robert. 1993. Root-induced irreversible transformation of a trioctahedral mica in the rhizosphere of rape. *Journal of Soil Science* 44: 535–545.

Hinsinger, P., and B. Jaillard. 1993. Root-induced release of interlayer potassium and vermiculitization of phlogopite as related to potassium depletion in the rhizosphere of ryegrass. *Journal of Soil Science* 44: 525–534.

Hinsinger, P., B. Jaillard, and J.E. Dufey. 1992. Rapid weathering of a trioctahedral mica by the roots of ryegrass. *Soil Science Society of America Journal* 56: 977–982.

Hong, H., F. Cheng, K. Yin, G.J. Churchman, and C. Wang. 2015. Three-component mixed-layer illite/smectite/kaolinite (I/S/K) minerals in hydromorphic soils, south China. *American Mineralogist* 100: 1883–1891.

Hubert, F., L. Caner, A. Meunier, and E. Ferrage. 2012. Unravelling complex <2 μm clay mineralogy from soils using x-ray diffraction profile modelling on particle-size subfractions: Implications for soil pedogenesis and reactivity. *American Mineralogist* 97: 384–398.

Hughes, J.C., and G. Brown. 1979. A crystallinity index for soil kaolins and its relation to parent rock, climate and soil maturity. *Journal of Soil Science* 30: 557–563.

Jaynes, W.F., J.M. Bigham, N.E. Smeck, and M.J. Shipitalo. 1989. Interstratified 1:1-2:1 mineral formation in a polgenetic soil from southern Ohio. *Soil Science Society of America Journal* 53: 1888–1894.

Johnson-Maynard, J.L., P.A. McDaniel, A.L. Falen, and D.E. Ferguson. 1997. Chemical and mineralogical conversion of andisols following invasion by bracken fern. *Soil Science Society of America Journal* 61: 549–555.

Jongkind, A.G., and P. Buurman. 2006. The effect of kauri (*Agathis australis*) on grain size distribution and clay mineralogy of andesitic soils in the Waitakere Ranges, New Zealand. *Geoderma* 134: 171–186.

Joussein, E., S. Petit, J. Churchman, B. Theng, D. Righi, and B. Delvaux. 2005. Halloysite clay minerals – A review. *Clay Minerals* 40: 383–426.

Kadir, S., and M. Eren. 2008. The occurrence and genesis of clay minerals associated with Quaternary caliches in the Mersin area, southern Turkey. *Clays and Clay Minerals* 56: 244–258.

Kanket, W., A. Suddhiprakarn, I. Kheoruenromne, and R.J. Gilkes. 2005. Chemical and crystallographic properties of kaolin from Ultisols in Thailand. *Clays and Clay Minerals* 53: 478–489.

Khadema, H., and J.M. Arocena. 2008. Kaolinite formation from palygorskite and sepiolite in rhizosphere soils. *Clays and Clay Minerals* 56: 429–436.

Khormali, F., and A. Abtahi. 2003. Origin and distribution of clay minerals in calcareous arid and semi-arid soils of Fars Province, southern Iran. *Clay Minerals* 38: 511–527.

Kleber, M., L. Schwendenmann, E. Veldkamp, J. Rößner, and R. Jahn. 2007. Halloysite versus gibbsite: Silicon cycling as a pedogenetic process in two lowland rain forest soils of La Selva, Coast Rica. *Geoderma* 138: 1–11.

Lanson, B. 1997. Decomposition of experimental X-ray diffraction patterns (profile fitting): A convenient way to study clays. *Clays and Clay Minerals* 45: 132–146.

Li, Z., B. Velde, and D. Li. 2003. Loss of K-bearing clay minerals in flood-irrigated, rice-growing soils in Jiangxi Province, China. *Clays and Clay Minerals* 51: 75–82.

Lin, C.-W., Z.-Y. Hseu, and Z.-S. Chen. 2002. Clay mineralogy of Spodosols with high clay contents in the subalpine forests of Taiwan. *Clays and Clay Minerals* 50: 726–735.

Loveland, P.J., and P. Bullock. 1975. Crystalline and amorphous components of the clay fractions in brown podzolic soils. *Clay Minerals* 10: 451–469.

Lowe, D.J. 1986. Controls on the rates of weathering and clay mineral genesis in airfall tephras: A review and New Zealand case study, p. 265–330. In: S.M. Colman, and D.P. Dethier (eds.), *Rates of Chemical Weathering of Rocks and Minerals*. Academic Press, Orlando, Florida.

Lowe, D.J. 1995. Teaching clays: from ashes to allophane, p. 19–23. In: G.J. Churchman, R.W. Fitzpatrick, and R.A. Eggleton (eds.), *Clays: Controlling the Environment*. Proceedings of the 10th International Clay Conference, Adelaide, Australia (1993). CSIRO Publishing, Melbourne.

Lundström, U.S., N. van Breemen, and D. Bain. 2000. The podzolisation process. A review. *Geoderma* 94: 91–107.

Ma, C., and R.A. Eggleton. 1999. Surface layer types of kaolinite: A high-resolution transmission electron microscope study. *Clays and Clay Minerals* 47: 181–191.

Melo, V.F., B. Singh, C.E.G.R. Schaefer, R.F. Novais, and M.P.F. Fontes. 2001. Chemical and mineralogical properties of kaolinite-rich Brazilian soils. *Soil Science Society of America Journal* 65: 1324–1333.

Mikutta, R., G.E. Schaumann, D. Gildemeister, S. Bonneville, M.G. Kramer, J. Chorover, O.A. Chadwick, and G. Guggenberger. 2009. Biogeochemistry of mineral-organic associations across a long-term mineralogical soil gradient (0.3–4100 kyr), Hawaiian Islands. *Geochimica et Cosmochimica Acta* 73: 2034–2060.

Mirabella, A., and M. Egli. 2003. Structural transformations of clay minerals in soils of a climosequence in an Italian alpine environment. *Clays and Clay Minerals* 51: 264–278.

Mortland, M.M., K. Lawton, and G. Uehara. 1956. Alteration of biotite to vermiculite by plant growth. *Soil Science* 82: 477–482.

Murakami.,T., H. Isobe, T. Sato, and T. Ohnuki. 1996. Weathering of chlorite in a quartz-chlorite schist: I. Mineralogical and chemical changes. *Clays and Clay Minerals* 44: 244–256.

Nakao, A., S. Funakawa, T. Watanabe, and T. Kosaki. 2009. Pedogenic alterations of illitic minerals represented by Radiocaesium Interception Potential in soils with different soil moisture regimes in humid Asia. *European Journal of Soil Science* 60: 139–152.

Ndayiragije, S., and B. Delvaux. 2003. Coexistence of allophane, gibbsite, kaolinite and hydroxyl-Al-interlayered 2:1 clay minerals in a perudic Andosol. *Geoderma* 117: 203–214.

Neaman, A., and A. Singer. 2004. The effects of palygorskite on chemical and physico-chemical properties of soils: A review. *Geoderma* 123: 297–303.

Newman, R.H., C.W. Childs, and G.J. Churchman. 1994. Aluminium coordination and structural disorder in halloysite and kaolinite by 27Al NMR spectroscopy. *Clay Minerals* 29: 305–312.

Norrish, K. 1973. Factors in the weathering of mica to vermiculite, p. 417–432. In: J.M. Serratosa (ed.), *Proceedings of the 1972 International Clay Conference, Madrid*. Div. de Ciencas, Madrid.

Norrish, K. 1995. An unusual fibrous halloysite, p. 275–284. In: G.J. Churchman, R.W. Fitzpatrick, and R.A. Eggleton (eds.), *Clays: Controlling the Environment*. Proceedings of the 10th International Clay Conference, Adelaide, Australia (1993). CSIRO Publishing, Melbourne.

Norrish, K., and J.G. Pickering. 1983. Clay minerals, p. 281–308. In: *Soils: An Australian Viewpoint*. CSIRO Australia, Melbourne/Academic Press, London.

Owliaie, H.R., A. Abtahi, and R.J. Heck. 2006. Pedogenesis and clay mineralogical investigation of soils formed on gypsiferous and calcareous materials, on a transect, southwestern Iran. *Geoderma* 134: 62–81.

Pai, C.W., M.K. Wang, and C.Y. Chiu. 2007. Clay mineralogical characterization of a toposequence of perhumid subalpine forest soils in northeastern Taiwan. *Geoderma* 138: 177–184.

Pai, C.W., M.K., Wang, H.B. King, C.Y. Chiu, and J.-L. Hwong. 2004. Hydroxy-interlayered minerals of forest soils in A-Li Mountain, Taiwan. *Geoderma* 123: 245–255.

Papoulis, D., P. Tsolis-Katagas, and C. Katagas. 2004. Progressive stages in the formation of kaolin minerals of different morphologies in the weathering of plagioclase. *Clays and Clay Minerals* 52: 275–286.

Paquet, H., and G. Millot. 1972. Geochemical evolution of clay minerals in the weathered products in soils of Mediterranean climate, p. 199–206. In: J.M. Serratosa (ed.), *Proceedings of the 1972 International Clay Conference, Madrid*. Div. de Ciencas, Madrid.

Parfitt, R.L. 1990. Allophane in New Zealand – A review. *Soil Research* 28: 343–360.

Parfitt, R.L. 2009. Allophane and imogolite: Role in soil biogeochemical processes. *Clay Minerals* 44: 135–155.

Parfitt, R.L., and G.J. Churchman. 1988. Clay minerals and humus complexes in five Kenyan soils derived from volcanic ash – A discussion. *Geoderma* 42: 365–366.

Parfitt, R.L., and T. Henmi. 1980. Structure of some allophanes from New Zealand. *Clays and Clay Minerals* 28: 285–294.

Parfitt, R.L., M. Russell, and G.E. Orbell. 1983. Weathering sequence of soils from volcanic ash involving allophane and halloysite, New Zealand. *Geoderma* 29: 41–57.

Parfitt, R.L., M. Saigusa, and J.D. Cowie. 1984. Allophane and halloysite formation in a volcanic ash bed under differing moisture conditions. *Soil Science* 138: 360–364.

Righi, D., K. Huber, and C. Keller. 1999. Clay formation and podzol development from postglacial moraines in Switzerland. *Clay Minerals* 34: 319–332.

Righi, D., S. Petit, and A. Bouchet. 1993. Characterization of hydroxy-interlayered vermiculite and illite/smectite interstratified minerals from the weathering of a chlorite in a Cryorthod. *Clays and Clay Minerals* 41: 484–495.
Robert, M., M. Hardy, and F. Elsass. 1991. Crystallochemistry, properties and organization of soil clays derived from major sedimentary rocks in France. *Clay Minerals* 26: 409–420.
Robertson, I.D.M., and R.A. Eggleton. 1991. Weathering of granitic muscovite to kaolinite and halloysite and of plagioclase-derived kaolinite to halloysite. *Clays and Clay Minerals* 39: 113–126.
Ross, G.J., C. Wang, A.I. Ozkan, and H.W. Rees. 1982. Weathering of chlorite and mica in New Brunswick podzol developed on till derived from chlorite-mica schist. *Geoderma* 27: 255–267.
Russell, E.W. 1973. *Soil Conditions and Plant Growth*, 10th edn. Longman, London.
Ryan, P.C., and F.J. Huertas. 2009. The temporal evolution of pedogenic Fe-smectite to Fe-kaolin via interstratified kaolin-smectite in a moist tropical soil chronosequence. *Geoderma* 151: 1–15.
Saigusa, M., S. Shoji, and T. Kato. 1978. Origin and nature of halloysite in Ando soils from Towada tephra, Japan. *Geoderma* 20: 115–129.
Sakharov, B.A., and V. Drits. 1973. Mixed-layer kaolinite-montmorillonite: A comparison of observed and calculated diffraction patterns. *Clays and Clay Minerals* 21: 15–17.
Sakharov, B.A., H. Lindgreen, A. Salyn, and V.A. Drits. 1999. Determination of illite-smectite structures using multispecimen X-ray diffraction profile fitting. *Clays and Clay Minerals* 47: 555–566.
Singer, A. 1989. Palygorskite and sepiolite group minerals, p. 829–872. In: J.B. Dixon, and S.B. Weed (eds.), *Minerals in Soil Environments*, 2nd edn. Soil Science Society of America, Madison, Wisconsin.
Singer, A. 2002. Palygorskite and sepiolite, p. 555–584. In: J.B. Dixon, and D.G. Schulze (eds.), *Soil Mineralogy with Environmental Applications*. Soil Science Society of America, Madison, Wisconsin.
Singer, A., and K. Norrish. 1974. Pedogenic palygorskite occurrences in Australia. *American Mineralogist* 59: 508–517.
Singh, B., and R.J. Gilkes. 1992a. Properties of soil kaolinites from south-western Australia. *Journal of Soil Science* 43: 645–667.
Singh, B., and R.J. Gilkes. 1992b. The electron-optical investigation of the alteration of kaolinite to halloysite. *Clays and Clay Minerals* 40: 212–229.
Skiba, M. 2007. Clay mineral formation during podzolisation in an alpine environment of the Tatra Mountains, Poland. *Clays and Clay Minerals* 55: 618–634.
Stevens, K.F., and C.G. Vucetich. 1985. Weathering of Upper Quaternary tephras in New Zealand, 2. Clay minerals and their climatic interpretation. *Chemical Geology* 53: 237–247.
Theng, B.K.G., G.J. Churchman, and R.H. Newman. 1986. The occurrence of interlayer clay-organic complexes in two New Zealand soils. *Soil Science* 142: 262–266.
Torn, M.S., S.E. Trumbore, O.A. Chadwick, P.M. Vitousek, and D.M. Hendricks. 1997. Mineral control of soil organic carbon storage and turnover. *Nature* 389: 170–173.
Utami, S.R., F. Mees, M. Dumon, N.P. Qafoku, and E. Van Ranst. 2019. Charge fingerprint in relation to mineralogical composition of Quaternary volcanic ash along a climatic gradient on Java Island, Indonesia. *Catena* 172: 547–557.
Velde, B. 1985. *Clay Minerals, a Physico-Chemcial Explanation of Their Occurrence*. Elsevier, Amsterdam.
Velde, B. 2001. Clay minerals in the agricultural surface soils in the Central United States. *Clay Minerals* 36: 277–294.
Velde, B., and P. Barré. 2010. *Soils, Plants and Clay Minerals: Mineral and Biologic Interactions*. Springer-Verlag, Berlin.
Velde, B., and A. Meunier. 2008. *The Origin of Clay Minerals in Soils and Weathered Rocks*. Springer-Verlag, Berlin.
Verrecchia, E.P., and M.-N. Le Coustumer. 1996. Occurrence and genesis of palygorskite and associated clay minerals in a Pleistocene calcrete complex, Sde Boqer, Negev Desert, Israel. *Clay Minerals* 31: 183–202.
Watanabe, T., Y. Sawada, J.D. Russell, W.J. McHardy, and M.J. Wilson. 1992. The conversion of montmorillonite to interstratified halloysite-smectite by weathering in the Omi acid clay deposit, Japan. *Clay Minerals* 27: 159–173.
Weaver, C.E. 1989. *Clays, Muds and Shales: Developments in Sedimentology*, vol. 44. Elsevier, Amsterdam.
White, G.N., and J.B. Dixon. 2002. Kaolin-serpentine minerals, p. 389–414. In: J.B. Dixon, and D.G. Schulze (eds.), *Soil Mineralogy with Environmental Applications*. Soil Science Society of America, Madison, Wisconsin.
Wilson, M.J. 1987. Soil smectites and related interstratified minerals: Recent developments, p. 167–173. In: L.G. Schultz, H. van Olphen, and F.A. Mumpton (eds.), *Proceedings of the International Clay Conference, Denver, 1985*. The Clay Minerals Society, Bloomington, Indiana.

Wilson, M.J. 1999. The origin and formation of clay minerals in soils: Past, present and future perspectives. *Clay Minerals* 34: 7–25.
Wilson, M.J. 2004. Weathering of the primary rock-forming minerals: Processes, products and rates. *Clay Minerals* 39: 233–266.
Wilson, M.J., D.C. Bain, and D.M.L. Duthie. 1984. The soil clays of Great Britain: II. Scotland. *Clay Minerals* 19: 709–735.
Wilson, M.J., W.A. Deer, R.A. Howie, and J. Zussman. 2013. *Rock-Forming Minerals,* vol. 3C, *Sheet Silicates: Clay Minerals*. The Geological Society, London.

FIGURE 1.3 Mosses and grass on a 16th century granite church wall northern France.

FIGURE 1.4 Alteration profile on granite (Massif Central France) showing soil, alterite (altered clay-rich zone) and altered rock basement.

FIGURE 3.1 Alteration of rocks in a wall of the Abbaye de Beauport, Brittany, France, showing differences in reaction rates between granite blocks and meta-basaltic rocks in a church wall 400 years old. Dissolution of the iron-rich basaltic minerals in dark-coloured rocks is far greater than that of granites (grey blocks to the left). Basalt mineral alteration is largely due to the effects of oxidation of iron in the basalt minerals. Reddish meta-sandstone (centre, right) is resistant to alteration.

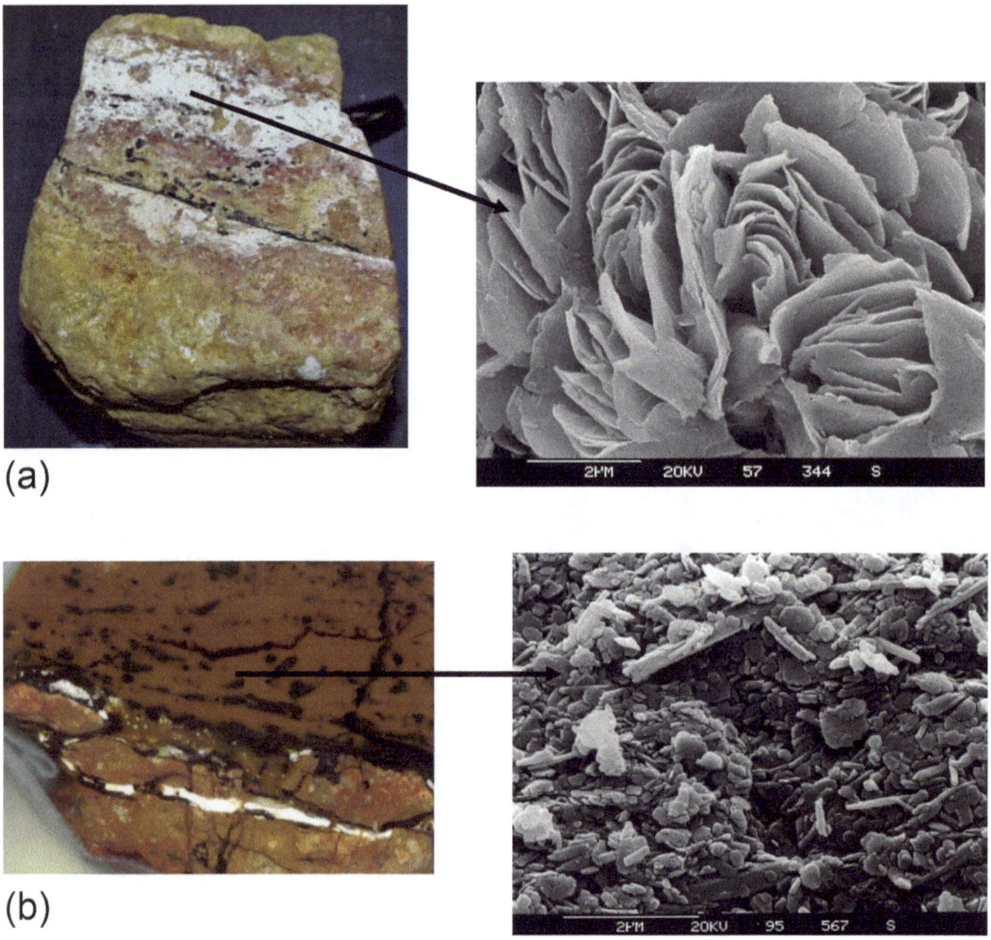

FIGURE 7.6 (A) Scanning electron micrograph (right) of kaolinite clay formed in the white infill in volcanic tuff (left), showing large plates. (With permission from Geotechnical Engineering Office, Civil Engineering and Development Department, Hong Kong.) (B) Scanning electron micrograph (right) of kaolinite clay formed in the reddish-brown infill in volcanic tuff (left), showing small plates (and some halloysite tubes). (With permission from Geotechnical Engineering Office, Civil Engineering and Development Department, Hong Kong.)

FIGURE 14.4 Examples of cracks developed in buildings in Australia as a result of the expansion and contraction of reactive soils. (Left) From https://www.abis.com.au/cracking-movement; (Right) From https://www.domain.com.au/news/is-your-house-cracking-up-what-to-do-about-cracks-in-the-home-20160420-goamq7/ (both accessed August 15, 2018).

FIGURE 14.9 A non-wetting soil in Western Australia with patchy plant growth (right) and improved growth through the addition of clay (at 300 t ha^{-1}) (left). (With thanks to David Hall, Western Australia Department of Primary Industries and Regional Development.)

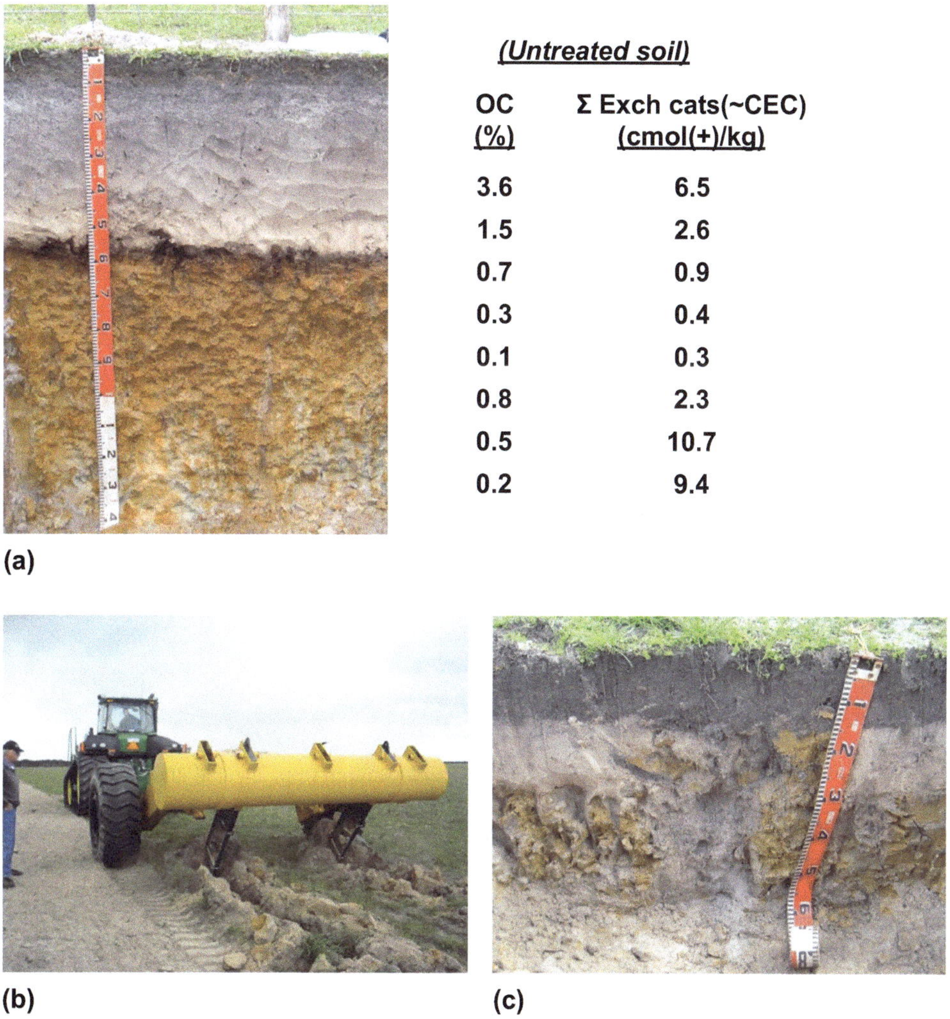

(Untreated soil)

OC (%)	Σ Exch cats(~CEC) (cmol(+)/kg)
3.6	6.5
1.5	2.6
0.7	0.9
0.3	0.4
0.1	0.3
0.8	2.3
0.5	10.7
0.2	9.4

FIGURE 14.10 (a) Profile of a duplex (sand over clay) soil, common in southern and western Australia. Organic carbon (OC) and exchangeable cation (ΣExch cats) values are low in all but the top of the A horizon, indicating low organic C and low clay except in the thin topsoil, while ΣExch cats values are higher in the B horizon, indicating higher clay content. (b) Clay from B horizons is moved into sandy A horizons by delving, giving the result in (c). The depth scale is in centimetres. (Photographs reproduced from Churchman, G.J., et al., 2014, Clay addition and redistribution to enhance carbon sequestration in soils, p. 327–335, in A.E. Hartemink and K. McSweeney (eds.), *Soil Carbon*, Progress in Soil Science, Springer, Switzerland. With permission from Springer Nature.)

FIGURE 14.12 The effect of adding bentonite clay to sandy soils in tropical Thailand upon the growth of forage sorghum. Crop under normal farming practice in foregrounds, crop on clay-amended soil in backgrounds. (Left) With average rainfall. (Right) In a dry year. (From Saleth, R.M., et al., 2009, Economic gains of improving soil fertility and water holding capacity with clay application: The impact of soil remediation research in Northeast Thailand, IWMI Research Report 130. International Water Management Institute, Colombo, Sri Lanka. With permission from International Water Management Institute.)

9 The Importance of Climate in the Formation of Soil Clays

> Science, my lad, is made up of mistakes, but they are mistakes which it is useful to make because they lead little by little to the truth.
>
> **Jules Verne (1828–1905)**

Of the various soil forming factors (Jenny, 1941), climate is a central driving force in the alteration of rock minerals and the formation of clays. It has been described as the most active of the soil factors (Critchfield, 1983). Precipitation is essential for the formation of soils – and clays. More important, however, is the throughflow of water, hence the influence of topography can modify that of climate. Temperature affects the rate of alteration and formation reactions. Biology, which has very much influence, is itself a function of climate. Parent materials provide the substrate for weathering and hence help provide the direction of the alteration and formation changes.

Time is another factor that needs consideration. Some (older) classification systems have included the concept of zonality to account for the effect of time. Zonal soils have well-formed horizons and are largely the effect of climatic and biological influences, rather than those from the parent materials. In azonal soils, by contrast, the time of alteration and formation has been too short for horizons to be well differentiated. Time is an important factor in these cases. There may also be intrazonal soils in which parent material, for one factor, has had a strong influence on the soils formed. The concept of zonality can also be applied to soil clays.

It is useful to examine the effect of climate upon soil clays following broad groupings, as defined by Griffiths (1976). Beginning with the coldest climate zones (the inverse of Griffiths' order), the groupings give rise to the following soil types:

1. Cold and cool zones
   a. Deserts – little organic matter (OM), lime near the surface. These mainly comprise largely unaltered minerals from the parent rocks such as micas, chlorites and feldspars. They may also contain some salts.
   b. Short summers (tundra) – anaerobic, little humus, very acid.
   c. Little rainfall (steppes) – deep humus layer, deep lime accumulation, retain water, very fertile.
   d. Medium rainfall (prairies) – high humus content, slight elevation, fertile.
   e. Wetter areas – podzolic soils (not quite podzols), thin layer of humus.
2. Warm zone
   a. Deserts – same as above (1a).
   b. Eastern regions – considerable laterisation, scanty OM.
   c. Mediterranean – reduced leaching, deeper lime accumulation.
3. Hot zone
   a. Deserts – same as in 1a and 2a.
   b. Tropical grassland – much humus, fertile but rapidly exhausted, dark colour.
   c. Rainforest and wet savannah – laterisation, alkalis eluviated to give acid soils, low humus, not very fertile, red in colour.

While strict adherence to the subdivisions in these groupings is difficult, we can examine soil clays by the broad division of soils into the three main climate zones.

9.1 COLD AND COOL ZONES

Soils in Antarctica serve as an extreme case of the minimum expression of climatic driving forces. Several studies have identified clays in and near to Antarctica (Blakemore and Swindale, 1958; Kelly and Zumberge, 1961; Claridge, 1965; Boyer, 1975; Claridge and Campbell, 1977; Simas et al., 2006; Claridge and Campbell, 2008). Common observations have included the evidence for early transformations of micas, with interstratified minerals forming. Several of the studies have also found evidence for the oxidation of iron, giving rise to iron stains on rocks. This is thought to come particularly from the breakdown of biotite and Fe sulphides (Kelly and Zumberge, 1961). Some smectites have been identified, but caution is noted (e.g. Simas et al., 2006) that some may originate from inheritance, possibly from hydrothermal alteration of parent rock and perhaps also wind transportation. Zeolites have been found (Claridge and Campbell, 2008), presumably from the deposition and subsequent alteration of volcanic ash in alkaline lakes. Salts also occur, consisting largely of the chlorides, nitrates and sulphates of sodium, potassium, calcium and magnesium (Campbell and Claridge, 1977).

Near Antarctica (Blume et al., 1997), some smectite was found, mostly on basalt. Podzolisation was seen to occur in the Falkland Islands (or Malvinas), and Fe was found to be organically bound. In another study on King George Island in the Antarctic area (Simas et al., 2006), soils were found to be basaltic, orthinogenic or acid sulphate types. Among them they showed smectitic interstratifications, allophane-like phases and ferrihydrite. However, soils in and around Antarctica are strictly not soils by some definitions because they contain no plants. Their clay contents are low. They are useful as indicators of early weathering of saprock/saprolite. In Canadian Arctic desert soils, berthierine, a ferroferric 1:1 mineral, was found (Kodama and Foscolos, 1981). This mineral is rare because it would be easily altered in warmer climates.

In cool climates away from Antarctica and the Arctic region, very early weathering changes have been detected. These include the release of Ca and Mg, mainly from biotite, anorthite and hornblende (Starr and Lindroos, 2006); the rapid development of podzol E horizons; and transformation of biotite and chlorite to vermiculite, with some smectite in E horizons (Burt and Alexander, 1996; Melkerud et al., 2000). In subarctic highlands in Iceland, sparse (<5 %) plant cover has led to soils that are "immature and retain mineralogical and physical properties of the volcanic parent material" (Mankasingh and Gísladóttir, 2019, p. 162). The severe climate largely accounts for the lack of plants, which in turn accounts for the paucity of weathering. In this latter respect, they resemble soils on newly erupted basalt on the island of Hawaii (Mikutta et al., 2009) (see also Chapter 8, Section 8.8), albeit that they are in quite different climatic zones. They lack the weathering effects of plants, in the Icelandic case, and that of the passage of sufficient time, in the Hawaiian situation.

There have been many studies of the minerals in soils forming out of retreating glaciers. In these, in the European Alps, smectite has been found to form by different pathways as well as to become decomposed during the soil formation process (Egli et al., 2001). Smectite is the end product of the alteration of chlorite, while micas have proceeded towards interstratified mica-smectite. At the same time, some smectite was identified as the product of hydrothermal alteration. This is decomposed by a process of 'retrograde podzolisation', according to Egli et al. (2001). In China, the development of soil properties following glacier retreat was found to be strictly time-dependant (He and Tang, 2008). The pH of the soil changed from 8.5 to 4.2 following the development of organic acids and its cation exchange capacity (CEC) changed from 0.4 to 19.8 $cmol_+kg^{-1}$ in the A horizon, likely due to the transformation of Fe- and Mg-containing biotite and the dissolution of carbonate. In Europe a study emphasising the development of organic matter after glacial retreat (Dümig et al., 2012) noted that in the first 700 years only very small amounts of a hydroxy-interlayered mineral was formed. The build-up of organic carbon was faster than the supply of mineral phases and the CEC developed was mainly due to the organic carbon. The development of organic matter began with the association of proteins with poorly crystalline iron oxides. In another European study (Egli et al., 2011), it was found that mineral formation started immediately at the beginning of soil formation, resulting in the transformation of mica to smectite. Vegetation accelerated the transformation process. Another study of glacial retreat in the Svalbard Archipelago, with a sub-zero mean annual temperature (Kabala and Zapart, 2012), produced a chronosequence which showed that carbonates were dissolved and leached out and that, while dithionite-soluble Fe was low, its oxalate-extractable form was high. There was a lack of crystalline secondary minerals.

Upland regions in temperate zones also constitute cool climates, at least seasonally. It is a feature of these that 2:1 phyllosilicates are transformed into interstratifications with vermiculite or smectite, that are often regular in layer superposition. These phases are 1:1 of either mica or chlorite to either vermiculite to smectite. They have been found in the upland regions of the former Yugoslavia (now Croatia) (Gjems, 1970), Scotland (Wilson, 1970), Norway (Kapoor, 1973) and South Island, New Zealand (Churchman, 1980). Because the climates in these regions are cool, alterations occur slowly, enabling subtle changes to occur where their rapidity would override them in warmer climates, in which interstratifications tend to be more random. In particular, when micas and chlorites in cool, upland climates lose an interlayer ion, whether potassium or hydroxy-Al, there is a realignment in the adjoining layer by which the adjoining interlayer holds its interlayer species more strongly. This likely occurs because of a shift of structural hydroxides in the adjoining layer towards the now-opened interlayer (Norrish, 1973) (see also Chapter 8, Section 8.4).

9.2 WARM ZONE

Though there have been innumerable studies of the clay mineralogy of individual soils or small sets of soils, only a few have attempted to give a nationwide summary of the types of clays occurring in soils. These include Folkoff and Meentemeyer (1985), Allen and Hajek (1989) and Velde (2001) for the United States; Loveland (1984) and Wilson et al. (1984) for Great Britain (respectively England and Wales, and Scotland); Norrish and Pickering (1983) for Australia; and Fieldes (1968) for New Zealand. Furthermore, Ito and Wagai (2017) have compiled a global dataset of clay-size minerals in topsoils and in subsoils from a survey of the literature.

A study of 99 A horizon soils from across the 48 adjoining states in the United States led Folkoff and Meentemeyer (1985) to find that there was a statistical (discriminant) relationship between climate data and clay mineral types. The climatic factors of the precipitation in the wettest month and a leaching index (Arkley's) accounted for most of the predictive power for clay minerals. Five groups were obtained for clay mineralogy of the soils. One with mica, montmorillonite and kaolinite was most prominent, followed by mica, vermiculite and kaolinite, then mica and kaolinite together, kaolinite and vermiculite together and a heterogeneous group. Allowing for variations in interpretations and naming of mineral types, it can be seen that either mica (probably illite) or its transformation products, vermiculite and some of the montmorillonite, were widespread, as was kaolinite. Although most soils in a compilation for the United States were of mixed mineralogy by Soil Taxonomy, Allen and Hajek (1989) found that Mollisols and Vertisols were dominantly smectitic and smectitic families were also prominent among Alfisols, Aridosols, Entisols and Inceptisols. However, it is noted by Allen and Hajek (1989) that the basis for allocating mineralogy families in the Soil Taxonomy requires revision if it is to be helpful. Velde (2001), sampling 86 surface soils in the Central United States, found that disordered (random) illite–smectites were the dominant type of clay mineral. Further, most of these were ~50% smectite, with another reasonable proportion having ~20% of smectite layers. The composition of the soils on sediments or sedimentary rocks largely reflects their origin. There was also kaolinite and mica-illite in lesser amounts than of illite-smectites but in approximately equal amounts of each (Velde, 2001). It is likely that Folkoff and Meentemeyer (1985) failed to observe the interstratified phases, more easily recognised since Lanson's (1997) work.

In England and Wales, the soil clays vary with the type of sedimentary rock (Loveland, 1984). Many are micaceous, including interstratifications with smectite. Kaolins are rarely dominant and some soils contain chlorite, sepiolite and palygorskite. In Scotland, which was largely glaciated some 10,000 years ago, clays in soils derive from three main sources (Wilson et al., 1984). Most are inherited from the parent material, whether sedimentary, igneous or metamorphic. This has given rise to a wide variety of clay minerals: illite, kaolinite, chlorite, smectite and interstratified minerals of various types. Some, especially kaolinite and halloysite, have derived from preglacial weathering in a warmer climate than at present. Recent (postglacial) weathering has also given rise to transformation of layer silicates, often with interlayering by aluminium. The poorly crystalline minerals, allophane and imogolite, have formed in podzolised soils.

Illite is the dominant clay mineral of the agricultural soils of Southern Australia (Norrish and Pickering, 1983). These occur in a Mediterranean (xeric) climate. Almost all Australian soils contain some kaolinite – it is dominant in soils in Northern Australia, as discussed under 'hot zone' soils (Section 9.3). Basalts often give rise to smectites. Dominantly kaolinitic soils ('krasnozems') also occur on basalts containing no layer silicates, even in Southern Australia. These soils also have high contents of iron oxides (hematite and/or goethite). The dominant illites there derive both from the transformation of primary micas and probably also chlorites and also by authigenesis (neoformation) in former lakes. Norrish and Pickering (1983) note that interstratifications are likely to be common in Australian soils. They make particular note of kaolin–smectite interstratifications, often ignored because of their weak peaks in X-ray diffraction, but important because they explain some anomalously high CECs for some clays. Churchman et al. (1994) have explored this possibility further on five widespread Australian soils, confirming the interstratification of kaolin and smectite in each of them and finding a pattern in each where the interstratification is more smectitic at depth and more kaolinitic towards the surface.

In many soils in New Zealand, where most of the weathering is postglacial, the clay fraction of zonal soils is mainly made up of micas and their transformation products (Fieldes, 1968). These are illites, vermiculites, smectites and hydroxy-interlayered expanded micas or illites (known as 'interlayered hydrous micas'). There can also be some kaolin. The clay fractions of intrazonal soils, so classified because they are formed on volcanic materials and this parent material is highly influential, are different. They contain allophane and/or kaolin, as halloysite. Some, especially those from basaltic materials, also contain a variety of oxides of aluminium, iron and titanium. Intrazonal soils also include those on carbonaceous parent materials, where smectites can form.

Globally, illite tends to be predominant in arid soils and in those of mid-latitude regions, although smectites are dominant in large areas of soils, apparently regardless of climate (Ito and Wagai, 2017). The common occurrence of smectites, especially in Alfisols, Aridosols, Mollisols and, of course, Vertisols, suggest the importance of other soil-forming factors besides climate, including parent materials (generally basalt, but including calcareous materials) and/or poor drainage, as in basins (Borchardt, 1989).

9.3 HOT ZONE

Soils in the hot – and often humid – zones can be regarded as zonal, where climate is the controlling factor, or intrazonal, where volcanic or other, e.g. serpentitic parent materials are the overriding influence on the results of weathering. In Australia, the hot and generally humid north of the continent gives rise to soils in which kaolin (usually as kaolinite) is dominant, but illite can also appear, as well as iron oxides and sometimes also smectite (Norrish and Pickering, 1983). In Southeastern Brazil, kaolinite and gibbsite occur together in a forested soil, with kaolinite more prominent in upper horizons and gibbsite in lower horizons (Furian et al., 2002). It is thought that Si and Al are brought to the surface by the forest, and this has enabled the formation and stability of kaolinite there. Kaolinite is the dominant mineral in Oxisols and Ultisols, which are usually found in the tropics where weathering is intense (Allen and Hajek, 1989). It is also found in laterites, including those in Western Australia, where it occurs in the pallid and mottled zone of lateritic profiles (Gilkes et al., 1973; Anand and Gilkes, 1984; Churchman and Gilkes, 1989). Typically, halloysite is found at the base of lateritic profiles, whether they are in tropical (Eswaran and Heng, 1976; Eswaran and

Bin, 1978), Mediterranean (xeric) (Churchman and Gilkes, 1989) or temperate climates (Calvert et al., 1980), with kaolinites towards the surface and sometimes also with aluminium oxides, both gibbsite and boehmite, on the surface. It may be inferred that this type of profile formed under tropical conditions even if now situated in another climate zone. Among iron oxides, hematite is most often found in soils in the tropics.

Studies in humid Asia have shown that while all 63 subsurface horizons of 44 upland soils (representing 240 sites sampled) from Japan, Thailand and Indonesia contained kaolinite, their clay mineralogy also entailed illite, and its transformation products provided that mica, and particularly muscovite, was present in the parent materials (Nakao et al., 2009). The products of the weathering of micas were vermiculitic when the particular climate regime was udic, where rainfall exceeds evapotransformation. On the other hand, in an ustic regime, with a distinct dry season, illite was less strongly altered. The soil pH also affected the products formed, with hydroxy-Al interlayers, and also gibbsite, forming when pHs were between 4.3 and 5.5, but not at higher or lower pHs (Watanabe et al., 2006). Analysing soil solutions and using thermodynamics, Watanabe et al. (2006) found that the composition of the clay fractions of the soils all tended towards kaolin as the stable product of weathering.

The different extents of alteration between the udic and ustic moisture regimes, essentially that between situations that are always wet (udic) and only wet intermittently (ustic), illustrate a general principle of the effects of moisture regime on weathering. When water is always present, more weathering occurs than when it is only present seasonally or intermittently. Even on seafloors, alteration occurs to give smectitic clays (Borchardt, 1989). In soils, plants can assist in weathering through their retention of water in contact with rocks. The important climate factor is the availability of water. Essentially water plus rock makes clay.

Almost all possible clay minerals may be formed in tropical soils, even if kaolinite is virtually ubiquitous (e.g. Ito and Wagai, 2017). Detailed analysis of the depth profile of a soil on greenschist in Zaire showed that chlorite weathered to chlorite–vermiculite, both regularly and irregularly interstratified, while kaolinite developed towards the surface (Herbillon and Makumbi, 1975). In Spodosols in Taiwan, the podzolisation process has given rise to vermiculite and vermiculite–illite minerals, with hydroxy-Al interlayers only in lower horizons where the pH is sufficiently high, as well as minor kaolinite (Lin et al., 2002).

Smectites also form in tropical soils. They are most likely to form from basaltic parent materials and smectitic layers often occur as kaolin–smectites in interstratifications with kaolinite. These have been recognised as intermediates in alterations from smectites to kaolinites occurring in catena in tropical as well as other climates (e.g., Herbillon et al., 1981, in Burundi; Bühmann and Grubb, 1991, in South Africa; Ryan et al., 2016, in Costa Rica) and also to halloysites (Delvaux et al., 1989, in Cameroon). In general, smectite is found at the base of a slope, where the appropriate ions accumulate, while the soils at the top of the slope are more kaolinitic. Kaolin-smectites are typically found part way along the slope. This change, including the formation of kaolin-smectite intermediates, can also occur within a profile (Churchman et al., 1994, in Australia; Vingiani et al., 2004, in Sardinia). In another situation, in Costa Rica, smectite appears to alter to kaolinite and halloysite with time without giving evidence for any intermediate phase (Ryan et al., 2016). Further, in Brazilian rainforest, a gibbsitic saprolite, denoting strong weathering from the loss of Si from saprolite and lower horizons, has nonetheless given rise to a topsoil containing kaolinite, absent from lower horizons (Furian et al., 2002). Uplift of Al (in organic complexes) and Si by the forest is invoked to explain this anomaly.

Generally speaking, the alterations occurring in tropical soils lead to different products on volcanic parent materials. Many give a range of contents of allophane and halloysite, one of these clay minerals occurring at the expense of the other. This is the case in soils formed at different elevations on an extinct volcano in Rwanda (Nizeyimana et al., 1997). The allophane to halloysite conversion is also shown in a sequence of age (0.3–4100 kyr) across the Hawaiian Islands (e.g. Torn et al., 1997; Mikutta et al., 2009). In equatorial Africa (Tanzania), a variety of climates – as a result of different altitudes of sampling sites on volcanoes, and also parent tephra compositions – enabled a study of these soil-forming factors on volcanic parent materials (Lyu et al., 2018). Differences in parent materials (0.06–0.45 Ma in age) had no perceptible effect compared with differences in both rainfall amount and patterns, and temperature. Both (relatively) low temperatures and low precipitation retarded the formation of nanocrystalline minerals like allophane and ferrihydrite, and their formation was also limited when precipitation increased as long as the temperature remained low. This mirrors the trend in the far more extreme cold temperature in Iceland (Mankasingh and Gísladóttir, 2019) (see also Section 9.1). When the sites had seasonally dry periods (in an ustic regime) kaolinite formed, rather than halloysite, echoing the findings of Churchman et al. (2010) in Hong Kong (see also Churchman et al., 2016; and Chapter 8, Section 8.7). When they were particularly wet, in an udic regime, gibbsite was formed. Soil solutions revealed that Si concentrations were the key factor governing the crystalline products, whether kaolinite, for high Si, or gibbsite, when Si concentration was low (Lyu et al., 2018) (see also Chapter 8, Section 8.10). However, not all clay minerals in soils on volcanic materials derive from weathering, with Van Ranst et al. (2008) finding that 2:1 minerals and also kaolinite in an Andisol in Indonesia was derived from inheritance from the parent rock material where they first formed hydrothermally. Quartz in Hawaiian soils was identified as an inheritance from dust carried hemispheric distances (Rex et al., 1969).

On serpentinite in Taiwan, a toposequence revealed that weathering has produced smectite and serpentine at the summit; chlorite and serpentine on the shoulder; smectite and chlorite-vermiculite on the backslope; and vermiculite on the footslope (Hseu et al., 2007). The soils were all of different types: an Entisol at the summit; a Vertisol on the shoulder; an Alfisol on the backslope; and an Ultisol on the footslope. As well, magnetite in the serpentinic rock was gradually replaced by pedogenic Fe oxides.

9.4 MECHANISMS OF CHANGE

Clearly, desilication is an important process occurring in all climates when enough water is provided in rainfall to effect significant leaching. This reflects the relative solubility of Si encompassing all pH conditions. By contrast, Al is only soluble at low and high pHs (e.g. Churchman, 2000). In general, smectite and also vermiculite and chlorite, as highly siliceous minerals, can undergo the changes on leaching that are shown in Figure 9.1.

These changes involve loss of interlayer potassium or hydroxyl followed by desilication of the layers. Not all of them necessarily occur in any of the steps shown (see e.g. Figure 8.10 in Chapter 8). Some of these steps involve transformations in the solid phase, while some involve dissolution and recrystallisation (see Figure 8.10 in Chapter 8).

Volcanic parent materials mostly tend to give rise to desilication alterations over the range encompassing kaolin minerals through to gibbsite. Resilication is also clearly possible, notably in

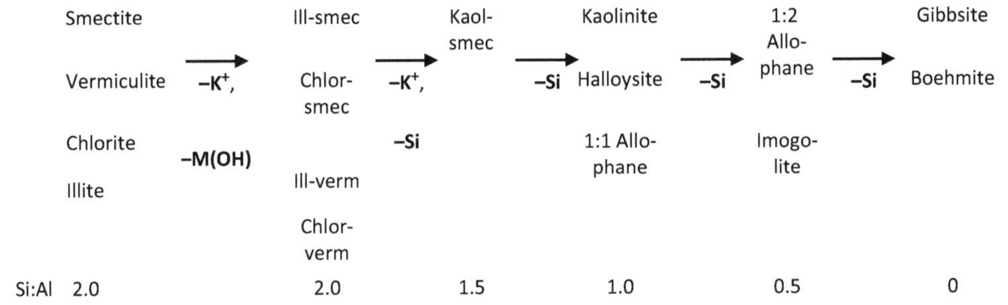

FIGURE 9.1 Possible mineral changes occurring on leaching soils. Where M(OH) = a hydroxyl interlayer; M = Mg, Al.

the change from allophane to halloysite in soils on volcanic parent materials (see Chapter 8, Section 8.8) and some of these materials can give rise to smectites where little leaching and/or poor drainage causes a build-up of Si (see Chapter 8, Section 8.3).

REFERENCES

Allen, B.L., and B.F. Hajek. 1989. Mineral occurrence in soil environments, p. 199–278. In: J.B. Dixon, and S.B. Weed (eds.), *Minerals in Soil Environments*, 2nd edn. Soil Science Society of America, Madison, Wisconsin.

Anand, R.R., and R.J. Gilkes. 1984. Weathering of hornblende, plagioclase and chlorite in meta-dolerite, Australia. *Geoderma* 34: 261–280.

Blakemore, L.C., and L.D. Swindale. 1958. Chemistry and clay mineralogy of a soil sample from Antarctica. *Nature* 182: 47–48.

Blume, H.P., L. Beyer, M. Bőlter, H. Erlenkauser, E. Kalk, S. Kneesh, U. Pfisterer, and D. Schneider. 1997. Pedogenic zonation in soils of the southern circum-polar region. *Advances in Geoecology* 30: 69–90.

Borchardt, G. 1989. Smectites. In: J.B. Dixon, and S.B. Weed (eds.), *Minerals in Soil Environments*, 2nd edn. Soil Science Society of America, Madison, Wisconsin. p. 675–727.

Boyer, S.J. 1975. Chemical weathering of rocks on the Lassiter Coast, Antarctic Peninsula, Antarctica. *New Zealand Journal of Geology and Geophysics* 18: 623–628.

Bühmann, C., and P.L.C. Grubb. 1991. A kaolin-smectite interstratification sequence from a red and black complex. *Clay Minerals* 26: 343–358.

Burt, R., and E.B. Alexander. 1996. Soil development on moraines of Mendenhall Glacier, southeast Alaska. 2. Chemical transformations and soil micromorphology. *Geoderma* 72: 19–36.

Calvert, C.S., S.W. Buol, and S.B. Weed. 1980. Mineralogical transformations of a vertical rock-saprolite-soil sequence in the North Carolina Piedmont. *Soil Science Society of America Journal* 44: 1096–1112.

Churchman, G.J. 1980. Clay minerals formed from micas and chlorites in some New Zealand soils. *Clay Minerals* 15: 59–76.

Churchman, G.J. 2000. The alteration and formation of soil minerals by weathering, p. F3–F76. In: M.E. Sumner (ed.), *Handbook of Soil Science*. CRC Press, Boca Raton, Florida.

Churchman, G.J., and R.J. Gilkes. 1989. Recognition of intermediates in the possible transformation of halloysite to kaolinite in weathering profiles. *Clay Minerals* 24: 579–590.

Churchman, G.J., P. Pasbakhsh, D.J. Lowe, and B.K.G. Theng. 2016. Unique but diverse: Some observations on the formation, structure and morphology of halloysite. *Clay Minerals* 51: 395–416.

Churchman, G.J., I.R. Pontifex, and S.G. McClure. 2010. Factors affecting the formation and characteristics of halloysites or kaolinites in granitic and tuffaceous saprolites in Hong Kong. *Clays and Clay Minerals* 58: 220–237.

Churchman, G.J., P.G. Slade, P.G. Self, and L.J. Janik. 1994. Nature of interstratified kaolin-smectites in some Australian soils. *Soil Research* 32: 805–822.

Claridge, G.G.C. 1965. The clay mineralogy and chemistry of some soils from the Ross Dependency, Antarctica. *New Zealand Journal of Geology and Geophysics* 8: 186–220.

Claridge, G.G.C., and I.B. Campbell. 1977. The salts in Antarctic soils, their distribution and relationships to soil processes. *Soil Science* 123: 377–384.

Claridge, G.G.C., and I.B. Campbell. 2008. Zeolites in Antarctic soils: Examples from Coombs Hills and Marble Point. *Geoderma* 144: 66–72.

Critchfield, H.J. 1983. *General Climatology*, 4th edn. Prentice-Hall, Englewood Cliffs, New Jersey.

Delvaux, B., A.J. Herbillon, and L. Vielvoye. 1989. Characterization of a weathering sequence of soils derived from volcanic ash in Cameroon. Taxonomic, mineralogical and agronomic implications. *Geoderma* 45: 375–388.

Dümig, A., W. Häusler, M. Steffens, and I. Kögel-Knabner. 2012. Clay fractions from a soil chronosequence after glacier retreat reveals the initial evolution of organo-mineral associations. *Geochimica et Cosmochimica Acta* 85: 1–18.

Egli, M., A. Mirabella, and P. Fitze. 2001. Clay mineral formation of two different chronosequences in the Swiss Alps. *Geoderma* 104: 145–175.

Egli, M., M. Wernli, C. Burga, C. Kneisel, C. Mavris, G. Valboa, A. Mirabella, M. Plötze, and W. Haeberli. 2011. Fast but spatially scattered smectite formation in the proglacial area Mortertasch: An evaluation using GIS. *Geoderma* 264: 11–21.

Eswaran, H., and W.C. Bin. 1978. A study of a deep weathering profile on granite in Peninsular Malaysia. Parts I, II, and III. *Soil Science Society of America Journal* 42: 144–158.

Eswaran, H., and Yeow Yew Heng. 1976. The weathering of biotite in a profile on gneiss in Malaysia. *Geoderma* 16: 9–20.

Fieldes, M. 1968. Clay mineralogy, p. 22–39. In: *Soils of New Zealand*, Part 2. New Zealand Soil Bureau Bulletin 26.

Folkoff, M.E., and V. Meentemeyer. 1985. Climatic control of the assemblages of secondary clay minerals in the A-horizon of United States soils. *Earth Surface Processes and Landforms* 10: 621–633.

Furian, S., L. Barbiéro, R. Boulet, P. Curmi, M. Grimaldi, and C. Grimaldi. 2002. Distribution and dynamics of gibbsite and kaolinite in an oxisol of Serra do Mar, southeastern Brazil. *Geoderma* 106: 83–100.

Gilkes, R.J., G. Scholz, and G.M. Dimmock. 1973. Lateritic deep weathering of granite. *Journal of Soil Science* 24: 523–536.

Gjems, O. 1970. Mineralogical composition and pedogenic weathering of the clay fraction in podzol weathering profiles in Zalesine, Yugoslavia. *Soil Science* 110: 237–243.

Griffiths, J.F. 1976. *Applied Climatology: An Introduction*, 2nd edn. Oxford University Press, Oxford.

He, L., and Y. Tang. 2008. Soil development along primary succession sequences on moraines of Hailuogou Glacier, Gongga Nountain, Sichuan, China. *Catena* 72: 259–269.

Herbillon, A.J., R. Frankart, and L. Vielvoye. 1981. An occurrence of interstratified kaolinite-smectite minerals in a red-black soil toposequence. *Clay Minerals* 16: 195–201.

Herbillon, A.J., and M.N. Makumbi. 1975. Weathering of chlorite in a soil derived from a chlorito-schist under humid tropical conditions. *Geoderma* 13: 89–104.

Hseu, Z.Y., H. Tsai, H.C. Hsi, and Y.C. Chen. 2007. Weathering sequences of clay minerals in soils along a serpentinitic toposequence. *Clays and Clay Minerals* 55: 389–401.

Ito, A., and R. Wagai. 2017. Global distribution of clay-size minerals on land surface for biogeochemical and climatological studies. *Scientific Data* 4: 170103.

Jenny, H. 1941. *Factors of Soil Formation: A System of Quantitative Pedology*. McGraw-Hill, New York.

Kabala, C., and J. Zapart. 2012. Initial soil development and carbon accumulation on moraines of the rapidly retreating Werenskiold Glacier, SW Spitsbergen, Svalbard archipelago. *Geoderma* 175–176: 9–20.

Kapoor, B.S. 1973. The formation of 2:1-2:2 intergrade clays in some Norwegian podzols. *Clay Minerals* 10: 79–86.

Kelly, W.C., and J.H. Zumberge. 1961. Weathering of a quartz diorite at Marble Point, McMurdo Sound, Antarctica. *The Journal of Geology* 69: 433–446.

Kodama, H., and A.E. Foscolos. 1981. Occurrence of berthierine in Canadian Arctic desert soils. *Canadian Mineralogist* 19: 279–283.

Lanson, B. 1997. Decomposition of experimental X-ray diffraction patterns (profile fitting): A convenient way to study clays. *Clays and Clay Minerals* 45: 132–146.

Lin, C.-W., Z.-Y. Hseu, and Z.-S. Chen. 2002. Clay mineralogy of Spodosols with high clay contents in the subalpine forests of Taiwan. *Clays and Clay Minerals* 50: 726–735.

Loveland, P.J. 1984. The soil clays of Great Britain: I. England and Wales. *Clay Minerals* 19: 681–707.

Lyu, H., T. Watanabe, M. Kilasara, and S. Funakawa. 2018. Effects of climate on distribution of soil secondary minerals in volcanic regions of Tanzania. *Catena* 166: 209–219.

Mankasingh, U., and G. Gísladóttir. 2019. Early indicators of soil formation in the Icelandic sub-arctic highlands. *Geoderma* 337: 152–163.

Melkerud, P.-A., D.C. Bain, A.G. Jongmans, and T. Tarvainen. 2000. Chemical, mineralogical and morphological characteristics of three podzols developed on glacial deposits in Northern Europe. *Geoderma* 94: 125–148.

Mikutta, R., G.E. Schaumann, D. Gildemeister, S. Bonneville, M.G. Kramer, J. Chorover, O.A. Chadwick, and G. Guggenberger. 2009. Biogeochemistry of mineral-organic associations across a long-term mineralogical soil gradient (0.3–4100 kyr), Hawaiian Islands. *Geochimica et Cosmochimica Acta* 73: 2034–2060.

Nakao, A., S. Funakawa, T. Watanabe, and T. Kosaki. 2009. Pedogenic alterations of illitic minerals represented by Radiocaesium Interception Potential in soils with different soil moisture regimes in humid Asia. *European Journal of Soil Science* 60: 139–152.

Nizeyimana, E., T.J. Bicki, and P.A. Agbu. 1997. An assessment of colloidal constituents and clay mineralogy of soils derived from volcanic materials along a toposequence in Rwanda. *Soil Science* 162: 361–371.

Norrish, K. 1973. Factors in the weathering of mica to vermiculite, p. 417–432. In: J.M. Serratosa (ed.), *Proceedings of the 1972 International Clay Conference, Madrid*. Div. de Ciencas, Madrid.

Norrish, K., and J.G. Pickering. 1983. Clay minerals, p. 281–308. In: *Soils: An Australian Viewpoint.* CSIRO, Melbourne/Academic Press, London.

Rex, R.W., J.K. Syers, M.L. Jackson, and R.N. Clayton. 1969. Eolian origin of quartz in soils of Hawaiian Islands and in Pacific pelagic sediments. *Science* 163: 277–279.

Ryan, P.C., F.J. Huertas, F.W.C. Hobbs, and L.N. Pincus. 2016. Kaolinite and halloysite derived from sequential transformation of pedogenic smectite and kaolinite-smectite in a 120 ka tropical soil chronosequence. *Clays and Clay Minerals* 64: 639–667.

Simas, F.N.B., C.E.G.R. Schaefer, V.F. Melo, M.B.B. Guerra, M. Saunders, and R.J. Gilkes. 2006. Clay-sized minerals in permafrost-affected soils (Cryosols) from King George Island, Antarctica. *Clays and Clay Minerals* 54: 721–736.

Starr, M., and A.-J. Lindroos. 2006. Changes in the rate of release of calcium and magnesium and normative mineralogy due to weathering along a 5,300-year chronosequence of boreal forest soils. *Geoderma* 133: 269–280.

Torn, M.S., S.E. Trumbore, O.A. Chadwick, P.M. Vitousek, and D.M. Hendricks. 1997. Mineral control of soil organic carbon storage and turnover. *Nature* 389: 170–173.

Van Ranst, E., S.R. Utami, A. Verdoodt, and N.P. Qafoku. 2008. Mineralogy of a perudic andosol in central Java, Indonesia. *Geoderma* 144: 379–386.

Velde, B. 2001. Clay minerals in the agricultural soils in the central United States. *Clay Minerals* 36: 277–294.

Vingiani, S., O. Righi, S. Petit, and F. Terribile. 2004. Mixed-layer kaolinite-smectite minerals in a red-black soil sequence from basalt in Sardinia (Italy). *Clays and Clay Minerals* 52: 473–483.

Watanabe, T., S. Funakawa, and T. Kosaki. 2006. Clay mineralogy and its relationship to soil solution composition in soils from different weathering environments in humid Asia: Japan, Thailand and Indonesia. *Geoderma* 136: 51–63.

Wilson, M.J. 1970. A study of weathering in a soil derived from a biotite-hornblende rock. *Clay Minerals* 8: 291–303.

Wilson, M.J., D.C. Bain, and D.M.L. Duthie. 1984. The soil clays of Great Britain: II. Scotland. *Clay Minerals* 19: 709–735.

10 Associations of Soil Clays

> Clay is moulded to form a cup, but it is on its non-being that the utility of the cup depends. Doors and windows are cut out to make a room, but it is on its non-being that the utility of the room depends. Therefore turn being into advantage, and turn non-being into utility.
>
> **Lao Tzu (600 BC–531 BC)**

One important feature of clay minerals in soils is that they are most often associated with other solid phases. These may be other aluminosilicates, and they almost invariably include oxides and oxyhydroxides of iron as well as those (along with hydroxides) of aluminium and manganese, but organic matter can be a key component in associations with minerals. This is a time of rapid acceleration in the number of research publications on soil organic carbon (SOC). Tracking two search engines (Scopus and Web of Science), McBratney et al. (2014) found that there had been a rise from ~200 publications on SOC per year in the year 2000 to around 1000 per year in 2012 and the trend is a rising one. It points to a paradigmatic shift having occurred in the topic of soil organic carbon. To understand how organic matter associations come about, it is useful to examine the interactions of minerals with biota and organic matter. First, however, we need to understand what is meant by 'organic matter' and what is known about it nowadays. In order to do this, it is helpful to put our current understanding of organic matter into context through a brief outline of the history of research into organic matter.

10.1 ORGANIC MATTER

Up until quite recently, organic matter in soils was thought to comprise humic substances and non-humic substances. The latter consisted of molecules with well-defined structures, such as carbohydrates, nucleic acids and proteins (e.g. Theng, 2012). Humic substances have always been less well defined. Indeed, they have traditionally been defined operationally, according to the results of a sequence of treatments with first sodium hydroxide, then hydrochloric or sulphuric acid (e.g. Russell, 1973). The material that is insoluble in caustic soda was defined as humin, while most interest lay in the material which remained soluble in acid (fulvic acid) and the product that is insoluble in acid (humic acid). While there was concern expressed among researchers that the products of the treatments, fulvic and humic acids and humin, may well be artefacts of the treatments with no real physical meaning, they often adopted the procedure. At the same time, efforts were made to deduce the structure of organic matter (e.g. Schnitzer and Khan, 1978). Eventually, however, it came to be recognised that soil organic matter is incredibly complex. And most significantly, it is probably the case that no two instances of soil organic matter are alike (Dubach and Mehta, 1963; Hayes et al., 1989). Hence there is little point in determining 'its' structure. Furthermore, a single structure for the products of alkali-acid treatments is unlikely.

Simply put, and following Kleber et al. (2015), organic matter is the sum of living and non-living biota and their breakdown products. The breakdown of dead plants, fauna and microorganisms occurs in an aqueous medium. Water is the solvent for the decomposition of the dead organic materials, and it also serves as a transport medium for microbes as active decomposition agents and for the organic materials in both dissolved and colloidal forms. Root exudates also promote microbial activity in the rhizosphere.

Microorganisms are the principal agents of decomposition of dead organic materials. They act to depolymerise larger biopolymers into smaller molecules that dissolve in water, and they oxidise and thereby functionalise the resulting molecules. Active microorganisms also exude a variety of substances including acids, chelators and so-called extracellular polymeric substances (EPS).

The latter, commonly polysaccharides, may be deposited on mineral surfaces. They can provide bridges between minerals and other organic molecules (e.g. Guckert, 1975; Foster, 1981; Chenu et al., 1987; Chenu, 1989, 1993; Robert and Chenu, 1992; Oades, 1993; Mills, 2003).

In lieu of a pursuit for a structure for organic matter, the great emphasis in studies of soil organic matter has been on the analysis of the occurrence of functional groups, originally in the operational fractions, humic and fulvic acids and then in whole soils. The main technique for these analyses, quantitative as well as qualitative, has been nuclear magnetic resonance (NMR) and its various manifestations, particularly magic angle spinning and most particularly with ^{13}C as the magnetic isotope. Other techniques have been used to also explore functional groups; these include Fourier transform infrared (FTIR) spectroscopy; pyrolysis, with gas chromatography or mass spectroscopy, for identification; and synchrotron-based near-edge X-ray absorption fine structure (NEXAFS). FTIR analyses can also benefit from the high energy in a synchrotron.

Nuclear magnetic resonance has revealed that functional groups in soil organic matter fit into four broad classes, namely, carbonyl, aromatic, O-alkyl and alkyl (Skjemstad et al., 1998). Their identification and quantification provide a more fruitful challenge than the search for a structure for organic matter or the products of its alkali-acid treatments. While some have proposed that humic substances comprise randomly coiled macromolecules, it may be that they consist of supramolecular aggregates of many molecules with relatively low molecular weights (Theng, 2012).

Undoubtedly, soil organic matter changes with time. The process of oxidative depolymerisation of the residue of plants and fauna by continuing microbial activity adds ionisable oxygen-containing functional groups to existing soil organic matter that leads to its decrease in molecular size and increase in aqueous solubility (Kleber et al., 2015). This process will continue to occur as long as the soil is moist and remnants of living biomass are present.

It is the carboxyl functional groups which are mainly responsible for both the aqueous solubility and also the chemical reactivity of soil organic matter. Following dissolution, organic matter coming into contact with mineral matter, e.g. within soil pores, is changed in composition and character to an extent dependent on the nature of the minerals contacted (Kleber et al., 2015). The interaction between minerals and organic matter in soils is symbiotic, with minerals affecting organic matter, while the latter greatly influences the nature and reactivity of minerals.

We have noted that scientists dealing with soil organic matter have largely discarded the search for well-defined compounds in favour of analyses revealing their functional groups. In contrast, the objects of soil clay mineralogists' studies have evolved in the opposite direction, namely from compounds found to be well-crystalline when X-ray diffraction was first applied to soil clays, through to a myriad of compounds with variable structures and characteristics. Thus, clay mineralogists working on soils find themselves dealing with poorly crystalline compounds hardly amenable to analysis by X-ray diffraction, such as allophane and ferrihydrite, a very large number of combinations of phyllosilicates in mixed-layered compounds, and many different varieties of interlayered hydrous aluminosilicates.

10.2 CHARGES ON CLAY PARTICLES

Soil mineral materials are to a large extent of two types, crystalline and 'amorphous' (or nanocrystalline), in their mineral structures. Amorphous materials are bound in normal chemical manners, usually by oxygen and metal covalent bonding, as are found in geological crystalline structures. However, the overall organisation of amorphous assemblages is such that little or no regularity is found, and bond orientations are more or less random. In crystalline materials the orientations of bonds are well defined, and the positions of the associated atoms are equally well defined. This leads to identifiable ionic structures which are visually distinct when seen under an electron microscope or with X-ray diffraction techniques. In the alteration zones there are two types of materials, as described earlier, where the crystalline materials are usually of crystal shapes that are tabular, i.e. flat in shape or book-shaped. In such structures the flat surfaces represent electronically

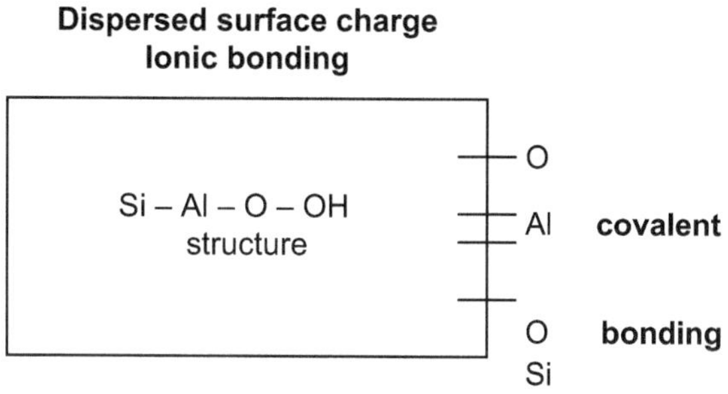

FIGURE 10.1 Sites for bonding on tabular phyllosilicates.

equilibrated surfaces or diffuse charge imbalances where all atoms are bonded in covalent structure attachments. The crystal edges are 'incomplete' in that the structure is not terminated in an equilibrated manner and hence unsatisfied charges, either negative or positive, are present on some ionic surfaces. At these points of the structure, free ions or ionic functions in assemblages present in the aqueous solutions can be attached to the clay crystal edges. The importance of bonding on crystal edges is that at these sites the covalent bonding structure of the silicate crystal is ruptured and the ions present would normally be fixed by strong covalent chemical bonding, but since the structure is discontinuous other ions are fixed on these sites (Figure 10.1). Here organic material is bonded to the clays. However, given the geometry of the clay crystals, generally tabular in form, the bonding with organic matter occurs on only specific sites. When the organic matter joins two phyllosilicate crystals by edge bonding, a specific type of microstructure begins to form. This is indicated in Figure 10.2. Edge-to-edge bonding leaves spaces between the crystals which are microvoids or pores. Here water is present in soils where it is fixed by capillary action which retains the water beyond that in free water under normal atmospheric conditions. The assemblage of these edge-bonded structures creates a special type of aggregate in soils which tends to retain soil water at times of drought. There is a strong tendency for the microaggregates to be joined together within macroaggregates which define soil structure.

Significant amounts of organic matter are bonded to diffuse charged clay surfaces by ionic attractions. This material is less strongly attached to the clays than that covalently bonded. It is less stable in the soil zone and is lost before the covalently bonded crystal edge material. This type of bonding is very important in the formation of macroaggregates.

In the case of amorphous solid materials in soils (oxides for the most part), most of the surfaces have residual ionic sites of either negative or positive charge which are distributed over the surface

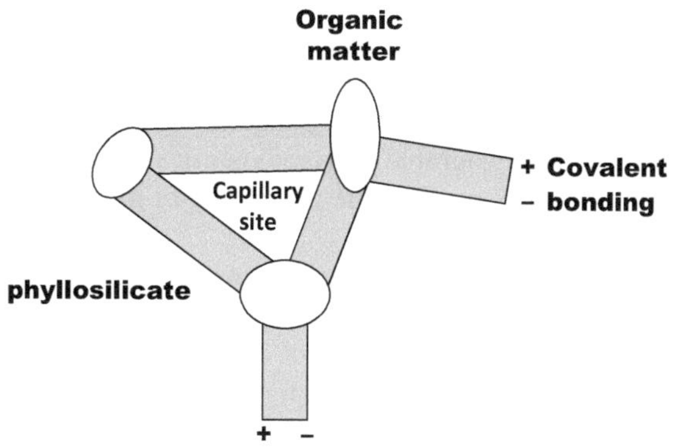

FIGURE 10.2 An example of phyllosilicate–organic bonding where organics join clay particles by covalent bonds on particle edges, often providing capillary pores for the retention of water.

of the particles. The sites of residual charge on silicate or oxide materials attract organic material which has the surface electronic configuration of an opposite charge. However, since there is no geometric restriction, there is no or little pore space between other inorganic material covered by oxide particles, and hence little micro (capillary) porosity.

10.3 ORGANO-MINERAL INTERACTIONS

In addition to playing a role in the formation of secondary carbonates, probably through processing by bacteria and fungi, biological agents can also bring about the formation of minerals with short-range order (or 'nanocrystalline' minerals; Churchman and Lowe, 2012) such as allophane and ferrihydrite, as part of their function in soil formation (Urrutia and Beveridge, 1995; Verboom and Pate, 2006). The formation of these minerals has often been observed in the process of podzolisation, and also in the formation of laterites (Verboom and Pate, 2006) as well as in the process of andisolisation from volcanic parent materials (Churchman and Lowe, 2012). As we will see later, biotic agents may be involved to a greater or lesser extent in these different processes. One of the main mechanisms for the formation of nanocrystalline forms of Fe and Al is through the extraction of these elements from the surfaces of primary minerals by means of their complexation by organic ligands of plant and/or microbial origin (Brantley et al., 2012). Typically, the organic ligands bind the metal cations via negatively charged oxygen atoms. Siderophores are among the most effective microbially derived compounds for the binding of Fe (III), which is very insoluble in noncomplexed form except at pHs well below those in almost all soils. Experiments have shown that, in abiotic conditions, so without organic ligands present, Fe and also Al are hardly released on weathering compared to Ca and Mg.

The growth of plants has often been seen to bring about the loss of potassium from the interlayers of micas to give more or less expandable 2:1 minerals including vermiculites and smectites by the process of 'transformation' in the solid state (Barré et al., 2009; Churchman and Lowe, 2012) (see also Chapter 14, Section 14.1).

In sediments, crystalline minerals have been seen to form through the actions of microorganisms (see also Chapter 7, Section 7.3.10). These include nontronite, which was found within polysaccharides originating from microbial cells in deep sea sediments (Ueshima and Tazaki, 2001). These authors also synthesised nontronite in groundwater containing Si-bearing Fe hydroxides that were seeded with polysaccharides. Halloysite was also formed through the incubation, over a period of months, of freshwater sediments containing kaolinite and feldspar, along with other, less labile, minerals (Tazaki, 2005). These minerals dissolved and then recrystallised ('neo-formed') halloysite. Kaolinite was also synthesised through the action of bacteria on solutions containing Si and Al together with complexing organic liquids (Fiore et al., 2011). Biota are capable of enabling the synthesis of all major types of minerals, whether by transformation or neoformation (Konhauser and Riding, 2012). They may perform a passive role by concentrating elements in order to catalyse the precipitation of minerals from supersaturated solutions. They may also play an active role via their physiology so that new minerals may form even though the contacting bulk solutions are not favourable thermodynamically for their formation and stability. These observations in sediments point to the possibility of microbial-mediated pathways for mineral formation in soils.

Microorganisms also play an active role in weathering through (1) physical disintegration of rocks; (2) production of acids, chelating ligands and oxidising agents to accelerate chemical alteration; (3) stabilisation of soils to increase their time of exposure to chemical agents; (4) production of extracellular polymers to moderate water potential, maintain diffusion channels and themselves act as ligands for chelation and also as nuclei for mineral neoformation; and (5) absorption of nutrients to lower solution concentrations and increase the chemical potential for weathering (Barker et al., 1997). Microorganisms can also act as a template in the formation of minerals. Thus, microbial synthesis (mineralisation) has been demonstrated of iron oxyhydroxides (Chan et al., 2009), short-range ordered aluminosilicates resembling allophane (Urrutia and Beveridge, 1995) and a halloysite-like mineral (Tazaki, 2005). The particular role played by minerals in the regulation of microbial communities and in their interactions with minerals has prompted the coinage of the term 'mineralosphere' for the active surrounds of minerals in the soils by analogy with the biologically active rhizosphere (Uroz et al., 2015).

From an evolutionary point of view, the plants producing soils some hundreds of millions of years ago more generally also resulted in a marked, and accelerating, increase in the rate of production of crystalline clay minerals (Kennedy et al., 2006). This is reflected in an increasing rate of development of the phyllosilicates, kaolinite and the expanding clays (notably vermiculite and smectite), and concomitant decreasing trends in illite (mica), chlorite and carbonates in shales, as common sedimentary rocks, with the passage of time over the past few hundred million years (Kennedy et al., 2006). The greatest rate of increase shown in North American shales is in smectites, for ages of around 500 million years, with the strongest increase in kaolinite occurring in shales from more recent times of around 200–300 million years (Verboom and Pate, 2013)

Undoubtedly, one of the most important effects of the onset of plants in the Earth's history was a great increase in the rate of production of carbon dioxide, CO_2, and its utilisation in the weathering of rocks via the agency of plants and associated biota (Berner, 1997). Although it may be cycled through the atmosphere, CO_2 derives ultimately from photosynthesis and consequent cellular respiration into the rhizosphere, where its acidifying effect (of up to two pH units; see Calvaruso et al., 2009) greatly promotes weathering of minerals from rocks. Weathering reactions of magmatic and high temperature metamorphic rocks are exemplified by the breakdown of the calcic and sodic plagioclases, anorthite $CaAl_2Si_2O_8$ and albite $NaAlSi_3O_8$ respectively, to yield 2:1 and 1:1 minerals (see e.g. Equations 7.1 and 7.2 in Chapter 7).

In the first place, CO_2 dissolves in water to give hydrogen ions:

$$CO_2 + H_2O = H^+ + HCO_3^-$$

It is the action of the hydrogen ions in water that affects the breakdown of the feldspars, with subsequent precipitation of phyllosilicates. The process can be seen as dissolution followed by recrystallisation, or neoformation, as in the case of the formation of the 1:1 mineral kaolinite:

$$CaAl_2Si_2O_8 + 2H^+ + H_2O = Al_2Si_2O_5(OH)_4 + Ca^{2+}$$

$$2NaAl_2Si_3O_8 + 2H^+ + 9H_2O = Al_2Si_2O_5(OH)_4 + 2Na^+ + 4H_4SiO_4$$

The chemical weathering of many – but not all – primary minerals (that of micas and chlorites, due to intracrystalline transformation, can be an exception) occurs by analogous dissolution and recrystallisation reactions initiated by the action of plant-derived carbon dioxide in water. Carbon dioxide from other sources such as volcanoes and other geothermal sources like hot springs and geysers had brought about some chemical weathering prior to the onset of vascular plants some hundreds of million years ago. Some weathering by incongruent dissolution had even resulted from the simple dissolution of primary minerals in non-acidified water. Nevertheless, chemical weathering of minerals from rocks accelerated when higher rooting plants evolved in the geological record.

Organic materials are key components of all soils, apart, perhaps, from those at the very early stages of weathering, those in especially cold environments, and in other environments hostile to most plant and animal life. They are also absent or rare in the deep abiotic zones in weathering profiles that are more properly described as regolith, saprock or saprolite, than as soil or alterite. These materials constitute a highly heterogeneous assortment of living organisms, including plants, macrofauna and microbial life, together with the dead detritus of all of these in varying stages of decomposition and also of transformation, largely as a result of the activities of the microbes. However, the non-living soil organic material, usually referred to as soil organic matter (SOM), per se may not be simply a product of these processes. It may also participate as an active agent in either (1) the decomposition of further dead organisms in and on soils, (2) as a source of slow release of nutrients for plants, (3) for the storage of water, for thermal insulation and solar heating of the soil, (4) as a pH buffer, (5) for the storage of some nutrients through cation exchange and (6) as a cement binding mineral particles together to build soil structure, help create pores and stabilise soils against erosion (e.g. Skjemstad et al., 1998). While it has long been known that most SOM is closely associated with minerals in soils – Greenland (1965) suggested that between 51% and 98% of all SOM was bound in complexes with clay minerals – it has been the recent thrust to maximise the proportion of carbon from rising atmospheric concentrations as gases that can be sequestered in soils which has led to much new research on the bonding between clay minerals and organic carbon as SOM. In turn, the insights being gained from the consequent and continuing explosion of research studies on the mechanism of association of SOM with minerals in soils has shone new light on those aspects of soil minerals which are most reactive towards SOM, as the other most reactive component in soils. It is clear that reactions are most rapid between small units of organic matter and clay minerals.

Provided plants and microbial agents that are capable of depolymerising plant biopolymers are present, as well as water, there is a strong tendency for the resulting monomers to associate with reactive mineral surfaces. Microorganisms can also provide organic acids, chelators and extracellular polymeric substances (EPS) by exudation and some of these compounds can be deposited on mineral surfaces, with some contributing to the dissolved pool of organic matter (Kleber et al., 2015).

The burgeoning spate of publications on soil organic matter includes many that have dealt with the associations of SOC/SOM with minerals. These associations have been studied by a number of different approaches. Albeit that some studies and publications are unable to be categorised by just one of the selected approaches, the various approaches, and some examples, with their outcomes are:

1. *Correlations between contents of SOC and various soil properties* (Nichols, 1984; Spain, 1990; Amato and Ladd, 1992; Arrouays et al., 1995; Saggar et al., 1999; Percival et al., 2000; Kahle et al., 2002; Kleber et al., 2005; Mikutta et al., 2005; Wiseman and Püttmann, 2006; Hernández et al., 2012; Rowley et al., 2018).

   Several studies, e.g. Nichols (1984) and Amato and Ladd (1992), have found good correlations between OC contents of soils and either their clay contents or properties which reflect clay content such as CECs. However, others have found to the contrary that there is no such correlation. There was a positive correlation between carbon and clay contents for most of a set of 72 soils in tropical Australia (Spain, 1990), but this relationship failed for soils derived from basalt, where interactions with free iron oxides better explained C contents. It may be that there is a correlation across the same soil type but not necessarily if several types are compared. In a particularly relevant study, Percival et al. (2000) attempted a correlation of this kind with 167 soils from around New Zealand and found no relationship at all between SOC and clay contents. Instead, SOC related best to oxalate-extractable Al and Si, which measure allophane content, and also with pyrophosphate-extractable Al, which measures Al associated with organic matter. Several of the other studies have found that SOC related best to measures of poorly crystalline Al and Fe, with surface areas of the material playing some role in affecting SOC content. However, the stabilising effects of Al and Fe in various forms upon SOC likely prevail only under acidic conditions, where most studies have been carried out, whereas good correlations between exchangeable Ca and SOM have been reported for soils under basic conditions (Rowley et al., 2018). It appears that pH governs the mechanism of stabilisation of SOC by minerals and ions.

2. *Indications from fractionations of soils.* Results have come from fractionations of soils by (a) particle density (Plante et al., 2006; Sollins et al., 2006, 2009), (b) selective chemical dissolutions (Eusterhues et al., 2003, 2005a, 2005b; Hobara et al., 2013; Wagai et al., 2013; Coward et al., 2017) and (c) physical disaggregation (Oades, 1989).

Fractionation by particle density enables separation of soils into light fractions comprising largely unbound organic material, and several increasingly denser fractions, comprising increasing proportions of mineral matter bound to organic materials, then unbound minerals. Of course, there is some gradation in densities according to mineral type, with a notable distinction between lighter smectites and heavier iron oxides. Identification of the functional groups of organic materials associated with each density fraction also enables the assignment of preferential associations of particular types of functional groups with particular types of mineral phases. As well, the mean residence time, or age, of the organic materials in each density fraction has sometimes been determined. Consideration of the types of minerals, given by e.g. X-ray diffraction analyses, the dominant organic functional groups and therefore their likely biological origin and also their age as an indicator of stability of association, can enable a reconstruction of the history of association of minerals and organic matter in soils.

Thus, it is indicated that peptidic groups, including on microbially derived proteins, are strongly associated with minerals (Sollins et al., 2006, 2009). Their strong bonding arises because of a combination of electrostatic and van der Waals forces (Theng, 2012). Carboxylic groups can also be strongly bound to minerals especially through hydroxyl groups linking with Fe oxyhydroxides and the nanocrystalline mineral ferrihydrite (Sollins et al., 2006, 2009). The initial, probably rapid and long-lasting, uptake of peptidic and/or carboxylic groups on mineral surfaces appears to condition the surfaces for the uptake of further organic molecules, including most likely by hydrophobic interactions. The result is an onion-like layering of organic molecules on mineral surfaces. A high concentration of fine particles (silt + clay), hence a high surface area, enhances protection of SOC in stable forms but not in the form of free particulate organic matter (Plante et al., 2006). Low energy physical dispersion (Plante et al., 2006) enabled separation of soils into aggregates of sizes that followed the classical hierarchical model of Tisdall and Oades (1982). As in this model, the smallest (micro)aggregates were the strongest and most long lasting. In temperate climate agricultural soils with little extractable Fe or Al, smectites, as the generally most reactive of the phyllosilicates, due to their high charge and ability to swell in water, tended to best stabilise microaggregates. Nitrogen-rich compounds, likely to be proteins, were concentrated in the smallest aggregates. Their mutual attraction means that smectite and proteins together help stabilise organic C (and N) in the microaggregates.

Chemical fractionation involving oxidation of easily accessible organic matter followed by dissolution of minerals (with hydrofluoric acid, HF) (Eusterhues et al., 2003, 2005a) revealed that some OM can be stabilised by minerals. Stabilisation of OM occurs to a significant extent by association with Fe oxides where these are common, but OM may also be stabilised by association with layer silicates if Fe oxides are less abundant. Furthermore, it has been found that not all of the SOC held by soils can be attributed to its adsorption by metal oxides and short-range order minerals (Wagai et al., 2013; Coward et al., 2017). Rather, a large part of the SOC may be incorporated in organo-mineral associations (Wagai et al., 2013), particularly those comprising Fe-bearing minerals (Coward et al., 2017). Stabilisation of OM by minerals occurs more commonly in subsoils than in topsoils. OM stabilised by minerals can be thousands of years older than in the bulk soil. Fe oxides play a strong role in the stabilisation, hence storage, of OM in acid soils because their positively charged singly coordinated hydroxyl groups link with carboxyl groups in OM. A study of organic soils in an Arctic ecosystem (Hobara et al., 2013) showed that OM in these non-mineral soils is strongly associated with Al oxides and hydroxides by 'organic chelation'.

Physical disaggregation has revealed different patterns of breakdown for two different broad groups of soils: leached soils and soils with high base status (Oades, 1989). Disaggregation of the leached soils, which are generally acid soils, leads to organo-mineral complexes of various kinds, including clay-encrusted organic matter clay and SOC bridged by cations (such as those of Fe and Al) and also polysaccharides and nitrogen compounds sorbed on clays. On the other hand, disaggregation of soils with high base status, while leading also to SOC adsorbed on clays by polysaccharides and N compounds, also reveals cation bridging between clays and SOC, with Ca^{2+} as the dominant cation in these soils. Bridging by Ca^{2+} can involve both inner-sphere and outer-sphere complexation of SOC (Rowley et al., 2018).

3. *Studies of incubations of soils* (Saggar et al., 1996; Rasmussen et al., 2007; Pronk et al., 2012; Vogel et al., 2014, 2015).

   Results of incubations in which either the decomposition of added organic matter, the priming effect of additions on existing OM or the formation of associations have been studied, generally bear out those from the correlations between SOC contents and various soil properties. Thus, associations with the most reactive of the inorganic components present had the strongest effects of suppressing decomposition of OM.

   In Saggar et al.'s (1996) study, these were those clay minerals – allophane and smectite – which had the greatest surface areas. In that of Rasmussen et al. (2007), it was the poorly crystalline Fe oxyhydroxides which most effectively restrained OM decomposition. Pronk et al.'s (2012) study of the formation of associations in artificial soils confirmed that clay minerals played active roles in enhancing the formation of aggregates. When Vogel et al. (2014) added additional OM to a European soil (a Luvisol), they observed that it was adsorbed quite selectively. The extra OM was attracted to rough surfaces and to sites where OM was already adsorbed. Many surfaces remained free of organic matter. This latter observation bears out those from surface area determinations, particularly by Mayer and Xing (2001), Lehmann et al. (2007), Wagai et al. (2009) and Heister et al. (2012).

4. *Studies of surface areas and energies* (Mayer and Xing, 2001; Kaiser and Guggenberger, 2003; Eusterhues et al., 2005b; Wagai et al., 2009).

   Measurements of the areas of surfaces before and after destruction of organic matter have shown that SOM tends to occur in patches on mineral surfaces, with most mineral surfaces not covered by SOM (Mayer and Xing, 2001; Wagai et al., 2009). Complete coverage of mineral surfaces occurred only at very high SOC contents. The most energetically favoured sites for SOM uptake are micropores, typically of ~10–1000 nm in diameter (Kaiser and Guggenberger, 2003). Otherwise SOC is more strongly held by Fe oxides, both crystalline and poorly crystalline, and, in some case, also by compounds of Al that are generally poorly crystalline (Eusterhues et al., 2005b).

5. *Direct studies of organo-mineral associations, microaggregates and pores.* These have been carried out with a wide variety of high-energy instruments. These have included: X-ray attenuation (Leifeld and Kögel-Knabner, 2003); solid-state ^{13}C nuclear magnetic resonance (NMR) (Schöning et al., 2005); transmission electron microscopy (TEM) (Chenu and Plante, 2006; Asano et al., 2018); scanning electron microscopy (SEM) (Asano et al., 2018); ultra-small X-ray scattering (McCarthy et al., 2008); a number of synchrotron radiation-based techniques, including Fourier transform infrared spectroscopy (FTIR), near-edge X-ray absorption fine structure spectroscopy (NEXAFS) and scanning transmission X-ray microscopy (STXM) (Lehmann et al., 2007; Wan et al., 2007; Solomon et al., 2012; Asano et al., 2018); and nanoscale secondary ion mass spectroscopy (NanoSIMS) (Heister et al., 2012; Steffens et al., 2017).

It has thereby been found that most fine material in soils occurs in microaggregates, with sizes that reflect those of their constituent mineral particles. This suggests the co-evolution of minerals and organic matter in soil development, and the protection of organic matter by clay minerals and/or by specific interactions of organic functional groups with Fe, Al and, in some cases, with Ca. The association between poorly crystalline Fe (hydrous) oxides and microbially derived sugars is particularly strong and long lasting. Even so, there is a considerable degree of heterogeneity in the nature of organic moieties within even small microaggregates. It is likely that the identity of the particular minerals associated with organic matter and the nature of their association depends on pH. However, at the finest (microdomain) level, in a common European soil there were limited types of associations of minerals and organic matter (Steffens et al., 2017). Using NanoSIMS with sophisticated image analyses, these authors found all microdomains belonged to one of only three types: Si mineral-rich with low OM, Al phyllosilicate-rich with high OM and an intermediate type. Pores in the first type are large, and they transport and exchange water, nutrients, contaminants and gases, while those in the other two groups are smaller and carry out storage functions. In a soil on volcanic ash, an aggregate isolated by ultrasonic dispersion showed that apparent heterogeneity in mineral-organic associations could be identified, using STXM, NEXAFS and both TEM and SEM as "a mosaic of two distinct regions" (Asano et al., 2018, p. 62). There were regions with smooth surfaces characterised by well-crystallised minerals and containing little C, and contrasting regions with rough surfaces that comprised X-ray amorphous minerals and high concentrations of C. These latter were labelled "organo-metal/mineral nanocomposites (OMN)" (Asano et al., 2018, p. 64) and their organic component thought to consist of microbially derived compounds that are predominantly amides and carboxyl C. These OMN are considered by Asano et al. (2018) to provide the binding agent linking organic carbon to minerals for the formation of micro-aggregates. These authors also consider that similar mechanisms apply to the binding of metals, organic matter and minerals in soils on other parent materials beside volcanic ash, albeit that allophane may not be present.

Consequences of microaggregation are very great, especially in soils in temperate climates. In these soils there is a direct relationship between clay minerals and organic matter, which creates the soil aggregates that are microporous and retain capillary water. This relationship is closely related to the water-holding capacity of soils (Chenu, 1993; Lal, 2000; Churchman et al., 2010) at low atmospheric water pressures (periods of dry climates). The construction of clay aggregates with edge-to-edge bonding through organic matter is the basis of this phenomenon. Churchman et al. (2010) show for example that the soils in a sequence with the highest content of organic material have the greatest microporosity (see Chapter 14, Figure 14.1). These relationships contribute to soil fertility as indicated by Murphy et al. (2015), due to the retention of water that is eventually available to plant roots and their mycorrhizae. The structure of clay aggregates and their water retention properties depend largely upon the shapes of the initial particles. Silicate clays (2:1 and 1:1 types) are especially suited to form microporous materials because they are joined by organic matter on their edges where charge variable sites are present (Figure 10.2). Iron oxides for example are more amorphous in shape and the resulting oxide–organic structures are more isotropic and less likely to create capillary porosity for containing water. Hence even though oxides form aggregates they are not necessarily useful for plant growth.

6. *Laboratory experiments.* Laboratory experiments related to the preservation of organic fossils (McMahon et al., 2016) have shown that, even without their formation into micro-aggregates, clay minerals are capable of preserving organic matter from predation and decomposition by bacteria. Their antibiotic effect is attributed to the toxicity of metal ions, particularly Al^{3+} and Fe^{3+} (McMahon et al., 2016). Al toxicity to bacteria has been well

demonstrated (McMahon et al., 2016), while the toxicity of Fe stems from the oxidative damage done to bacterial cells (e.g., Morrison et al., 2014). As also discussed in Chapter 7, Section 7.3.10, it has thus been illustrated again how clay minerals may play an active part in the ecology of soils.

The apparently unrelenting development of increasing sophisticated analytical instruments with higher energies and greater spatial resolution will continue to ensure more advances in our understanding of links between minerals and organic matter in soils. Advanced instrumental approaches to the associations of minerals and organic matter in soils offer the considerable advantage of enabling exploration of the soil and its components *in situ* and thereby avoid the need to mentally reconstruct the real soil from artefacts of the analytical procedure.

10.4 COMBINED INDICATIONS ON LINKS BETWEEN MINERALS AND SOM

Together, these and some other relevant studies indicate that SOC can be associated with almost any (secondary and also altered primary) mineral phase, but shows a preference for poorly crystalline oxides and also silicates of Fe and Al, if they are present in large concentrations. If not, SOC becomes bound to phyllosilicates according to their surface reactivities. Smectites (2:1 permanent charge phyllosilicates) are generally the most reactive of these. The relative concentrations and also contributions of poorly crystalline Fe and Al minerals versus those of phyllosilicates may often be related to soil pH, with a low pH favouring the Fe and Al compounds and higher pHs favouring phyllosilicates. The nature of the association between organic matter and mineral surfaces depends upon the nature of the surfaces and also the pH. At low pHs, organic matter, on the one hand, and particularly the variably charged oxides, oxyhydroxides and hydroxides and nanocrystalline silicates of Fe and Al, on the other, tend to form inner sphere associations, hence are stronger than the outer-sphere complexes formed by organic matter at higher pHs and with phyllosilicates (Kleber et al., 2015). Furthermore, the association of minerals and organic matter tend to be through outer-sphere complexing in arid soils, even if the minerals are hydrous oxides or hydroxides.

Different types of OM can be bound to minerals in a series of layers, with proteins and molecules containing carboxyl groups often – although not invariably – near mineral surfaces, and those with more hydrophobic groups in outer layers. Some organic functional groups have a strong attraction to minerals and form associations early in soil development that tend to be long lasting. Chief among these are proteins and also carboxyl and O/N-alkyl groups (Schöning et al., 2005). Generally, these are aliphatic and derived from microbial action.

SOM is held on mineral surfaces in patches rather than as an evenly spread layer. Many surfaces in soils have little or no associated SOM. Whereas organic matter is largely associated with minerals, minerals often occur devoid of any organic matter. SOM appears to be preferentially attracted, within nanocomposites formed with non-crystalline minerals, to rough surfaces. When it comes to their affinity for organic matter, not all minerals are equal.

SOM is most likely to occur in biological 'hot spots' such as in rhizospheres, where exudation of organic molecules and microbial activity is most intense. Much SOC is held within micropores resulting from the formation of microaggregates. Minerals play a role in the formation of all sizes of aggregates, especially microaggregates. Often, but not always, associations of the minerals themselves surround organic materials, e.g. polysaccharides, or even microorganisms to form microaggregates (see Figures 11.3 and 11.5 in Chapter 11). Minerals can also have an antibacterial action, regardless of whether they are in microaggregates.

Associations between minerals and organic matter do not always occur by surface adsorption but can also take place through co-precipitation with species of Fe and Al (Kleber et al., 2015). The resulting co-precipitates often comprise a type of aggregate (a nano- or microaggregate) in which metal oxides, oxyhydroxides and hydroxides are embedded in a matrix of organic matter.

Minerals, organic matter, their associations together and microaggregates are all highly heterogeneous both within soils and between soils. Compounds of Fe, especially, and also Al, are almost ubiquitous in soils, probably as a result of oxidation occurring at the initial stages of weathering of primary minerals.

These generalisations can be explained when it is appreciated that minerals and organic matter and their associations evolved together in the formation of soils. Their co-evolution was suggested by Chenu and Plante (2006) and has been explicitly pursued in some studies. Initial changes following glacial retreat has shown that while organic matter and minerals may co-evolve, the two processes can occur at different rates. Generally, plants are a key factor in the early development of soils, with the onset in geological time of vascular plants leading to the development of most soils and of hydrous secondary minerals. Biotite weathering and the formation of poorly crystalline Fe oxides and Al phases give rise to new reactive mineral surfaces for association with, and the stabilisation of, organic molecules. These organic molecules formed at a faster rate than that of the increase in mineral surfaces (Dümig et al., 2012).

In the Hawaiian Islands, organic matter has resulted across the island chain from the growth and decomposition of a common suite of plants, while minerals show a development with time that reflects the volcanic history of the island chain (Mikutta et al., 2009) (see also Chapter 8, Section 8.8). Over the time sequence represented by the island chain, a rapidly increasing accumulation of organic matter occurred in association with the increasingly dominant poorly crystalline minerals of Fe and Al, but these gave way with time to more well-crystallised mineral phases, particularly halloysite, and also well-crystallised oxyhydroxides. Less OM was associated with the crystalline minerals than with those of poor crystallinity and, in addition, the different mineral assemblages favoured the association of different types of OM. This particular situation illustrates the general principle that minerals can have both quantitative and qualitative effects on OM in soils.

The nature of the organic matter and the form of its interactions with clay minerals is affected by the climate, which is reflected in the degree of alteration of the primary minerals. Degree of alteration also affects the nature of the secondary minerals, whether those from a low intensity of alteration (2:1 minerals) or from a higher intensity of alteration (1:1 minerals), while other secondary minerals reflect particular conditions of formation. For example, palygorskite and sepiolite reflect alkaline environments, while oxides, hydroxides and oxyhydroxides of iron and aluminium reflect a strong intensity of weathering, and those of iron and manganese reflect highly oxidising conditions.

Often, there is an initial association of organic matter with minerals, layers of which may well be associated together in 'tactoids' (see e.g. Figure 11.2, Chapter 11). Organo-mineral associations can then form the basis for further incorporation of minerals and organic matter to form microaggregates.

10.5 FORMATION AND STABILISATION OF MICROAGGREGATES

Microaggregates can be regarded as the fundamental units of soils. In philosophical terms, soils are best explained by reduction down to the level of microaggregates (Churchman, 2010a, 2010b). Further reduction, to the level of individual clay particles, is not useful, except for explaining 'bulk' soil properties.

Microaggregates have been distinguished by size from macroaggregates but often the distinction is arbitrary and pragmatic (Totsche et al., 2018). Many (e.g. Beare et al., 1994; Six et al., 1999; Denef et al., 2004) have denoted microaggregates as the 53–250 μm size fraction of soils after minimal physical disruption. Others (Tisdall and Oades, 1982; Paradelo et al., 2016) have described one level of microaggregation occurring at 53–250 μm and another at silt size (~2–50 μm). Using X-ray attenuation, Leifeld and Kögel-Knabner (2003) measured the mean weight diameter of microaggregates <63 μm in a range of soils to be from 11.8 to 15.6 μm, with that of the primary particles constituting them ranging from 7.4 to 15.6 μm. Much of the understanding of organo-mineral associations and the microaggregates that often contain them has been obtained by transmission electron microscopy of ultra-thin slices of soil that is first stained to identify individual organic components, e.g.

polysaccharides, polyphenols and proteins (e.g. Foster and Martin, 1981), hence examined without mechanical disruption of soils and of associations within them. Using this approach, Churchman (2000), Chenu and Plante (2006) and Churchman et al. (2010) have discovered microaggregates involving mineral and organic associations in the size range of 1 to 5 μm.

Generally, all such studies have shown that, at the scale of a few micrometres, soils include associations of organic matter and minerals, both 2:1 and 1:1 phyllosilicates, together with oxides, especially of iron. These microaggregates show a variety of composition and shapes. Organic components are often enclosed within an arrangement of minerals (see e.g. Figures 11.3 and 11.5, Chapter 11). Many such arrangements are likely to constitute two-dimensional slices through pores filled with, in these cases, organic matter. Microaggregates provide the basis of the structure of soils, acting to ensure the stability of soils against disruption through agricultural management practices, especially cultivation, as well as against erosion, by both water and wind. Microaggregates have different compositions in different soils. They constitute a level of association of soils that is difficult to break down, either in the field through e.g. cultivation (e.g. Tisdall and Oades, 1982; Ramsay et al., 1986; Watteau et al., 2012) or in the laboratory through disruption treatments (Churchman and Tate, 1986, in a variety of soil types; and Oades and Waters, 1991, in an Oxisol dominated by kaolinite and iron oxides; Fernández-Ugalde et al., 2013, in some temperate soils; and Asano and Wagai, 2014, in an allophanic Andisol). In the examples given, associations of fine materials with and without organic matter in microaggregates occurred through oxides in the Oxisols, was favoured by interstratified minerals in the temperate soils and by minerals with short-range order, particularly allophane, in the Andisol. The inherent stability of microaggregates in some soils, at least, has been shown by estimations of the duration of the organic matter in microaggregates and that of the microaggregates themselves. These have been shown to be similar to each other, with microaggregates and their constituent organic matter having a life of about one century in some cultivated soils, whereas aggregates coarser than 1 mm last for only a few years (Balesdent, 1996; Puget et al., 2000). Average turnover of C in microaggregates was calculated to be 412 years, whereas it was 140 years in macroaggregates (Jastrow et al., 1996). Density fractionation revealed that the densest fraction of organic matter, hence mineral-associated OM, had a turnover time of 945 years (Sollins et al., 2009). Furthermore, a study using plant residues labelled with ^{14}C showed that there was a vast difference between the calculated lifetimes of C in larger fragments of soils, including macroaggregates, which showed C lifetimes of no more than 10 years, and silt- and clay-sized material, including microaggregates, with lifetimes for C of ~400 years and ~1000 years respectively (Buyanovsky et al., 1994). This indicates the difference in intensity of chemical bonding (covalent compared to ionic) in microaggregates compared to macroaggregates. Chronosequence studies have shown that the accretion of newly formed soil components, such as microbial residues or hydrous Fe oxides, seems to take place within about 200 years and it may be supposed that the formation of soil microaggregates follows a similar timescale (Totsche et al., 2018).

It appears that the amount of organic matter that is associated with clay minerals in soils does not always depend on the nature of the minerals (Wattel-Koekkoek et al., 2001). However, the type of organic compounds that are associated with clay minerals shows a dependence on the type of clay minerals providing the sorptive surfaces (Blanco-Canqui and Lal, 2004). Fractionation of minerals in soils by their densities showed that the densest material included the oldest carbon in soils, which had a long residence time, as noted (Sollins et al., 2009). The densest material contained the most iron.

An effect of mineralogy on associated types of organic molecules was indicated when more SOM was extracted from smectitic soils by pyrophosphate, which preferentially extracts polyvalent cations such as Al and Ca, than was extracted by NaOH, which extracts polysaccharides, among other compounds (Wattel-Koekkoek et al., 2001). The opposite was the case for kaolinitic soils. Thus, there appears to be different mechanisms for binding OM by kaolinite and by smectite. Kaolinite preferentially binds polysaccharides, while smectite binds to organic matter containing aromatic compounds through polyvalent cation bridges.

It is important to note that clay minerals do not behave towards organic matter simply as inert materials. The reaction of carbon with clay minerals involves (1) adsorption of OM, probably selectively by type, on clays; (2) polymerisation of OM by clay surfaces; and (3) sequestration of polymerised organic compounds so that they are inaccessible to soil organisms (Laird et al., 2001; Blanco-Canqui and Lal, 2004). In a review article, Six et al. (2002, p. 161) state that "since the physical protection of POM (particulate organic matter) seems to be mostly determined by microaggregation, we hypothesise that the maximum physical protection capacity for SOM is determined by the maximum microaggregation, which is in turn determined by clay content, clay type". In microaggregates, minerals and organic matter both play active functions in their stability; minerals, and especially carbonates and short-range ordered phases of Fe, Mn and Al, can be seen as acting as 'cementing' agents, while organic materials of various types act as 'gluing' agents (Totsche et al., 2018).

Mechanisms of binding of organic matter to minerals may differ between topsoils and subsoils. This may reflect a greater role of iron oxyhydroxides in binding to OM at depth than in surface horizons (Eusterhues et al., 2005b). In acid forest soils, binding of clay minerals to OM occurred primarily through short-range order Al silicates in topsoils but through Fe oxyhydroxides in subsoils. A multiple correlation across samples from 167 pedons from varied soils throughout New Zealand showed that pyrophosphate-extractable Al and also allophane, when present in low concentrations (so as not to saturate soils), gave the best predictions of SOC (Percival et al., 2000). Other studies have indicated the strong links that can form between SOM and iron oxyhydroxides.

In general, it appears that organic matter is associated differently and with different strengths in soils in which poorly crystalline minerals dominate in comparison with soils in which the dominant minerals are well-crystallised layer silicates. The main mechanisms by which minerals and organic molecules are bound together in soils are (1) via singly coordinated surface hydroxyl groups (especially on ferrihydrite and allophane); (2) on minerals via negatively charged organic functional groups, particularly carboxyl and alcoholic/phenolic OH groups (Kleber et al., 2005, 2015); and (3) via polyvalent cations bridging negatively charged minerals, i.e. all clay minerals at soil pHs, and negatively charged organic matter (Wattel-Koekkoek et al., 2001). The two former mechanisms, involving inner-sphere complexation, tend to be stronger than the latter, as outer-sphere complexation. It has been demonstrated (Kahle et al., 2004) that the interactions of dissolved organic carbon (DOC) with siloxane groups may be weaker than those with hydroxyl groups. More broadly, the first mechanism listed above can be seen to include a continuum of complex mixtures of Fe and Al oxides ranging from "low-polymeric metal-organic complexes to well-crystalline phases with surface attached OM" (Kleber et al., 2015, p. 68) or, in other words, from co-precipitation of OM with Al and Fe to adsorption of OM by oxides, oxyhydroxides and hydroxides of Fe and Al. These evolve together (Chenu and Plante, 2006). The mechanism that is adopted depends on the path and state of evolution of both the minerals and the organic molecules in SOM, and also on the chemical composition of the soil solution.

Organic matter has been found to be distributed in a heterogeneous fashion in soils. Most surfaces are mineral rather than organic (Mayer and Xing, 2001). Microorganisms and organic matter occupy only a small fraction of soil surfaces ($\ll$1% according to Kleber et al., 2015). Soil surfaces have been described as "a huge desert" in relation to the occurrence of living organisms (Hinsinger et al., 2009), and this is presumably also the case for the breakdown products of living organisms, i.e. organic matter. The 'hot spots' for microorganisms, most likely including the rhizospheres of plants (Hinsinger et al., 2009), are also likely to give rise to the highest turnover rates for metal-organic associations (Kleber et al., 2015). With additions of iron oxyhydroxides to various soils leading to increases in both their specific surface areas and also in their capacities for organic matter, it is likely that micropores formed by the Fe compounds led to the preferential uptake of SOM at edges, rough surfaces and, particularly, at the mouths of these micropores (Kaiser and Guggenberger, 2003). Pedogenic oxides are especially reactive towards organic matter (Kögel-Knabner et al., 2008). They are particularly dominant in mineral-organic matter interactions in

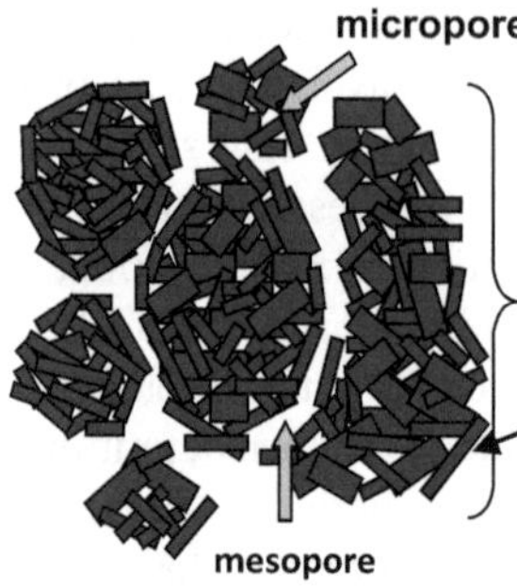

FIGURE 10.3 Schematic diagram of a possible disposition of minerals with and without associated organic matter into microaggregates leading to micropores and mesopores. Mesopores can contain water or liquid organic material.

subsoils. Organic carbon in subsoils is consistently older when occluded within minerals than when free so that "association with minerals is the most important factor in stabilisation of OC in soils, irrespective of vegetation, soil type and land use" (Schrumpf et al., 2013, p. 1675).

Attachment of OM to mineral surfaces besides those of Fe oxides where these are less reactive, namely at neutral and high pHs, is also patchy rather than as a homogeneous spread (Heister et al., 2012). Whereas soils with a high loading of organic matter could have all mineral surfaces covered by OM, there was a threshold (3 mg-OC m^{-2}) below which coverage of mineral surfaces by OM was patchy (Wagai et al., 2009). The energetics of sorption of N_2 gas indicated that the OM below this threshold of coverage was occluded by minerals, including, most likely, those forming micropores.

More often than not, soil structure has been discussed in the literature in relation to aggregates. However, aggregates are just one aspect of soil structure. They are important for stabilising soils against erosion and perhaps also against complete disruption by agricultural practices such as cultivation. The other aspect of soil structure that is most important for plant growth is that of pores (Figure 10.3). In many ways, aggregation describes the architecture of soils viewed as buildings. The pores are then the rooms in the buildings, and these are where people live and carry out useful and interesting activities. The core role played by clays in the formation and stability of aggregates and especially microaggregates means that clays are also important in the formation and stability of pores. This aspect is discussed next.

10.6 CLAY AND ORGANIC AGGREGATION AND SOIL STRUCTURE

10.6.1 Organic Particle Size and Fixation on Different Clay-Sized Particles (OM Content and Clay-Size Fraction)

Organic material in soils generally combines with mineral soil clay particles, silicates and oxides. Organic material can be of quite various sizes, from root particles to root exudates in the micrometre scale (Kleber et al., 2015). All of the organic material is unstable under surface conditions and especially in the biosphere of microorganisms of soils. The residence time of organic material in soils varies from less than a year to the order of hundreds to thousands of years. The chemical binding of organics and mineral materials is often strong. It is likely to be responsible for the extension of the residence time of organic material into hundreds to thousands of years. It creates a soil structure (aggregates) that is stable for agricultural uses of soils, even if unstable over the short term in geological time.

The site of fixation depends on the clay structure (chemical bonding and structure) where chemical bonding intensity can be quite variable. In a smectite 2:1 mineral with a diffuse surface ionic charge and a strong covalent charge of crystal edges, the bonding or chemical attraction of organic material will be much more intense on crystal edge sites than crystal surface sites (Barré et al., 2014). However, large amounts of organic matter will be fixed on 2:1 mineral surfaces. The 2:1 minerals attract organics more than the 1:1 minerals due to mineral surface charges. However, oxides have a strong affinity for organic material, especially iron oxides (Kleber et al., 2015), and they fix relatively large amounts of organic matter on high-charge crystal edge sites.

The 2:1 minerals (phyllosilicates), through strong interaction by covalent chemical bonding at crystal edges and a less strong ionic bonding on particle surfaces, lead to the formation of clay particle-organic networks, which form organo-mineral associations, which then can combine, along with unassociated minerals, to form microaggregates with capillary mesoporosities (Figure 10.3). Capillary water is held here during dry periods at the surface of the soil. The 1:1 minerals have strong covalent chemical attraction for organic materials at crystal edges and little interaction on crystal surfaces, and they also form microaggregate structures. Oxides, with bonding to organics occurring in variable sites but with no structural preferences, do not form micronetworks and hence have little or no microporosity. Laboratory experiments indicate that 2:1 smectites form honeycomb structures; 1:1 kaolinites form more flat and parallel grained structures (Chenu, 1993; Dorioz et al., 1993). This indicates the more complex interactions of organic matter with charged surface 2:1 minerals compared to 1:1 minerals. These materials are important in their capillary water retention and their tendency to form microaggregates where micro- or mesoporosity holds water by capillary action (Smagin et al., 2004).

Hence, as clays show less silica content (2:1 to 1:1, oxides), the capillary water content due to the formation of micronetworks is smaller. The change in clay soil structure is due to the presence of high water content in the soil environment, and hence the retention of soil water is less important in that water is present much of the year. The retention of capillary water is very important for 2:1 dominated soil clays because it is in the temperate climates where these minerals form, and it is here that dry periods occur for parts of the year and continued chemical reaction can be affected by capillary water in the soil structures.

Given the structure of organo-mineral associations, it is evident that the size of organic particulate matter must be of the same order of magnitude as that of the soil clay mineral particles. Larger organic matter particles can be incorporated in the microaggregates (see Figures 11.3 and 11.5, Chapter 11).

10.6.2 Formation of Macroaggregates

Larger units (macroaggregates) are formed through wetting and drying and plant growth, with associated faunal and microbial activities. Microaggregates are initially formed within the faster-forming macroaggregates (Oades, 1984; Six et al., 2004). The mechanisms of binding of macroaggregates differ from those of microaggregates, generally involving plant roots and fungal hyphae (Tisdall and Oades, 1982; Oades, 1984). Subsequent to their formation, small microaggregates may combine into larger units. The initial smaller units may be brought into contact during the drying cycle. In most soils rainfall is not continuous but periodic, and the surface soils are saturated and subsequently dried where inter-unit pore space increases. The wetting cycle approaches the smaller units through swelling of clays associated with water and the smaller aggregates come into contact. This contact approaches the clay-organic particles allowing a new set of chemical bonding conditions, while plant growth allows physical binding to occur through entanglement of pre-formed microaggregates and other soil particles, including primary particles by root hairs and fungal hyphae. Larger organic particles (such as root debris and extracellular polysaccharides; Dinel et al., 1991; Dorioz et al., 1993; Churchman, 2000) can be incorporated adding to the organic content (Six et al., 2000) and overall macroaggregate stability (Arthur et al., 2012) (see also Figures 11.3, 11.4 and 11.5 in Chapter 11). This material can evolve chemically and physically in the soil horizon re-adapting to the local soil clay environment (Puget et al., 2000) forming and re-forming macroaggregate structures.

Overall clay mineralogy can be important in the formation of macroaggregates in soils (Denef and Six, 2005). These authors demonstrated experimentally that a kaolinitic soil formed macroaggregates abiotically from mineral–mineral interactions faster but less effectively than an illitic soil, which formed aggregates by interaction with plant root and plant residues. The different mechanisms for the soils with different dominant clay minerals reflects the results of disaggregating treatments (Oades and Waters, 1991), whereby soils with 2:1 minerals yielded a hierarchy of aggregates by size, but those with 1:1 kaolinite (and oxides) showed no hierarchy but rather many small particles.

10.6.3 Evolution of Aggregates with Rainfall Episodes (Wetting and Drying)

We see in Figure 10.4 that the smaller aggregates associate to form larger units and this leaves a certain amount of larger pore spaces between some of the units. The formation of new pore structures is not homogeneously distributed in the soils where organics are present to bind clay particles together, but it becomes more homogeneous as a function of depth. At greater depth there is more homogeneous confinement and a tendency to maintain larger aggregate units. Further, the amount of dissolved and hence renewable and mobile organic matter decreases greatly with depth (Kleber et al., 2015) depending upon the soil type, as does the larger-sized organic material such as root debris.

The effect of depth in a soil can be seen in the example in Figure 10.5 where a Mollisol from the midwest of the United States sampled in September, after numerous cycles of rain and drying, shows a concentration of macropores into cracks forming at depth compared to the surface.

10.6.4 Macroaggregates and Soil Structure: The Formation of Fractures

As macroaggregates increase in proportion in a soil, porosity develops at their margins eventually forming cracks or long zones of porosity. This phenomenon can be seen in Figure 10.6. The cracks are interconnected eventually and form a network that transmits rainwater and air to different levels of the soil. When organic matter becomes rarer, the macroaggregates are rarer forming large units and strong cracking is apparent. This occurs with depth in the soil profiles.

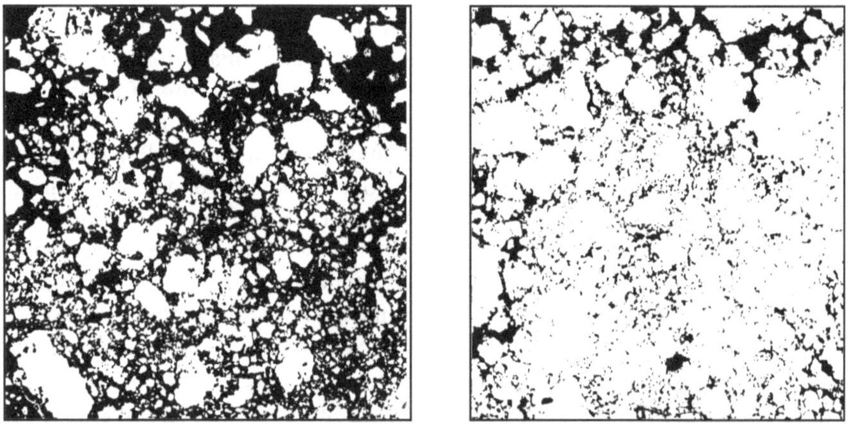

LaTouche Rendzina 3 cycles
3 cm depth

La Touche Redzina 9 cycles
3 cm depth

FIGURE 10.4 Results of laboratory experiment of wetting and drying of a cultivated soil in samples at the same depth (3 cm).

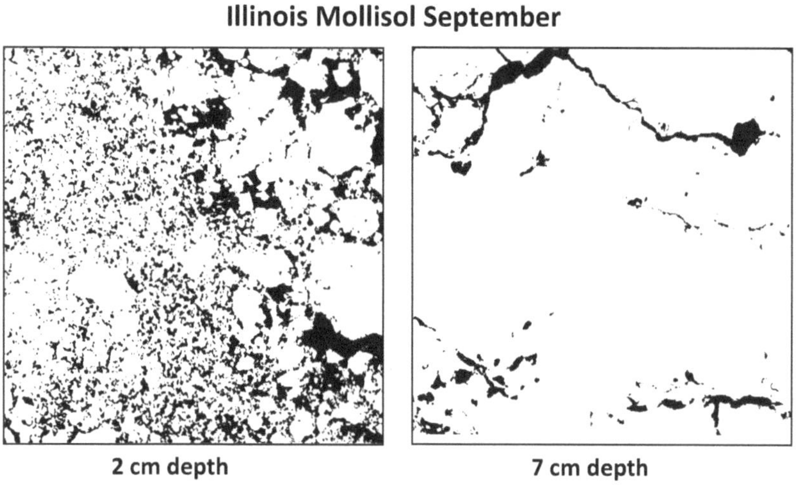

FIGURE 10.5 A Mollisol from the midwest of the United States sampled at different depths in September, after many cycles of rain and drying.

10.6.5 Change of Structure with Depth: Aggregation and Fractures

Fractures are rare but ubiquitous in these materials as is the case in most soils. If there is little organic matter present in the soil, cracks will be found at the surface.

In general, the formation of macroaggregates determines to a large extent the fertility of a soil for plants with significant root systems near the surface. The aggregation is the basis for the agricultural term 'tilth' which dates from the 12th century.

10.6.6 Organic Matter and Clay Structures and Retention of Capillary Water

Organic matter and clay minerals (silicates) form complex structures depending upon the tendency to chemical bonding of these components on crystal edges or on charged clay surfaces (see Figure 10.1).

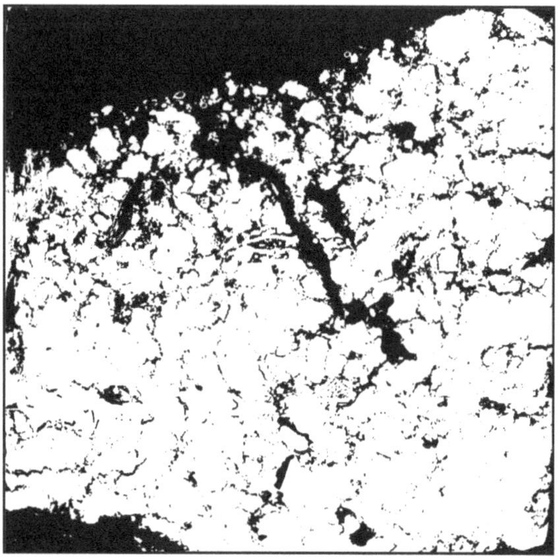

FIGURE 10.6 Profile of prairie soil. Total depth is 10 cm.

The bonding on clay surfaces leaves little free water present, but the enclosure of water in microcapillary sites within clays and organic edge site bonded structures leaves free water in the soil structure. Smagin et al. (2004) show that the smaller size fractions in soils (higher clay-sized content) tend to contain more water. However, in general the coarser size fractions of soils contain more organic matter (Six et al., 2000). Thus, it is not the relative amounts of the two components that influence the water holding capacity but the manner in which the organic and clay fractions are bound together. Organic matter with smaller molecular sizes is more likely to form capillary microaggregate materials (Dorioz et al., 1993) that will form capillary zones for water retention. Churchman (2012) indicates relations between carbon sequestration (strong chemical bonding between clays and organic material) in soils, on the one hand, and water retention, on the other.

In numerous studies where 2:1 clays are a significant part of the soil clays (in temperate climates), the relationship between small organic molecules formed from plant exudates and 2:1 clays indicates higher water retention as a result. This is an important aspect of clay-organic bonding.

REFERENCES

Amato, M., and J.N. Ladd. 1992. Decomposition of 14C-labelled glucose and legume material in soils: Properties influencing the accumulation of organic residue C and microbial biomass C. *Soil Biology and Biochemistry* 24: 455–464.

Arrouays, D., I. Vion, and J.L. Kicin. 1995. Spatial analysis and modeling of topsoil carbon storage in temperate forest humic loamy soils of France. *Soil Science* 159: 191–198.

Arthur, E., P. Schjønning, P. Moldrup, and L.W. de Jonge. 2012. Soil resistance and resilience to mechanical stress for three differently managed sand loam soils. *Geoderma* 173–174: 50–60.

Asano, M., and R. Wagai. 2014. Evidence of aggregate hierarchy at micro- to submicron scales in an allophanic Andisol. *Geoderma* 216: 62–74.

Asano, M., R. Wagai, N. Yamaguchi, Y. Takeichi, M. Maeda, H. Suga, and Y. Takahashi. 2018. In search of a binding agent: Nano-scale evidence of preferential carbon associations with poorly-crystalline mineral phases in physically-stable, clay-sized aggregates. *Soil Systems* 2: 32.

Balesdent, J. 1996. The significance of organic separates to carbon dynamics and its modelling in some cultivated soils. *European Journal of Soil Science* 47: 485–493.

Barker, W.W., S.A. Welch, and J.F. Banfield. 1997. Biochemical weathering of silicate minerals, p. 391–428. In: J.F. Banfield, and K.H. Nealson (eds.), *Geomicrobiology: Interactions between Microbes and Minerals.* Mineralogical Society of America, Washington DC.

Barré, P., G. Berger, and B. Velde. 2009. How element translocation by plants may stabilize illite clays in the surface of temperate soils. *Geoderma* 151: 22–30.

Barré,P., O. Fernandez-Ugalde, I. Virto, B. Velde, and C. Chenu. 2014. Impact of phyllosilicate mineralogy on organic carbon stabilization in soils: Incomplete knowledge and exciting prospects. *Geoderma* 235–236: 382–395.

Beare, M.H., P.F. Hendrix, M.L. Cabrera, and D.C. Coleman. 1994. Aggregate-protected and unprotected organic matter pools in conventional and no-tillage soils. *Soil Science Society of America Journal* 58: 787–795.

Berner, R.A. 1997. Geochemistry and geophysics: The rise of plants and their effect on weathering and atmospheric CO_2. *Science* 276: 544–546.

Blanco-Canqui, H., and R. Lal. 2004. Mechanisms of carbon sequestration in soil aggregates. *Critical Reviews in Plant Sciences* 23: 481–504.

Brantley, S.L., M. Lebedeva, and E.M. Hausrath. 2012. A geobiological view of weathering and erosion, p. 205–227. In: A.H. Knoll, D.E. Canfield, and K.O. Konhauser (eds.), *Fundamentals of Geobiology.* Blackwell Publishing, Chichester, United Kingdom.

Buyanovsky, G.A., M. Aslam, and G.H. Wagner. 1994. Carbon turnover in soil physical fractions. *Soil Science Society of America Journal* 58: 1167–1173.

Calvaruso, C., L. Mareschal, M.-P. Turpault, and E. Leclerc. 2009. Rapid clay weathering in the rhizosphere of Norway spruce and oak in an acid forest ecosystem. *Soil Science Society of America Journal* 73: 331–338.

Chan, C.S., S.C. Fakra, D.C. Edwards, D. Emerson, and J.F. Banfield. 2009. Iron oxyhydroxide mineralization on microbial extracellular polysaccharides. *Geochimica et Cosmochimica Acta* 73: 3807–3818.

Chenu, C. 1989. Influence of a fungal polysaccharide, scleroglucan, on clay microstructures. *Soil Biology and Biochemistry* 21: 299–305.

Chenu, C. 1993. Clay- or sand-polysaccharide associations as models for the interface between micro-organisms and soil: Water related properties and microstructure. *Geoderma* 56: 143–156.

Chenu, C., and A.F. Plante. 2006. Clay-sized organo-mineral complexes in a cultivation chronosequence: Revisiting the concept of the 'primary organo-mineral complex'. *European Journal of Soil Science* 57: 596–607.

Chenu, C., C.H. Pons, and M. Robert. 1987. Interaction of kaolinite and montmorillonite with neutral polysaccharides, p. 375–381. In: L.G. Schultz, H. van Olphen, and F.A. Mumpton (eds.), *Proceedings of the International Clay Conference, Denver, 1985*. The Clay Minerals Society, Bloomington.

Churchman, G.J. 2000. The alteration and formation of soil minerals by weathering, p. F3–76. In: M.E. Sumner (ed.), *Handbook of Soil Science*. CRC Press, Boca Raton, Florida.

Churchman, G.J. 2010a. The philosophical status of soil science. *Geoderma* 157: 214–221.

Churchman, G.J. 2010b. Is the geological concept of clay minerals appropriate for soil science? A literature-based and philosophical analysis. *Physics and Chemistry of the Earth, Parts A/B/C* 35: 927–940.

Churchman, G.J. 2012. Small heterogeneous associations and water retention link soil quality to carbon sequestration in soils: Philosophical and practical implications, p. 134–137. In: L. Burkitt, and L. Sparrow (eds.), *Proceedings of the 5th Australia and New Zealand Soil Science Conference, Hobart*. Soil Science Australia, Adelaide, Australia.

Churchman, G.J., R.C. Foster, L.P. D'Acqui, L.J. Janik, J.O. Skjemstad, R.H. Merry, and D.A. Weissmann. 2010. Effect of land-use history on the potential for carbon sequestration in an Alfisol. *Soil and Tillage Research* 109: 23–35.

Churchman, G.J., and D.J. Lowe. 2012. Alteration, formation and occurrence of minerals in soils, p. 20.1–20.72. In: P.M. Huang, Y. Li, and M.E. Sumner (eds.), *Handbook of Soil Sciences: Properties and Processes*, 2nd edn. CRC Press, Boca Raton, Florida.

Churchman, G.J., and K.R. Tate. 1986. Aggregation of clay in six New Zealand soil types as measured by disaggregation procedures. *Geoderma* 37: 207–220.

Coward, E.K., A.T. Thompson, and A.F. Plante. 2017. Iron-mediated mineralogical control of organic matter accumulation in tropical soils. *Geoderma* 306: 206–216.

Denef, K., and J. Six. 2005. Clay mineralogy determines the importance of biological versus abiotic processes for macroaggregate formation and stabilization. *European Journal of Soil Science* 56: 469–479.

Denef, K., J. Six, R. Merckx, and K. Paustian. 2004. Carbon sequestration in microaggregates of no-tillage soils with different clay mineralogy. *Soil Science Society of America Journal* 68: 1935–1944.

Dinel, H., G. Mehuys, and M. Levesque. 1991. Influence of humic fibric materials on the aggregation and aggregate stability of lacustrine silty clay. *Soil Science* 151: 141–158.

Dorioz, J.M., M. Robert, and C. Chenu. 1993. The role of roots, fungi and bacteria on clay particle organization. An experimental approach. *Geoderma* 56: 179–194.

Dubach, P., and N.C. Mehta. 1963. The chemistry of soil humic substances. *Soils and Fertilisers* 26: 293–300.

Dümig, A., W. Häusler, M. Steffens, and I. Kögel-Knabner. 2012. Clay fractions from a soil chronosequence after glacier retreat reveal the initial evolution of organo-mineral associations. *Geochimica et Cosmochimica Acta* 85: 1–18.

Eusterhues, K., C. Rumpel, M. Kleber, and I. Kögel-Knabner. 2003. Stabilisation of organic matter by interactions with minerals as revealed by mineral dissolution and oxidative degradation. *Organic Geochemistry* 34: 1591–1600.

Eusterhues, K., C. Rumpel, and I. Kögel-Knabner. 2005a. Stabilization of soil organic matter isolated via oxidative degradation. *Organic Geochemistry* 36: 1567–1575.

Eusterhues, K., C. Rumpel, and I. Kögel-Knabner. 2005b. Organo-mineral associations in sandy acid forest soils: Importance of specific surface area, iron oxides and micropores. *European Journal of Soil Science* 56: 753–763.

Fernández-Ugalde, O., P. Barré, F. Hubert, I. Virto, C. Girardin, E. Ferrage, L. Caner, and C. Chenu. 2013. Clay mineralogy differs qualitatively in aggregate-size classes: Clay-mineral-based evidence for aggregate hierarchy in temperate soils. *European Journal of Soil Science* 64: 410–422.

Fiore, S., S. Dumontet, F.J. Huertas, and V. Pasquale. 2011. Bacteria-induced crystallization of kaolinite. *Applied Clay Science* 53: 566–571.

Foster, R.C. 1981. Polysaccharides in soil fabrics. *Science* 214: 665–667.

Foster, R.C., and J.K. Martin. 1981. In situ analysis of soil components of biological origin, p. 75–110. In: E.A. Paul, and J.N. Ladd (eds.), *Soil Biochemistry*, vol. 5. Marcel Dekker, New York.

Greenland, D.J. 1965. Interactions between clays and organic compounds in soils. Part 1. Mechanisms of interaction between clays and defined organic compounds. *Soils and Fertilisers* 28: 415–425.

Guckert, A. 1975. Interface sol-racine-I. Etude au microscope electronique des relations mucigel-argile-microorganisms. *Soil Biology and Biochemistry*: 241–250.

Hayes, M.H.B., P. MacCarthy, R.L. Malcolm, and R.S. Swift. 1989. The search for structure: Setting the scene, p. 3–31. In: M.H.B. Hayes, P. MacCarthy, R.L. Malcolm, and R.S. Swift (eds.), *Humic Substances II*. John Wiley & Sons, Chichester, United Kingdom.

Heister, K., C. Höschen, G.J. Pronk, C.W. Mueller, and I. Kögel-Knabner. 2012. NanoSIMS as a tool for characterizing soil model compounds and organomineral associations in artificial soils. *Journal of Soils and Sediments* 12: 35–47.

Hernández, Z., G. Almendros, P. Carral, A. Álvarez, H. Knicker, and J.P. Pérez-Trujillo. 2012. Influence of non-crystalline minerals in the total amount, resilience and molecular composition of the organic matter in volcanic ash soils (Tenerife Island, Spain). *European Journal of Soil Science* 63: 603–615.

Hinsinger, P., A.G. Bengough, D. Vetterlein, and I.M. Young. 2009. Rhizosphere: Biophysics, biogeochemistry and ecological relevance. *Plant and Soil* 321: 117–152.

Hobara, S., K. Koba, N. Ae, A.E. Giblin, K. Kushida, and G.R. Shaver. 2013. Geochemical influences on solubility of soil organic carbon in arctic tundra ecosystems. *Soil Science Society of America Journal* 77: 473–481.

Jastrow, J.D., R.M. Miller, and T.W. Boutton. 1996. Carbon dynamics of aggregate-associated organic matter estimated by carbon-13 natural abundance. *Soil Science Society of America Journal* 60: 801–807.

Kahle, M., M. Kleber, and R. Jahn. 2002. Predicting carbon content in illitic clay fractions from surface area, cation exchange capacity and dithionite-extractable iron. *European Journal of Soil Science* 53: 639–644.

Kahle, M., M. Kleber, and R. Jahn. 2004. Retention of dissolved organic matter by phyllosilicate and soil clay fractions in relation to mineral properties. *Organic Geochemistry* 35: 269–276.

Kaiser, K., G. Guggenberger. 2003. Mineral surfaces and soil organic matter. *European Journal of Soil Science* 54: 219–236.

Kennedy, M., M. Droser, L.M. Mayer, D. Pevear, and D. Mrofka. 2006. Late Precambrian oxygenation; inception of the clay mineral factory. *Science* 311: 1446–1449.

Kleber, M., K. Eusterhues, M. Keiluweit, C. Mikutta, R. Mikutta, and P.S. Nico. 2015. Mineral-organic associations: Formation, properties, and relevance in soil environments. *Advances in Agronomy* 130: 1–140.

Kleber, M., R. Mikutta, M.S. Torn, and R. Jahn. 2005. Poorly crystalline mineral phases protect organic matter in acid subsoil horizons. *European Journal of Soil Science* 56: 717–725.

Kögel-Knabner, I., G. Guggenberger, M. Kleber, E. Kandeler, K. Kalbitz, S. Scheu, K. Eusterhues, and P. Leinweber. 2008. Organo-mineral associations in temperate soils: Integrating biology, mineralogy, and organic matter chemistry. *Journal of Plant Nutrition and Soil Science* 171: 61–82.

Konhauser, K., and R. Riding. 2012. Bacterial biomineralization, p. 105–130. In: A.H. Knoll, D.E. Canfield, and K.O. Konhauser (eds.), *Fundamentals of Geobiology*. Wiley-Blackwell, Chichester, United Kingdom.

Laird, D.A., D.A. Martens, and W.L. Kingery. 2001. Nature of clay-humic complexes in an agricultural soil. *Soil Science Society of America Journal* 65: 1413–1418.

Lal, R. 2000. Physical management of soils of the tropics: priorities for the 21st century. *Soil Science* 165: 191–207.

Lehmann, J., J. Kinyangi, and D. Solomon. 2007. Organic matter stabilization in soil microaggregates: Implications from spatial heterogeneity of organic carbon contents and carbon forms. *Biogeochemistry* 85: 45–57.

Leifeld, J., and I. Kögel-Knabner. 2003. Microaggregates in agricultural soils and their size distribution determined by X-ray attenuation. *European Journal of Soil Science* 54: 167–174.

Mayer, L.M., and B. Xing. 2001. Organic matter-surface area relationships in acid soils. *Soil Science Society of America Journal* 65: 250–258.

McBratney, A.B., U. Stockmann, D.A. Angers, B. Minasny, and D.J. Field. 2014. Challenges for soil organic carbon research, p. 3–16. In: A.E. Hartemink, and K. McSweeney (eds.), *Soil Carbon*. Springer, Cham, Switzerland.

McCarthy, J.F., J. Ilavsky, J.D. Jastrow, L.M. Mayer, E. Perfect, and J. Zhuang. 2008. Protection of organic carbon in soil microaggregates via restructuring of aggregate porosity and filling of pores with accumulating organic matter. *Geochimica et Cosmochimica Acta* 72: 4725–4744.

McMahon, S., R.P. Anderson, E.E. Saupe, and D.E.G. Briggs. 2016. Experimental evidence that clay inhibits bacterial decomposers: Implications for preservation of organic fossils. *Geology* 44: 867–870.

Mikutta, R., M. Kleber, and R. Jahn. 2005. Poorly crystalline minerals protect organic carbon in clay subfractions from acid subsoil horizons. *Geoderma* 128: 106–115.

Mikutta, R., G.E. Schaumann, D. Gildemeister, S. Bonneville, M.G. Kramer, J. Chorover, O.A. Chadwick, and G. Guggenberger. 2009. Biogeochemistry of mineral–organic associations across a long-term mineralogical soil gradient (0.3–4100 kyr), Hawaiian Islands. *Geochimica et Cosmochimica Acta* 73: 2034–2060.

Mills, A.L. 2003. Keeping in touch: Microbial life on soil particle surfaces. *Advances in Agronomy* 78: 1–43.

Morrison, K.D., J.C. Underwood, D.W. Metge, D.D. Eberl, and L.B. Williams. 2014. Mineralogical variables that control the antibacterial effectiveness of a natural clay deposit. *Environmental Geochemistry and Health* 36: 613–631.

Murphy, C.J., E.M. Baggs, N. Morley, D.P. Wall, and E. Paterson. 2015. Rhizosphere priming can promote mobilisation of N-rich compounds from soil organic matter. *Soil Biology and Biochemistry* 81: 236–243.

Nichols, J.D. 1984. Relation of organic carbon to soil properties and climate in the Southern Great Plains. *Soil Science Society of America Journal* 48: 1382–1384.

Oades, J.M. 1984. Soil organic matter and structural stability: Mechanisms and implications for management. *Plant and Soil* 76: 319–337.

Oades, J.M. 1989. Introduction to organic matter in mineral soils, p. 89–159. In: J.B. Dixon, and S.B. Weed (eds.), *Minerals in Soil Environments*, 2nd edn. Soil Science Society of America, Madison, Wisconsin.

Oades, J.M. 1993. The role of biology in the formation, stabilization and degradation of soil structure. *Geoderma* 56: 377–400.

Oades, J.M., and A.G. Waters. 1991. Aggregate hierarchy in soils. *Soil Research* 29: 815–828.

Paradelo, R., F. van Oort, P. Barré, D. Billiou, and C. Chenu. 2016. Soil organic matter stabilization at the pluri-decadal scale: Insight from bare fallow soils with contrasting physicochemical properties and macrostructures. *Geoderma* 275: 48–54.

Percival, H.J., R.L. Parfitt, and N.A. Scott. 2000. Factors controlling soil carbon levels in New Zealand grasslands: Is clay content important? *Soil Science Society of America Journal* 64: 1623–1630.

Plante, A.F., R.T. Conant, C.E. Stewart, K. Paustian, and J. Six. 2006. Impact of soil texture on the distribution of soil organic matter in physical and chemical fractions. *Soil Science Society of America Journal* 70: 287–296.

Pronk, G.J., K. Heister, G.-C. Ding, K. Smalla, and I. Kögel-Knabner. 2012. Development of biogeochemical interfaces in an artificial soil incubation experiment; aggregation and formation of organo-mineral associations. *Geoderma* 189–190: 585–594.

Puget, P., C. Chenu, and J. Balesdent. 2000. Dynamics of soil organic matter associated with particle-size fractions of water-stable aggregates. *European Journal of Soil Science* 51: 595–605.

Ramsay, A.J., R.E. Stannard, and G.J. Churchman. 1986. Effect of conversion from ryegrass pasture to wheat cropping on aggregation and bacterial populations in a silt loam soil in New Zealand. *Soil Research* 24: 253–264.

Rasmussen, C., N. Matsuyama, R.A. Dahlgren, R.J. Southard, and N. Brauer. 2007. Soil genesis and mineral transformation across and environmental gradient on andesitic lahar. *Soil Science Society of America Journal* 71: 225–237.

Robert, M., and C. Chenu. 1992. Interactions between soil minerals and microorganisms, p. 307–404. In: G. Stotzky, and J.-M. Bollag (eds.), *Soil Biochemistry*, vol. 7. Marcel Dekker, New York.

Rowley, M.C., S. Grand, and É.P. Verrecchia. 2018. Calcium-mediated stabilization of soil organic carbon. *Biogeochemistry* 137: 27–49.

Russell, E.W. 1973. *Soil Conditions and Plant Growth*, 10th edition. Longman, London.

Saggar, S., A. Parshotam, C. Hedley, and G. Salt. 1999. 14C-labelled glucose turnover in New Zealand soils. *Soil Biology and Biochemistry* 31: 2025–2037.

Saggar, S., A. Parshotam, G.P. Sparling, C.W. Feltham, and P.B.S. Hart. 1996. 14C-labelled ryegrass turnover and residence times in soils varying in clay content and mineralogy. *Soil Biology and Biochemistry* 28: 1677–1686.

Schnitzer, M., and S.U. Khan (eds.). 1978. *Soil Organic Matter.* Elsevier, New York.

Schöning, I., H. Knicker, and I. Kögel-Knabner. 2005. Intimate association between O/N-alkyl carbon and iron oxides in clay fractions of forest soils. *Organic Geochemistry* 36: 1378–1390.

Schrumpf, M., K. Kaiser, G. Guggenberger, T. Persson, I. Kögel-Knabner, and E.-D. Schulze. 2013. Storage and stability of organic carbon in soils as related to depth, occlusion within aggregates, and attachment to minerals. *Biogeosciences* 10: 1675–1691.

Six, J., H. Bossuyt, S. Degryze, and K. Denef. 2004. A history of research on the link between (micro)aggregates, soil biota, and soil organic matter dynamics. *Soil and Tillage Research* 79: 7–31.

Six, J., R.T. Conant, E.A. Paul, and K. Paustian. 2002. Stabilization mechanisms of soil organic matter: Implications for C-saturation of soils. *Plant and Soil* 241: 155–176.

Six, J., E.T. Elliott, and K. Paustian. 1999. Aggregate and soil organic dynamics under conventional and no-tillage systems. *Soil Science Society of America Journal* 63: 1350–1359.

Six, J., K. Paustian, E.T. Elliott, and C. Combrink. 2000. Soil structure and organic matter: I. Distribution of aggregate size classes and aggregate-associated carbon. *Soil Science Society of America Journal* 64: 681–689.

Skjemstad, J.O., L.J. Janik, and J.A. Taylor. 1998. Non-living soil organic matter: What do we know about it? *Australian Journal of Experimental Agriculture* 38: 667–680.

Smagin, A., N. Sadovnikova, T. Nazarova, A. Kiryushova, A. Mashika, and A. Eremina. 2004. The effect of organic matter on water retention capacity of soils. *Eurasian Soil Science* 37: 267–275.

Sollins, P., M.G. Kramer, C. Swanston, K. Lajtha, T. Filley, A.K. Aufdenkampe, R. Wagai, and R.D. Bowden. 2009. Sequential density fractionation across soils of contrasting mineralogy: Evidence for both microbial- and mineral-controlled soil organic stabilization. *Biogeochemistry* 96: 209–231.

Sollins, P., C. Swanston, M. Kleber, T. Filley, M. Kramer, S. Crow, B.A. Caldwell, K. Lajtha, and R. Bowden. 2006. Organic C and N stabilization in a forest soil: Evidence from sequential density fractionation. *Soil Biology and Biochemistry* 38: 3313–3324.

Solomon, D., J. Lehmann, J. Harden, J. Wang, J. Kinyangi, K. Heymann, C. Karunakaran, Y. Lu, S. Wirick, and C. Jacobsen. 2012. Micro- and nano-environments of carbon sequestration: Multi-element STXM–NEXAFS spectromicroscopy assessment of microbial carbon and mineral associations. *Chemical Geology* 329: 53–73.

Spain, A.V. 1990. Influence of environmental conditions and some soil chemical properties on the carbon and nitrogen contents of some tropical Australian rainforest soils. *Soil Research* 28: 825–839.

Steffens, M., D.M. Rogge, C.W. Mueller, C. Höschen, J. Lugmeier, A. Kölbl, and I. Kögel-Knabner. 2017. Identification of distinct functional domains controlling C storage in soils. *Environmental Science and Technology* 51: 12182–12189.

Tazaki, K. 2005. Microbial formation of a halloysite-like mineral. *Clays and Clay Minerals* 53: 224–233.

Theng, B.K.G. 2012. *Formation and Properties of Clay-Polymer Complexes*, vol. 4. Elsevier, Amsterdam.

Tisdall, J.M., and J.M. Oades. 1982. Organic matter and water-stable aggregates in soils. *Journal of Soil Science* 33: 141–163.

Totsche, K.U., W. Amelung, M.H. Gerzabek, G. Guggenberger, E. Klumpp, C. Knief, E. Lehndorff, R. Mikutta, S. Peth, A. Prechtel, N. Ray, and I. Kögel-Knabner. 2018. Microaggregates in soils. *Journal of Plant Nutrition and Soil Science* 181: 104–136.

Ueshima, M., and K. Tazaki. 2001. Possible role of microbial polysaccharides in nontronite formation. *Clays and Clay Minerals* 49: 292–299.

Uroz, S., L.C. Kelly, M.P. Turpault, C. Lepleux, and P. Frey-Klett. 2015. The mineralosphere concept: Mineralogical control of the distribution and function of mineral-associated bacterial communities. *Trends in Microbiology* 23: 751–762.

Urrutia, M.M., and T.J. Beveridge. 1995. Formation of short-range ordered aluminosilicates in the presence of a bacterial surface (*Bacillus subtilis*) and organic ligands. *Geoderma* 65: 149–165.

Verboom, W.H., and J.S. Pate. 2006. Bioengineering of soil profiles in semiarid ecosystems: The 'phytotarium' concept. A review. *Plant and Soil* 289: 71–102.

Verboom, W.H., and J.S. Pate. 2013. Exploring the biological dimension to pedogenesis with emphasis on the ecosystems, soils and landscapes of southwestern Australia. *Geoderma* 211–212: 154–183.

Vogel, C., K. Heister, F. Buegger, I. Tanuwidjaja, S. Haug, M. Schloter, and I. Kögel-Knabner. 2015. Clay mineral composition modifies decomposition and sequestration of organic carbon and nitrogen in fine soil fractions. *Biology and Fertility of Soils* 51: 427–442.

Vogel, C., C.W. Mueller, C. Höschen, F. Buegger, K. Heister, S. Schulz, M. Schloter, and I. Kögel-Knabner. 2014. Submicron structures provide preferential spots for carbon and nitrogen sequestration in soils. *Nature Communications* 5: 2947.

Wagai, R., L.M. Mayer, and K. Kitayama. 2009. Extent and nature of organic coverage of soil mineral surfaces assessed by a gas sorption approach. *Geoderma* 149: 152–160.

Wagai, R., L.M. Mayer, K. Kitayama, and Y. Shirato. 2013. Association of organic matter with iron and aluminum across a range of soils determined via selective dissolution techniques coupled with dissolved nitrogen analysis. *Biogeochemistry* 112: 95–109.

Wan, J., T. Tyliszczak, and T.K. Tokunaga. 2007. Organic carbon distribution, speciation, and elemental correlations within soil microaggregates: Applications of STXM and NEXAFS spectroscopy. *Geochimica et Cosmochimica Acta* 71: 5439–5449.

Watteau, F., G. Villemin, F. Bartoli, C. Schwartz, and J.L. Morel. 2012. 0–20 μm aggregate typology based on the nature of aggregative organic materials in a cultivated silty topsoil. *Soil Biology and Biochemistry* 46: 103–114.

Wattel–Koekkoek, E.J.W., P.P.L. van Genuchten, P. Buurman, and B. van Lagen. 2001. Amount and composition of clay-associated organic matter in a range of kaolinitic and smectitic soils. *Geoderma* 99: 27–49.

Wiseman, C.L.S., and W. Püttmann. 2006. Interactions between mineral phases in the preservation of soil organic matter. *Geoderma* 134: 109–118.

11 Occurrence and Extraction of Soil Clays

To try to understand the soil by taking a few trowelfuls and submitting them to chemical tests is like trying to understand the human body by cutting off the finger, grinding it to paste and performing the same tests. You may learn a lot about the chemistry of pastes, but about the intricate anatomical linkages – and about the body's function as a whole – you will learn nothing at all.

William Bryant Logan, 1995

11.1 CLAY ASSOCIATIONS

As already stated, clay minerals as components of soils are of interest from two main points of view, either for explaining the origin of a soil or else for explaining and/or predicting its properties. From the first viewpoint, clay minerals most often occur in soils as the principal inorganic products of the pedogenic processes involved in the formation of the soil. From the second, clay minerals are of particular interest because they provide the most reactive surfaces in soils. Clay mineral particles provide the inorganic source of the electrical charge and the predominant area of surfaces for reactions with other entities in soils, including organic constituents, both living and dead, and also other clay mineral particles. Thus Greenland (1965) found that organic matter was largely found associated with minerals in soils and Tisdall and Oades (1982) posited that the association of minerals together and also with processed organic matter is the basis for the strongest and most persistent associations found in soil, i.e. those constituting microaggregates.

As a result of the great reactivity of their surfaces, clay minerals rarely occur in soils as separate submicron particles. This is just as well for the stability of soils as very small particles would be easily removed by the erosive forces of water and wind. However, although *in situ* studies of clays show promise, as we shall explain, clay minerals most commonly need to be disassociated from their attachments in soils in order to be able to characterise them. Nonetheless, all such dissociations are in danger of producing artefacts by changing the essential nature of the clays that are inherent in their *in situ* associations within soils. We will first examine examples of associations of clays within soils that have been recorded in the literature.

Electron microscopy has provided some of the most useful techniques for examining associations within soils *in situ*. Scanning electron microscopy (SEM) undoubtedly has a role to play, and some work has been done using this technique, often with modifications to avoid changes in soil fabric and interparticle associations that invariably occur upon drying in the essentially aqueous soil system, whether prior to or during analysis. Thus, for example, samples have been frozen by cryofixation in liquid coolants prior to examination by SEM, by the technique of low temperature scanning electron microscopy, or LTSEM (Chenu and Tessier, 1995). Alternatively, they have been purposely kept wet, as sampled using environmental scanning electron microscopy, or ESEM (Foster, 1994; Churchman et al., 2010).

Important information has also been obtained using transmission electron microscopy (TEM), which has been capable of high resolution for some time. This technique is commonly used in clay mineral studies where the common procedure involves prior dispersion of the sample and then sedimentation of its constituent small particles onto a flat surface for examination by an electron beam upon its transmission through the thin sedimented particles. For examination of undisturbed soil samples *in situ*, an approach for the TEM analysis of biological specimens has been employed in which small samples of soil were first fixed chemically immediately after sampling to kill biota. In order to both distinguish organic matter from minerals and also to identify major types of organic groups present in the samples, they are stained with appropriate reagents. Then they are dehydrated and embedded in a resin. The resulting solid subsamples are then sliced to give ultra-thin sections which can be examined by TEM. The procedure is detailed by Foster and Martin (1981), with particular reagents defined for killing biota, staining to identify organic components and resin impregnation. This approach has been used in many subsequent studies by R.C. Foster in Australia and also by C. Chenu in France. Examples of work by these people and their associates using this approach are given here in micrographs showing clays and their associations *in situ* in soils (e.g. Figure 11.1).

Among the most common associations involving clays, as fine mineral material, in soils are their associations with other particles of clay. These associations may themselves constitute microaggregates. This is shown in Figure 11.2.

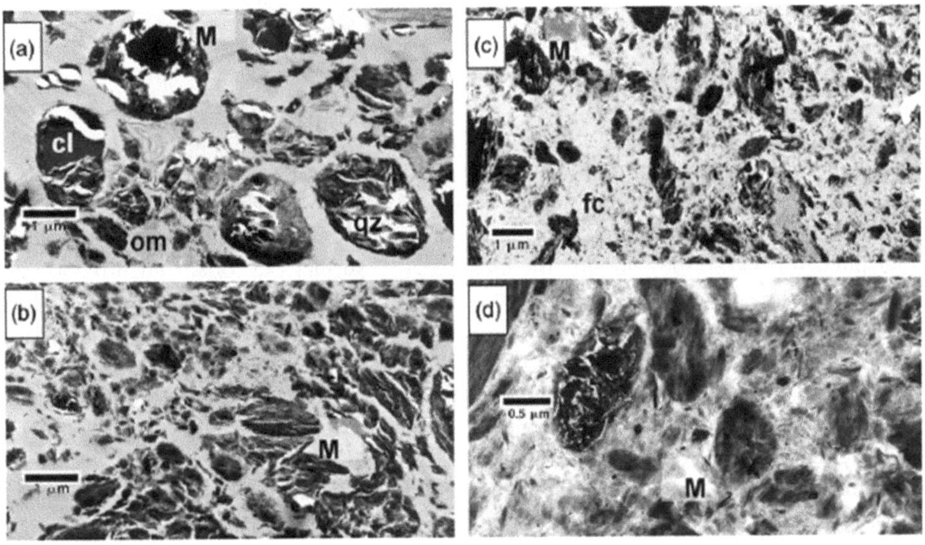

FIGURE 11.1 Transmission electron micrographs of ultra-thin sections of an Alfisol, both uncultivated (a) and also after short-term (b) and also long-term cultivation (c and d). In (a), microaggregates (M) are shown that are comprised of fine mineral matter (clay cl), often with organic matter (om) and others including shards from the breakage of quartz (qz) during thin sectioning. There is dispersed fine clay material (fc) surrounding microaggregates (M) in (b), (c) and (d), but not in (a). Note that the scale for (d) is different from that for (a), (b) and (c). (From Churchman, G. J., et al., 2010, *Soil and Tillage Research* 109: 23–35. Thanks to Dr. Ralph Foster. With permission from Elsevier.)

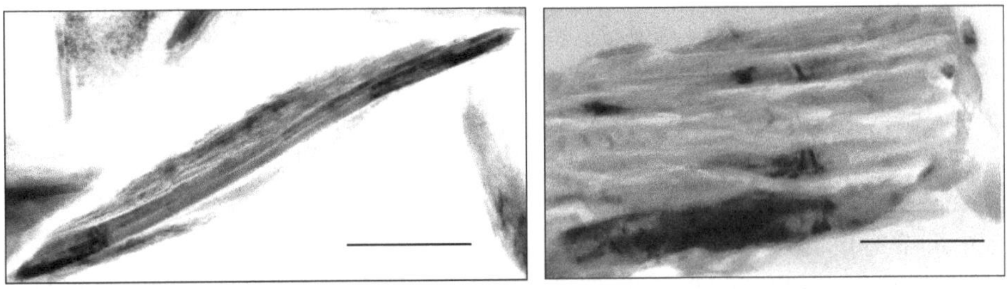

FIGURE 11.2 Transmission electron micrographs of ultrathin sections showing an illite crystal (left) and a kaolinite crystal (right). The scale bar is 0.5 µm. (From Chenu, C., and A.F. Plante, 2006, *European Journal of Soil Science* 57: 596–607. With permission from John Wiley & Sons.)

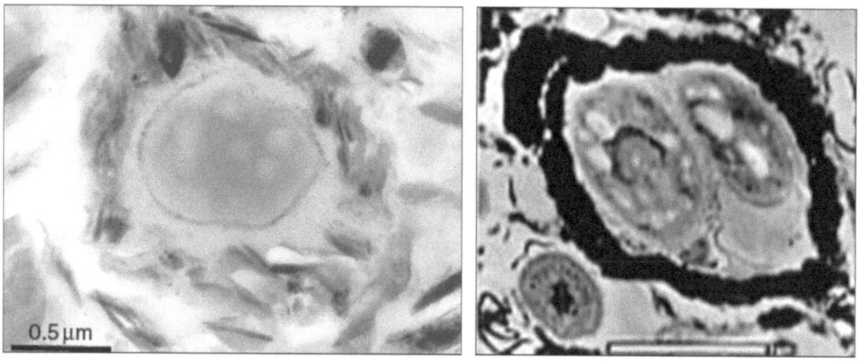

FIGURE 11.3 Transmission electron micrographs of ultra-thin sections of soils showing denser clay material surrounding bacteria. (Left) A Hamplumbrept (or Luvisol). (From Chenu, C., and A.F. Plante, 2006, *European Journal of Soil Science* 57: 596–607. With permission from John Wiley & Sons.) (Right) An Alfisol. (From Churchman, G.J., 2000, The alteration and formation of soil minerals by weathering, p. F3–F76, in M.E. Sumner (ed.), *Handbook of Soil Science*, CRC Press, Boca Raton, Florida. With thanks to Dr. Ralph Foster. With permission from CRC Press/Taylor & Francis.)

Quite often, however, there is a close association of clays with organic components of soils. These include situations where clays surround bacteria (Figure 11.3).

In other cases, clays are seen to be closely associated with other forms of biota, including fine roots and a variety of living and remnant species (Figure 11.4).

Clays can also be seen to surround other organic materials. These include extracellular polysaccharides (Figure 11.5).

Clays – or organo-clays – may also be associated with one another to mainly constitute a microaggregate, as seen (albeit with a small organic core) in an ultra-thin section of a Vertisol by Jocteur Monrozier et al. (1991, Figure 8). In this case, they closely resemble the schematic representation in Chapter 10, Figure 10.3.

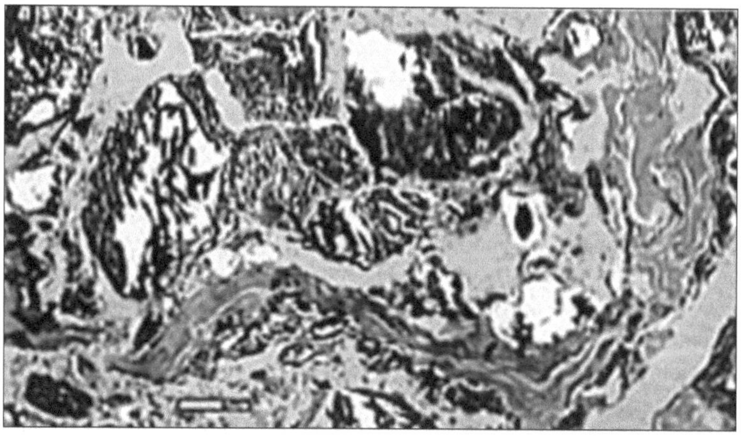

FIGURE 11.4 Transmission electron micrographs of ultra-thin section of an Alfisol showing clay material surrounding fine roots and other biological materials. (From Churchman, G.J., et al., 2014, Clay addition and redistribution to enhance carbon sequestration in soils, p. 327–335, in A.E. Hartemink and K. McSweeney (eds.), *Soil Carbon*, Springer, Cham, Switzerland. With thanks to Dr. Ralph Foster. With permission from Springer Nature.)

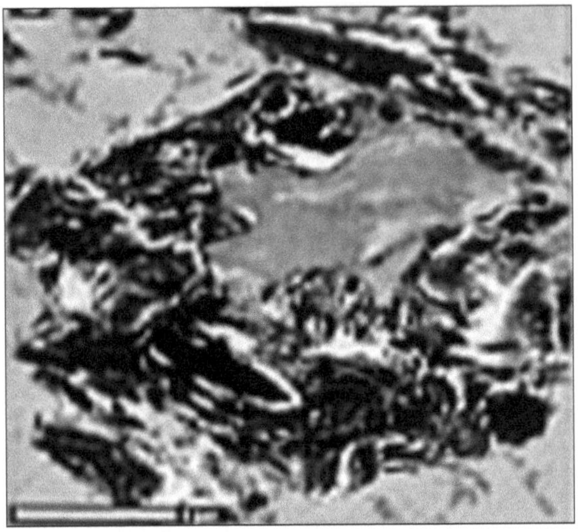

FIGURE 11.5 Transmission electron micrographs of ultra-thin section of an Alfisol showing clay material surrounding extracellular polysaccharides. (From Churchman, G.J., 2000, The alteration and formation of soil minerals by weathering, p. F3–F76, in M.E. Sumner (ed.), *Handbook of Soil Science*, CRC Press, Boca Raton, Florida. With thanks to Dr. Ralph Foster. With permission from CRC Press/Taylor & Francis.)

11.2 EXTRACTION OF SOIL CLAYS

There are strong implications for the extraction of clays arising from the observations of close associations between minerals and organic matter. For one, these associations are often very intimate. They arise from the operation of a variety of heterogeneous binding mechanisms including ion exchange, hydrogen bonding and hydrophobic bonding. In particular, for example, there are close chemical interactions between clay minerals and polysaccharides (e.g. Chenu, 1989; Chenu and Stotzky, 2002) (see also Chapter 10, Section 10.3). Even the associations of minerals with one another (Figure 11.2) may occur through heterogeneous bonds, involving metal oxides as well as well-processed organic matter (e.g. Tisdall and Oades, 1982). Procedures for the separation of soils into particle size fractions almost certainly change the nature of the constituent materials and especially the fine materials.

Attempts to maximise the yields of fine material from soils involve two main types of treatment: (1) disaggregation, and (2) dispersion or peptisation. Disaggregation involves the disruption of bonds between particles, which may be physical but are commonly also chemical. Physical bonding includes the results of close associations between particles from surface tension on drying and can also be physico-chemical in nature, e.g. reflecting the tendency of polyvalent cations to link together different anionic species – which include most aluminosilicate clay minerals and also organic matter – more strongly than monovalent cations. Dispersion or peptisation involves the maximisation of repulsive forces between fine particles in suspension to prevent their prolonged associations in suspension and hence ensure maximum dispersion.

The most common approaches to disaggregation have involved the removal of organic matter and of oxides, hydroxides and oxyhydroxides of Fe, Al and Mn, all of which are considered to play a role in cementing together particles in most soils. In some soils, particularly alkaline soils that are commonly found in more arid zones, disaggregation procedures also aim to remove carbonate cements. As well, salts may need to be removed from soils to aid their separation into fine particles and their subsequent dispersion. This usually occurs upon repeated washing with water.

Organic matter is removed from soils by oxidation. Hydrogen peroxide treatments, usually with heating, are most often used for this purpose, but other oxidants, such as sodium hypochlorite, sodium hyprobromite and potassium permanganate, may also be used. Carbonate cements are removed by acid treatments, with care being taken, e.g. through the use of indicators, to ensure that the pH does not drop so low that soil particles are dissolved. Soil clays are stable in the acid range at pH >4.5 due to the insolubility of aluminium at these pHs, and the use of sodium acetate is safe because the pH remains above 5. Even so, dissolution is a kinetic process, and a pH >3.5 is safe for the short duration of most laboratory preparation work.

The sodium ion (added as a salt, with excess sodium being removed, or via a resin) is often used to enable dispersion through peptisation but may also play a role in disrupting polyvalent ionic links between negatively charged entities. In many laboratories, the main chemical employed for peptisation is sodium hexametaphosphate at an alkaline pH obtained with sodium hydroxide or carbonate. pH elevation by alkali addition alone can aid dispersion. The effect of the hexametaphosphate, commercially known as 'Calgon' (standing for 'calcium gone') is to increase the negative charge of the edges of clay particles, which increases their mutual repulsion. This is a permanent effect. Addition of alkalis alone similarly increases the negative charges on particles, but their effect is only temporary.

Figure 11.6 shows the proportions of clay-size (<2 μm) material yielded by various disaggregation procedures designed to disrupt different types of associations occurring in a variety of soils, all of them from New Zealand. The nine treatments applied to each of the six soils in the study comprised five (I, S, Na, U(wet), U(dry)) that were physical only and four that were chemical, three of which were designed to remove either different types of organic matter (AcAc, P-B) or the total organic component (H2O2) and one of which (H2O2/CDB) was designed to chemically destroy all likely links, both organic and inorganic.

The set of soils ranged over a variety of textures, from sandy to clayey; represented a variety of geological and weathering origins, from alluvial to deeply weathered *in situ* and from volcanic ash and pumice; and covered a variety of dominant clay minerals, from micas through smectites to halloysite and allophane. A notable feature of the results for all of them was that a minimal disruption, consisting of an inversion of the cylinders containing the soils in aqueous suspension, produced almost no clay (<2 μm) material in every case. In four soils, the strongest physical disruption (ultrasonification of field-moist soil) was just as disruptive or nearly as disruptive as the extreme chemical treatment involving oxidation with hydrogen peroxide and reduction with dithionite, along with chelation using citrate.

Even physical treatments can change the nature of the fine material that is extracted. This was shown by the effects of shaking two different soils for increasing times upon yield of fine clay (<0.2 μm) relative to that of coarse clay (0.2–2 μm), as well as that of the total clay fraction (<2 μm) (Chittleborough, 1982). In both, a fine sandy loam and a medium clay soil, there was a continued increase in yield of fine clay with time of shaking even after 14 h. In the fine sandy loam, the yields of both fine and coarse clay, hence of total clay, continued to increase with prolonged shaking, so that clay material, both fine and coarse, was produced at the expense of coarser fractions such as silt. In the medium clay soil, by contrast, the total clay fraction yield reached a maximum after shaking for only 100 s, and hence fine clay was produced at the expense of coarse clay in this soil. Nonetheless, the nature of the fine material obtained clearly depended on time of shaking for each soil.

As seen in Churchman and Tate's (1986) results for soils from volcanic materials from Egmont, Taupo and Puniu (Figure 11.6), clays in some soils are strongly held by chemical bonds. For these types of soils, and also for some soils with high contents of metal oxides (generally, Oxisols or Ultisols according to Soil Taxonomy), particularly strong disruption is required to achieve the disaggregation of fine particles (e.g. Silva et al., 2015). Working with soils from Hawaii, Silva et al. (2015) found that ultrasonication at increasingly high energies was required to maximise the yield of clay from a variety of Andisols, Oxisols and Ultisols.

Soils that are difficult to disperse are by no means confined to those from volcanic materials or highly oxidic soils. It has been known for a long time that there are soils in Australia which are especially difficult to disaggregate, hence apparently coarse-textured but which display properties,

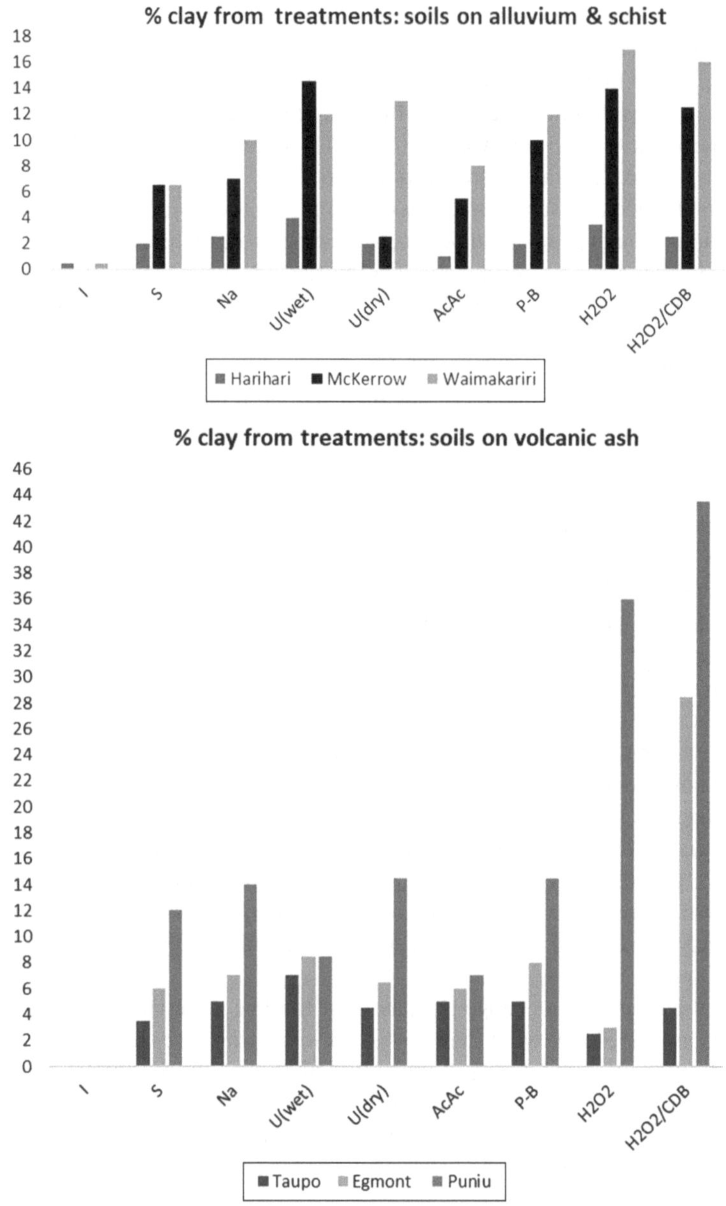

FIGURE 11.6 Percentage of each soil sample in the clay (<2 μm) fraction following disaggregation treatments: I, inversion; S, prolonged shaking ; Na, shaking with Na resin; U(wet), 3 min ultrasonics of field-moist soil; U(dry), 3 min ultrasonics of air-dried soil ; AcAc, acetylacetone soak; P-B, periodate–borate; H_2O_2, exhaustive (×10) peroxidation; H_2O_2/CDB, hydrogen peroxide + citrate–dithionite–bicarbonate (each twice). Clay minerals are predominantly mica (Harihari and Waimakariri) mica–smectite (McKerrow); allophane (Taupo and Egmont); or halloysite (Puniu). (Re-plotted from data in Churchman, G.J., and K.R. Tate, 1986, *Geoderma* 37: 207–220.)

especially water-holding capacities and cation exchange capacities, that indicate they are likely to have high contents of clay-size particles, albeit in strong associations. When a selection of soils of this type, known as 'sub-plastic' because of these characteristics, was treated with sodium hexametaphosphate and prolonged shaking in pastes with water, designed to enhance disruptive collisions between particles (Norrish and Tiller, 1976), it was first found that the clay yield increased

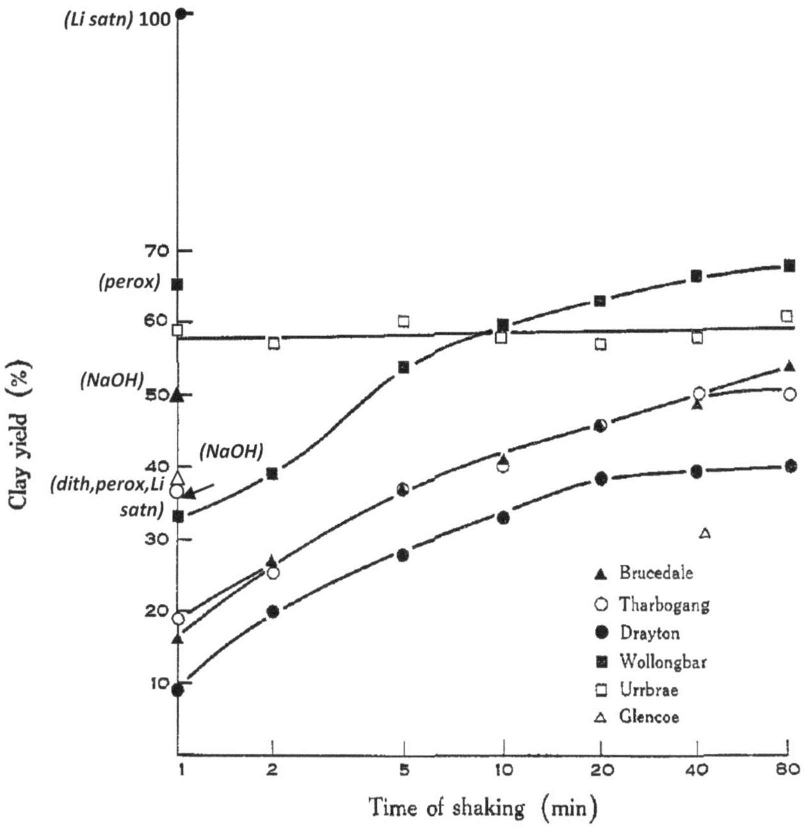

FIGURE 11.7 The effect of time of shaking on the yield of clay from so-called sub-plastic soils (and another soil; Urrbrae) and the yield after one minute shaking following the chemical pre-treatments which gave the maximum yield for each soil. Li saturation gave the maximum yield (of 100%) from the Drayton soil, while peroxidation gave the maximum yield (66%) from the Wollongbar soil. NaOH produced the most clay from the Brucedale and Tharbogang soils (50% and 37% respectively), while it required a dithionite treatment together with peroxidation and Li saturation to give the most clay (38%) from the Glencoe soil. (Based on Norrish, K., and K.G. Tiller, 1976, *Soil Research* 14: 273–289.)

progressively with time of shaking up to and beyond 80 minutes (Figure 11.7). This was in contrast to a soil ('Urrbrae') without sub-plastic characteristics, which dissociated into its apparent maximum yield after just 1 minute's shaking. However, targeted chemical treatments increased clay yield of all of the sub-plastic soils, albeit to different extents with each treatment. The treatments attempted included saturation with Li^+, oxidation with hot hydrogen peroxide, reduction with dithionite, chelation of Al and Fe with citrate, acidification (with M HCl) and an alkaline treatment (with NaOH of different concentrations) and various combinations of these. Figure 11.7 also shows the maximum yield that was achieved for each treated soil after 1 minute's shaking and identifies the chemical agents or their combination leading to these maxima (along the y-axis of the plot). The observation that different agents led to maximum production of clay in different soils illustrated that the forces and agents responsible for aggregating and cementing fine material in soils may differ with each type of soil. The sub-plastic soils studied by Norrish and Tiller (1976) included one ('Drayton') in which electrostatic forces between particles dominated (these were overcome by saturation with Li^+); one ('Wollongbar') in which organic matter was the predominant aggregation agent (this was destroyed by peroxidation); two ('Brucedale' and 'Tharbagong') in which interparticle intergrowths apparently provided links between them (this effect was destroyed by alkali (NaOH) dissolution); and one soil ('Glencoe') wherein particles were held together by a combination of electrostatic forces and cements of organic matter and also iron oxides (destroyed by dithionite, then Li^+ saturation, then alkali dissolution).

11.3 RECOMMENDED PROCEDURES

However, recognition that there is no such thing as an absolute value of clay content nor is there one procedure that can achieve it with all soils is recorded in several protocols that are recommended for the particle size separation of soils. An authoritative chapter on the topic states that "depending on the method of chemical treatment and physical dispersion used, measured clay contents can vary by factors of two to four or more" (Gee and Bauder, 1986, p. 397). Furthermore, one volcanic ash soil, by way of example of a type of soil that is difficult to disperse, especially following drying, gave rise to clay contents varying from 1% to 56% (Gee and Bauder, 1986). This drying effect is also shown for the two allophanic soils on volcanic ash in Figure 11.6. For this type of soil, it is recommended that drying from field moisture be avoided. A pH of 3 or above 9 has been recommended for their dispersion (Warkentin et al., 1980). A more general recommendation has been made that adequate dispersion of soils with variable charge needs to occur in a solution with a pH well above or else below the zero point of charge (Tama and El-Swaify, 1978; El-Swaify, 1980). Lessons learnt from these soils which offer particular challenges for their dispersion, i.e. avoiding drying and maximising net charge (in order to aid the mutual repulsion of particles), apply to soils generally. In a sense, all determinations of the clay content of soils are arbitrary, as we have seen. Since clay particles in soils generally occur in associations giving rise to larger entities, the clay content depends upon the efficiency of the disaggregating and subsequent peptisation processes. "The method that produces the most complete dispersion of a soil sample is the more acceptable method", according to Gee and Bauder (1986, p. 387). In other words, if clay content is to provide a meaningful index or surrogate for such important properties of a soil as its ability to retain and supply water or nutrients, it is important that the method used for each particular soil provides the most complete dispersion possible for that soil, even if the procedure employed differs from that employed for other types of soils. Otherwise, apparently low clay contents resulting from incomplete disaggregation and/or poor peptisation may hide a high potential capacity for the retention and supply of either or both water or nutrients by presenting a strongly aggregated or poorly dispersed soil as apparently coarser than is represented by its maximum achievable clay content.

For each soil, then, the aim of obtaining the maximum achievable clay content involves the removal of all cements to maximise the disaggregation of particles and also the repulsive forces between the resulting set of particles to ensure maximum dispersion. Cements to be removed include organic matter, which is destroyed by oxidation, with hydrogen peroxide being the most efficient oxidant for most soils (Day, 1965; Gee and Bauder, 1986). They also include iron oxides, which are removed by reduction in a pH-buffered solution, and the subsequent chelation of dissolved Fe. A sodium bicarbonate-buffered sodium dithionite–citrate system devised by Mehra and Jackson (1958) is considered to be the most effective for this purpose while minimising the destruction of silicates which contain iron. Nonetheless, some silicates may be dissolved by this treatment.

Carbonates, where they occur, also need to be removed to enhance disaggregation and dispersion. This is achieved by acidification, but it is recommended that the weakly acidic sodium acetate be used for this purpose rather than a strong acid like HCl. The use of sodium acetate ensures that the pH does not go below 5, and the treatment may be only slowly or partly effective in highly calcareous soils. To ensure maximum dispersion of particles it is also necessary to remove soluble salts, which act as flocculants. This is done by washing or leaching samples with de-ionised or distilled water.

It has been illustrated in comparative studies by, e.g. Norrish and Tiller (1976) and Churchman and Tate (1986) that the choice of a disaggregation or dispersion method to achieve the maximum possible clay content can be different for different soils. Generally speaking, the most effective methods involve dispersion with either a chemical peptiser such as alkaline sodium hexametaphosphate, together with physical disruption such as prolonged shaking, or physical disruption alone, mostly by ultrasound, but also through sodium saturation, followed by a physical disruption treatment (Mikhael and Briner, 1978; Churchman et al., 1999).

A note of caution concerns the common hydrogen peroxide and citrate–bicarbonate–dithionite (CDB) pretreatments of soils. They may each have effects beyond those respectively of simply oxidising organic matter external to mineral structures, and reducing and extracting Fe that is extraneous to silicates as oxides and oxyhydroxides. In particular there may be effects on the nature and extent of hydroxyl interlayers in 2:1 expandable layer silicates such as smectites and vermiculites.

For a series of agricultural surface soils in the Central United States, Velde (2001) found that treatment with hydrogen peroxide led to these interlayers becoming more expandable, probably because organic matter had helped to close them down in the soils. In a series of soils on loess in Southern New Zealand, Churchman and Bruce (1988) found that CDB treatment of some soils containing hydroxy-interlayered minerals also led to more expandable 2:1 layer silicates. The reason for the change arises because citrate strongly complexes aluminium as well as iron. However, as previously discussed (Rich, 1968; Barnhisel, 1977) the extent of removal of interlayer material that was responsible for the expansion depended on the degree of development and composition of the interlayer material.

REFERENCES

Barnhisel, R.I. 1977. Chlorites and hydroxy interlayered vermiculite and smectite, p. 331–357. In: J.B. Dixon, and S.B. Weed (eds.), *Minerals in Soil Environments*, 1st edn. Soil Science Society of America, Madison, Wisconsin.

Chenu, C. 1989. Influence of a fungal polysaccharide, scleroglucan, on clay microstructures. *Soil Biology and Biochemistry* 21: 299–305.

Chenu, C., and A.F. Plante. 2006. Clay-sized organo-mineral complexes in a cultivation chronosequence: Revisiting the concept of the 'primary organo-mineral complex'. *European Journal of Soil Science* 57: 596–607.

Chenu, C., and G. Stotzky. 2002. Interactions between microorganisms and soil particles: An overview, p. 1–40. In: P.M. Huang (ed.), *Interactions Between Soil Particles and Microorganisms: Impact on the Terrestrial Ecosystem*. John Wiley & Sons, Manchester.

Chenu, C., and D. Tessier. 1995. Low temperature scanning electron microscopy of clay and organic constituents and their relevance to soil microstructures. *Scanning Microscopy* 9: 989–1010.

Chittleborough, D.J. 1982. Effect of the method of dispersion on the yield of clay and fine clay. *Soil Research* 20: 339–346.

Churchman, G.J. 2000. The alteration and formation of soil minerals by weathering, p. F3–F76. In: M.E. Sumner (ed.), *Handbook of Soil Science*. CRC Press, Boca Raton, Florida.

Churchman, G.J., F. Bartoli, G. Burtin, J. Rouiller, and D. Weissmann. 1999. Comparison of methods using sodium for the size fractionation of soils, p. 331–338. In: H. Kodama, A.R. Mermut, and J.K. Torrance (eds.), *Clays for Our Future*. Proceedings of the 11th International Clay Conference, Ottawa, Canada 1997. ICC97 Organising Committee, Ottawa, Canada.

Churchman, G.J., and J.G. Bruce. 1988. Relationships between loess deposition and mineral weathering in some soils in Southland, New Zealand, p. 11–31. In: D.N. Eden, and R.J. Furkert (eds.), *Loess, Its Distribution, Geology and Soils*. Balkema, Rotterdam.

Churchman, G.J., R.C. Foster, L.P. D'Acqui, L.J. Janik, J.O. Skjemstad, R.H. Merry, and D.A. Weissmann. 2010. Effect of land-use history on the potential for carbon sequestration in an Alfisol. *Soil and Tillage Research* 109: 23–35.

Churchman, G.J., A. Noble, G. Bailey, D. Chittleborough, and R. Harper. 2014. Clay addition and redistribution to enhance carbon sequestration in soils, p. 327–335. In: A.E. Hartemink, and K. McSweeney (eds.), *Soil Carbon*. Springer, Cham, Switzerland.

Churchman, G.J., and K.R. Tate. 1986. Aggregation of clay in six New Zealand soil types as measured by disaggregation procedures. *Geoderma* 37: 207–220.

Day, P.R. 1965. Particle fractionation and particle-size analysis, p. 545–567. In: C.A. Black (ed.), *Methods of Soil Analysis*. Part 1. *Physical and Mineralogical Properties, Including Statistics of Measurement and Sampling*. American Society of Agronomy and Soil Science Society of America, Madison, Wisconsin.

El-Swaify, S.A. 1980. Physical and mechanical properties of Oxisols, p. 303–324. In: B.K.G. Theng (ed.), *Soils With Variable Charge*. New Zealand Society of Soil Science, Lower Hutt.

Foster, R.C. 1994. Microorganisms and soil aggregates, p. 144–155. In: C.E. Pankhurst, B.M. Doube, V.V.S.R. Gupta, and P.R. Grace (eds.), *Soil Biota*. CSIRO Publications, Melbourne.

Foster, R.C., and J.K. Martin. 1981. In situ analysis of soil components of biological origin, p. 75–110. In: E.A. Paul, and J.N. Ladd (eds.), *Soil Biochemistry*, vol. 5. Marcel Dekker, New York.

Gee, G.W., and J.W. Bauder. 1986. Particle-size analysis, p. 383–411. In: A. Klute (ed.), *Methods of Soil Analysis*. Part I. *Physical and Mineralogical Methods*. Agronomy Monograph No. 9, 2nd edn. American Society of Agronomy and Soil Science Society of America, Madison, Wisconsin.

Greenland, D.J. 1965. Interactions between clays and organic compounds in soils. Part 1. Mechanisms of interaction between clays and defined organic compounds. *Soils and Fertilisers* 28: 415–425.

Jocteur Monrozier, L., J.N. Ladd, R.W. Fitzpatrick, R.C. Foster, and M. Rapauch. 1991. Components and microbial biomass content of size fractions in soils of contrasting aggregation. *Geoderma* 50: 37–62.

Mehra, O.P., and M.L. Jackson. 1958. Iron oxide removal from soils and clays by a dithionite-citrate system buffered with sodium bicarbonate. *Clays and Clay Minerals* 7: 317–327.

Mikhael, E.H., and G.P. Briner. 1978. Routine particle size analysis of soils using sodium hypochlorite and alkaline dispersion. *Soil Research* 16: 241–244.

Norrish, K., and K.G. Tiller. 1976. Subplasticity in Australian soils. V. Factors involved and techniques of dispersion. *Soil Research* 14: 273–289.

Rich, C.I. 1968. Hydroxy interlayers in expansible layer silicates. *Clays and Clay Minerals* 16: 15–30.

Silva, J.H.S., J.L. Deenik, R.S. Yost, G.L. Bruland, and S.E. Crow. 2015. Improving clay content measurement in oxidic and volcanic ash soils of Hawaii by increasing dispersant concentration and ultrasonic energy levels. *Geoderma* 237–238: 211–223.

Tama, K., and S.A. El-Swaify. 1978. Charge, colloidal and structural stability interrelationships for oxidic soils, p. 41–49. In W.W. Emerson, R.D. Bond, and A.R. Dexter (eds.), *Modification of Soil Structure*. Wiley, London.

Tisdall, J.M., and J.M. Oades. 1982. Organic matter and water-stable aggregates in soils. *Journal of Soil Science* 33: 141–163.

Velde, B. 2001. Clay minerals in the agricultural surface soils in the central United States. *Clay Minerals* 36: 277–294.

Warkentin, B.P., T. Maeda, and B.K.G. Theng. 1980. Soils with variable charge. In: B.K.G. Theng, *Soils with Variable Charge*. New Zealand Society of Soil Science, Lower Hutt.

12 Identification and Quantification of Clay Minerals in Soils

Names and attributes must be accommodated to the essence of things, and not the essence to the names, since things come first and names afterwards.

Galileo Galilei (1564–1642)

12.1 IDENTIFICATION OF SOIL CLAYS BY X-RAY DIFFRACTION

X-ray diffraction (XRD) remains the main technique for identification of soil clays. Other techniques, e.g. infrared spectroscopy, differential thermal analysis and electron microscopy, are mainly useful for confirmation of initial identification by XRD.

Table 12.1 gives the diagnostic peaks in XRD for the main minerals comprising soil clays. Ideally, clay fractions extracted from soils are examined in two main forms: (1) as oriented samples deposited from a suspension onto a flat surface (usually a glass or silicon slide), which enhances peaks from basal planes (00l peaks); and (2) as random samples. Often random samples are supposed to be obtained by back-filling powders into a well in a sample holder. However, it is doubtful that complete randomisation is reliably obtained in this manner. Other methods have been suggested to obtain random samples, with spray-drying being a promising approach (e.g. Hillier, 1999). Peaks for phyllosilicates dominate the patterns from oriented samples, while those for primary minerals such as quartz, feldspars and gibbsite are strong in the patterns from random samples. These latter patterns also show peaks for iron and aluminium oxides, oxyhydroxides and hydroxides, provided their concentrations and crystallinities are high enough.

12.2 IDENTIFICATION AND ANALYSES OF SOIL CLAYS BY CHEMICAL EXTRACTIONS

X-ray diffraction, while the main identification technique for crystalline minerals, is unsuitable for poorly crystalline or nanocrystalline minerals such as allophane and ferrihydrite. However, imogolite gives a number of broad peaks, with the strongest at or near 16, 7.9, 5.6, 3.3 and 2.25 Å, and some allophane samples can give exceptionally broad reflections around 3.3 and 2.25 Å (Wada, 1989). Instead, allophane and ferrihydrite are identified and measured by the yield of Si from the dissolution of the soil or soil clay fraction in 0.15 M ammonium oxalate in the dark for 4 hours (Parfitt, 1990). The multiplication factor used to measure the content of allophane (%) depends on the Al:Si ratio in the dissolution product, so oxalate-extractable Al is also measured. The multiplication factor is 7 when the ratio of Al:Si is 2.0, but varies linearly from 5 when the ratio is 1.0, to 16 when it is 3.5 (Parfitt, 1990). The Al:Si ratio identifies the type of allophane present. The per cent content of ferrihydrite can be estimated as 1.7 times the content of oxalate-extractable Fe (Childs et al., 1991).

Iron oxides, oxyhydroxides and hydroxides give rise to peaks in XRD if they are crystalline (Table 12.1b). However, these peaks are typically weak and, in any case, many metal 'oxides' (inclusive term) are poorly crystalline. The total 'free' Fe, which includes oxides as well as ferrihydrite,

TABLE 12.1
Diagnostic Peaks in XRD

(a) Layer Silicates, including Results of Treatments Where Applicable

Name	Spacing (Å)					
	Mg- or Ca-Saturated, Air-Dry	**+ Glycerol, Ethylene Glycol**	**K-Saturated, 110°C**	**K-Saturated, 350°C**	**K-Saturated, 550°C**	**+ Formamide (<2 h)**
Chlorite-smectite (regular)	24	31–32	24	24	24	n.a.
Mica*-smectite (regular)	20–25	28	10	10	10	n.a.
Mica*-vermiculite (regular)	20–24	24	10	10	10	n.a.
Smectite	10–15	17–18	10	10	10	n.a.
Vermiculite	10–14	14	10	10	10	n.a.
Illite-vermiculite (variable)	10–14	10–14	10–14	10–12	10–12	n.a.
Smectite-illite (variable)	12–13	13–14	10	10	10	n.a.
Illite-smectite (variable)	10–12	12–13	10	10	10	n.a.
Smectite-chlorite (variable)	14–15	16–17	11–12	11–12	11–12	n.a.
Chlorite-smectite (variable)	14–15	15–16	12–13	12–13	12–13	n.a.
Mg, Fe-Chlorite	14	14	14	14	14	n.a.
Al-Chlorite (= HIV or HIS)	14	14	12	10	10	n.a.
Poorly crystallised illite	10.2-10.9	10.2–10.9	~10	~10	~10	n.a.
Mica* (well crystallised)	10.0	10.0	10.0	10.0	10.0	n.a.
Kaolin-smectite	15#, 7.2–7.9#	18#, 7.2–7.9#	10, 7.2	10, 7.2	10	n.a.
Halloysite	7.2 and/or 10#	7.2 and/or 10.9	7.2	7.2	—	10.1
Kaolinite	7.1	7.1	7.1	7.1	—	7.1

(b) Other Minerals (Air-Dry Sample Only)

Name	Spacing (Å)
Sepiolite	12.2
Palygorskite	10.4
Talc	9.3
Hornblende	8.38, 3.12
Gibbsite	4.85
Gypsum	4.28
Goethite	4.16
Cristobalite	4.15
Opal-CT	4.10
Feldspars	6.3–6.6, 4.02–4.04, 3.70–3.83, 3.62–3.67
Hematite	3.67
Anatase	3.52
Quartz	4.26, 3.34
Calcite	3.04

n.a., Not applicable.
* Or illite.
\# Often broad or indistinct.
— No basal spacing peaks.

can be measured by chemical extraction, with heating, using sodium dithionite (added as a solid) as a reductant, sodium citrate (typically 22% in solution) as a complexing reagent, and usually also sodium bicarbonate (1 M) as a pH buffer (Mehra and Jackson, 1958; Holmgren, 1967; Parfitt and Childs, 1988). The extraction treatment is often denoted as DC or DCB. In order to separately estimate crystalline Fe oxides such as (haematite + goethite) from ferrihydrite, the amount of Fe extracted by ammonium oxalate is subtracted from that extracted by DC or DCB.

12.3 IDENTIFICATION OF SOIL CLAYS BY INFRARED SPECTROSCOPY

Infrared spectroscopy has sometimes been used to identify clays in soils (Farmer, 1974; Russell and Fraser, 1994). Probably the main use for this technique has been to identify the presence of kaolinite and halloysite. These give distinctive peaks in the –OH stretching region of mid-infrared (MIR) spectra. For kaolinite, there are four peaks (at 3697, 3669, 3652 and 3620 cm^{-1}), with just three for disordered kaolinites (they give no peak at 3669 cm^{-1}). For halloysite, there are two, generally broader peaks, at 3695 and 3620 cm^{-1}. These peaks may be used to confirm the presence of kaolinite and/or halloysite, and may also be used to identify these minerals when the sample is too small for use in XRD analyses (e.g. Collyer et al., 1984).

Visible–near infrared (vis-NIR) spectra have been shown to display individual peaks for different clay minerals (e.g. kaolinite, illite and smectite), and these can be used to give estimates of their composition in soil samples (Viscarra Rossel et al., 2009). These estimates showed good agreement with those from XRD (Viscarra Rossel et al., 2009). The spectrophotometers for this technique are available in portable form so that they can be used for analyses *in situ* as well as in the laboratory (Viscarra Rossel et al., 2009).

12.4 IDENTIFICATION OF SOIL CLAYS BY THERMAL ANALYSES

Differential thermal analysis, together with thermo-gravimetric analysis, gives rise to mainly endothermic peaks at distinctive temperatures for different minerals (Mckenzie, 1957). These destructive analyses may be most useful for semi-quantitative analysis of kaolin minerals, which yield endotherms near 600°C. However, the temperature and the size of the peaks are subject to various factors relating to the instrument, the nature of the sample, and experimental factors, including heating rate (Holdridge and Vaughan, 1957).

12.5 IDENTIFICATION OF SOIL CLAYS BY ELECTRON MICROSCOPY

Electron microscopy is one of the most widely used techniques for identifying and characterising clays, including soil clays. Many studies of soil clays include electron micrographs of the clays, from transmission and/or scanning electron microscopy. Whereas micrographs of clays from geological deposits, so-called standard clays, often show distinctive shapes, those from soils are often poorly distinguished. For example, kaolinites are often distinguished by hexagonal-shaped plates, while halloysites are generally distinguished by non-platy shapes, often tubes or spheres. In soils, however, kaolinites may be more amorphous in shape (e.g. Singh and Gilkes, 1992; also see Figure 7.5 herein), and halloysites similarly have less distinctive particles in soils (e.g. Churchman, 1990). Furthermore, halloysites have been found to occur in platy shapes (Tazaki, 1979; Noro, 1986; Joussein et al., 2005) and even more recently in book-like arrangements of platy particles (Cunningham et al., 2016). Hence, particle shape by electron microscopy may be unreliable for identifying minerals, even if an unusual particle shape can give an indication that further distinguishing tests, such as the formamide expansion test for halloysite (Churchman et al., 1984), are required for confirmation.

Allophane and, particularly, imogolite have been found to have distinctive shapes with transmission electron microscopy (see Figure 8.8).

12.6 ANALYSIS OF SOIL CLAYS BY OTHER TECHNIQUES

Analysis of clays may be conducted with a variety of other techniques. These include Mössbauer spectroscopy (Childs et al., 1979; Goodman, 1980); nuclear magnetic resonance (Fripiat, 1980; Stone and Sanz, 1980; Goodman and Chudek, 1994); electron paramagnetic resonance or electron spin resonance (McBride, 1980; Pinnavaia, 1980; Vadrine, 1980; Paterson and Swaffield, 1994); neutron scattering (Ross and Hall, 1980; Hall et al., 1985; Bordallo et al., 2008; Cavallaro et al., 2018); and photoacoustic spectroscopy (Schmidt, 1980; Churchman et al., 2010a). Not all have been applied – or reports published – of their application to soil clays. In any case, none is satisfactory for the complete mineralogical analysis of a sample but each may provide useful additional information to that obtained from X-ray diffraction, together with that obtained from infrared spectroscopy and thermal analysis methods. For example, iron oxides in soils have been characterised by Goodman and Berrow (1976) and Childs et al. (1979) using Mössbauer spectroscopy, and photoacoustic spectroscopy was employed to measure surface quartz signals, hence the extent of covering materials on the quartz crystals by Churchman et al. (2010a).

12.7 QUANTITATIVE ANALYSES

Accurate quantitative analyses by all techniques founder on the problem that minerals in soils are different from those in 'pure' mono-mineral deposits (Churchman, 2010). As a result, the choice of standards for establishing comparisons with responses from known quantities of material for quantitative analyses is fraught with difficulties. This applies as much for X-ray diffraction as for any other method. An additional complication in the use of X-ray diffraction for quantitative analyses lies in the nature of X-rays (e.g. Klug and Alexander, 1954). The intensities of reflections of X-rays depend on (1) the angle of diffraction, where intensity is a wave with angular variation, and (2) the so-called structure factor that reflects the composition of the material under investigation and is also dependent on the angle of diffraction. As a result, analyses should factor in these functions for calculating the response of X-rays to the angle of diffraction and the specific mineral composition. It has been suggested that divisors be applied to compare either peak heights or peak areas with others at different angles of diffraction. In particular, peaks are often ratioed to the size (height or area) of a peak at 10 Å (for illite, or hydrated halloysite). Peaks at 18 Å (for expanded smectite) are divided by 4, those at 14 Å (for vermiculite, unexpanded smectite and chlorite) by 3, those at 12 Å (some mixed-layer minerals) by 2, and those at 7 Å (kaolinite and dehydrated halloysite) by 1.5 (Johns et al., 1954; Churchman, 1980). Even so, with the crystallinities of the minerals being diverse and generally poor, the resulting analyses can only be regarded as semi-quantitative and are often given in ranges rather than precise percentages.

It is nevertheless possible to make numerical measurements on phyllosilicate clay mineral XRD spectra that allow comparisons of relative abundances of different phases in different samples and hence observe changes due to different histories of the soils. Curve decomposition of the contributing peaks of the phases allows these comparisons (see Annex). Using peak height and width at half height, multiplied one by the other, normalised to a per cent of the total clay peak surfaces, gives a reasonable idea of relative abundance of the different phases in a sample by the relative surface area of the peaks. Comparing the relative percentage of each phase between different samples (for example smectite compared to illite) enables observation of the mineralogical differences on a comparative basis as a function of the chemical histories of the soils. Thus, absolute abundance is difficult to obtain but comparisons of changes in relative mineral abundances using X-ray diffraction data and curve decomposition methods can be made.

Some improvement of the analysis may be obtained following measurement of the cation exchange capacity (CEC) of the soil or its clay fraction. When it is assumed that all of the CEC derives from the clay fraction, an algorithm can be built from the contribution that each mineral component makes to the CEC value, measured, for example using XRF analyses of a Ba^{2+}-exchanged sample where samples contain no Ba in their structures (Churchman et al., 2010b). This algorithm may be used as a simultaneous equation together with equations for the different clay mineral phases from XRD and then iterated to provide the best fit. CECs of the individual minerals are mid-point values of those given in Table 8.1 (Chapter 8). Typically, CECs employed for common soil clay minerals (in cmol(+) kg^{-1}) are 100 for smectites, 25 for illites and 5 for kaolinites.

Analyses are most reliable when the soils or soil clays form a series, e.g. a chronosequence, or when examining the effect of a particular treatment on a soil or its clays. One of the features of soil clays is that, for 2:1 clays, peaks for related phases often overlap. Peaks can encompass the whole range from 14 to 10 Å for untreated clays and from 18 to 10 Å for clays treated with a polar liquid, typically glycerol of ethylene glycol. These collections of peaks can be separated into their constituents by a computer program (see Annex).

Nonetheless, the analysis of clay minerals in soils remains only semi-quantitative, and this is the case even for full profile fitting approaches such as the Rietveldt method (Bish, 1993). There are limitations to the application of the Rietveldt method to soil clays owing to their general, and variable, disorder (Ufer and Raven, 2017). However, profile fitting using appropriate software (Reynolds and Reynolds, 1996) is useful for distinguishing and quantifying mixed-layer phases which are not easily recognised visually.

Some positive indications of the future for quantitative analysis comes from the recent development of infrared diffuse reflectance spectroscopy for the prediction of useful soil properties such as pH, organic carbon, clay, silt and sand contents, CEC, exchangeable Al and K, available phosphorus, and electrical conductivity of soils. The procedure involves correlating the mid-infrared (MIR) or near-infrared (NIR) spectra of whole soils with their chemical and physical properties in training sets appropriate for the soils under analysis (e.g. Janik, 1998; Viscarra Rossel et al., 2006). Statistical techniques (principal components analysis, Janik, 1998; then, later, principal least-squares regression, Viscarra Rossel et al., 2006) are used to provide models matching spectra and properties. The spectra are not interpreted for components but used as data in themselves. A recent development in this area has been to set up a global spectral library in lieu of the need to provide local training sets (Viscarra Rossel et al., 2016). A similar approach can be used with whole profiles from X-ray diffraction in place of the infrared spectra. Recent developments of this approach have used a machine learning algorithm called 'Cubist' to predict and interpret a range of soil properties from X-ray powder diffraction (XRPD) patterns of micronised whole soil (i.e. <2 mm sieved) samples (Butler et al., 2018; Hillier and Butler, 2018). The predictions thus obtained for a set of 1246 Scottish soils were superior to those from (near) infrared spectroscopy for total carbon, total nitrogen, sand, silt and clay contents, and, notably also – in terms of Loveland et al.'s (1999) negative conclusions about the value of clay mineralogy for predictions of K at that time (see also Chapter 1, Section 1.8) – a measure of plant-available potassium. It was less successful than NIR for predictions of soil pH (in water) and CEC. The mineral phases responsible for each soil property of interest can be identified using this approach. As a useful example, the measure of plant-available K was shown to be highly sensitive to phyllosilicate K rather than to the often more-abundant K-feldspar (Butler et al., 2018; Hillier and Butler, 2018). An approach using cluster analysis more similar to that used in the infrared spectroscopic methods for predicting soil properties (e.g. Janik, 1998; Viscarra Rossel et al., 2006) has also been shown to be useful with whole-profile XRPD patterns as input data (Hillier, 2018). The successful launch of these procedures augurs well for an emerging field of 'digital soil mineralogy' (Hillier and Butler, 2018).

REFERENCES

Bish, D.L. 1993. Studies of clays and clay minerals using x-ray powder diffraction and the Rietveld method, p. 79–122. In: R.C. Reynolds Jr., and R.E. Ferrell Jr. (eds.), *Computer Applications to X-Ray Powder Diffraction Analysis of Clay Minerals*, CMS Workshop Lectures 5, Clay Minerals Society, Bloomington.

Bordallo, H.N., L.P. Aldridge, G.J. Churchman, W.P. Gates, M.T.F. Telling, K. Kiefer, P. Fouquet, T. Seydel, and S.A.J. Kimber. 2008. Quasi-elastic neutron scattering studies on clay interlayer-space highlighting the effect of the cation in confined water dynamics. *The Journal of Physical Chemistry C* 112: 13982–13991.

Butler, B.M., S.M. O'Rourke, and S. Hillier. 2018. Using rule-based regression models to predict and interpret soil properties from X-ray powder diffraction data. *Geoderma* 329: 43–53.

Cavallaro, G., L. Chiappisi, P. Pasbakhsh, M. Gradzielski, and G. Lazzara. 2018. A structural comparison of halloysite nanotubes of different origin by small-angle neutron scattering (SANS) and electric birefringence. *Applied Clay Science* 160: 71–80.

Childs, C.W., B.A. Goodman, and G.J. Churchman. 1979. Application of Mössbauer spectroscopy to the study of iron oxides in some red and yellow-brown soil samples from New Zealand, p. 555–565. In: M.M. Mortland, and V.C. Farmer (eds.), *Proceedings of the International Clay Conference, Oxford, 1978.* Developments in Sedimentology, vol. 27. Elsevier, Amsterdam.

Childs, C.W., N. Matsue, and N. Yoshinaga. 1991. Ferrihydrite in volcanic ash soils of Japan. *Soil Science and Plant Nutrition* 37: 299–311.

Churchman, G.J. 1980. Clay minerals formed from micas and chlorites in some New Zealand soils. *Clay Minerals* 15: 59–76.

Churchman, G.J. 1990. Relevance of different intercalation tests for distinguishing halloysite from kaolinite in soils. *Clays and Clay Minerals* 38: 591–599.

Churchman, G.J. 2010. Is the geological concept of clay minerals appropriate for soil science? A literature-based and philosophical analysis. *Physics and Chemistry of the Earth, Parts A/B/C* 35: 927–940.

Churchman, G.J., R.C. Foster, L.P. D'Acqui, L.J. Janik , J.O. Skjemstad, R.H. Merry, and D.A. Weissmann. 2010a. Effect of land-use history on the potential for carbon sequestration in an Alfisol. *Soil and Tillage Research* 109: 23–35.

Churchman, G.J., I.R. Pontifex, and S.G. McClure. 2010b. Factors affecting the formation and characteristics of halloysites or kaolinites in granitic and tuffaceous saprolites in Hong Kong. *Clays and Clay Minerals* 58: 220–237.

Churchman, G.J., J.S. Whitton, G.G.C. Claridge, and B.K.G. Theng. 1984. Intercalation method using formamide for differentiating halloysite from kaolinite. *Clays and Clay Minerals* 32: 241–248.

Collyer, F.X., B.G. Barnes, G.J. Churchman, T.S. Clarkson, and J.T. Steiner. 1984. A trans-Tasman dust transport event. *Weather and Climate* 4: 42–46.

Cunningham, M.J., D.J. Lowe, J.B. Wyatt, V.G. Moon, and G.J. Churchman. 2016. Discovery of halloysite books in altered silicic Quaternary tephras, northern New Zealand. *Clay Minerals* 51: 351–372.

Farmer, V.C. 1974. *Infrared Spectra of Minerals.* Mineralogical Society, London.

Fripiat, J.J. 1980. The application of NMR to the study of clay minerals, p. 245–315. In: J.W. Stucki, and W.L. Banwart (eds.), *Advanced Chemical Methods for Soil and Clay Minerals Research.* D Reidel, Dordrecht.

Goodman, B.A. 1980. Mössbauer spectroscopy, p. 1–92. In: J.W. Stucki, and W.L. Banwart (eds.), *Advanced Chemical Methods for Soil and Clay Minerals Research.* D Reidel, Dordrecht.

Goodman, B.A., and M.L. Berrow. 1976. The characterization by Mössbauer spectroscopy of the secondary iron in pans formed in Scottish podzolic soils. *Le Journal de Physique, Colloques* 37(C6): C6-849–C6-855.

Goodman, B.A., and J.A. Chudek. 1994. Nuclear magnetic resonance spectroscopy, p. 120–172. In: M.J. Wilson (ed.), *Clay Mineralogy: Spectroscopic and Chemical Determinative Methods.* Chapman & Hall, London.

Hall, P.L., G.J. Churchman, and B.K.G. Theng. 1985. Size distribution of allophane unit particles in aqueous suspensions. *Clays and Clay Minerals* 33: 345–349.

Hillier, S. 1999. Use of an air brush to spray dry samples for X-ray powder diffraction. *Clay Minerals* 34: 127–135.

Hillier, S. 2018. "Digital soil mineralogy": New approaches to old problems, p. 108 [abstract]. In: Program and Abstracts, 55th Annual Meeting, The Clay Minerals Society, University of Illinois at Urbana-Champaign, June 11–14, 2018. The Clay Minerals Society, Chantilly, Virginia.

Hillier, S., and B. Butler. 2018. New XRD-based approaches to soil mineralogy. *Spectroscopy* 33: 34–37.

Holdridge, D.A., and F. Vaughan. 1957. The kaolin minerals (kandites), p. 98–139. In: R.C. Mckenzie (ed.), *The Differential Thermal Investigation of Clays.* Mineralogical Society, London.

Holmgren, G.G.C. 1967. A rapid citrate-dithionite-bicarbonate extractable iron procedure. *Soil Science Society of America Proceedings* 31: 210–211.

Janik, L.J., R.H. Merry, and J.O. Skjemstad. 1998. Can infrared diffuse reflectance analysis replace soil extractions? *Australian Journal of Experimental Agriculture* 38: 681–696.

Johns, W.F., R.E. Grim, and W.F. Bradley. 1954. Quantitative estimation of clay minerals by diffraction methods. *Journal of Sedimentary Petrology* 24: 242–251.

Joussein, E., S. Petit, J. Churchman, B. Theng, D. Righi, and B. Delvaux. 2005. Halloysite clay minerals – A review. *Clay Minerals* 40: 383–426.

Klug, H.P., and L.E. Alexander. 1954. *X-Ray Diffraction Procedures for Polycrystalline and Amorphous Materials*. Wiley, New York.
Loveland, P.J., I.G. Wood, and A.H. Weir. 1999. Clay mineralogy at Rothamsted: 1934–1988. *Clay Minerals* 34: 165–183.
McBride, M.B. 1980. Application of spin probes to ESR studies of organic-clay systems, p. 423–450. In: J.W. Stucki, and W.L. Banwart (eds.), *Advanced Chemical Methods for Soil and Clay Minerals Research*. D Reidel, Dordrecht.
Mckenzie, R.C. (ed.). 1957. *The Differential Thermal Investigation of Clays*. Mineralogical Society, London.
Mehra, O.P., and M.L. Jackson. 1958. Iron oxide removal from soils and clays by a dithionite-citrate system buffered with sodium bicarbonate. *Clays and Clay Minerals* 7: 317–327.
Noro, H. 1986. Hexagonal platy halloysite in an altered tuff bed, Komaki city, Aichi prefecture, Central Japan. *Clay Minerals* 21: 401–415.
Parfitt, R.L. 1990. Allophane in New Zealand – A review. *Soil Research* 28: 343–360.
Parfitt, R.L., and C.W. Childs. 1988. Estimation of forms of Fe and Al – A review, and analysis of contrasting soils by dissolution and Mossbauer methods. *Soil Research* 26: 121–144.
Paterson, E., and R. Swaffield. 1994. X-ray photoelectron spectroscopy, p. 226–259. In: M.J. Wilson (ed.), *Clay Mineralogy: Spectroscopic and Chemical Determinative Methods*. Chapman & Hall, London.
Pinnavaia, T.J. 1980. Applications of ESR spectroscopy to inorganic-clay systems, p. 391–421. In: J.W. Stucki, and W.L. Banwart (eds.), *Advanced Chemical Methods for Soil and Clay Minerals Research*. D Reidel, Dordrecht.
Reynolds, R.C. Jr., and R.C. Reynolds III. 1996. *NEWMOD II, A Computer Program for the Calculation of the Basal Diffraction Intensities of Mixed-Layer Clay Minerals*. R.C. Reynolds, Hanover, New Hampshire.
Ross, D.K., and P.L. Hall. 1980. Neutron scattering methods of investigating clay systems, p. 93–168. In: J.W. Stucki, and W.L. Banwart (eds.), *Advanced Chemical Methods for Soil and Clay Minerals Research*. D Reidel, Dordrecht.
Russell, J.D., and A.R. Fraser. 1994. Infrared methods, p. 11–67. In: M.J. Wilson (ed.), *Clay Mineralogy: Spectroscopic and Chemical Determinative Methods*. Chapman & Hall, London.
Schmidt, R.L. 1980. Applications of photoacoustic spectroscopy to the study of soils and clay minerals, p. 451–465. In: J.W. Stucki, and W.L. Banwart (eds.), *Advanced Chemical Methods for Soil and Clay Minerals Research*. D Reidel, Dordrecht.
Singh, B., and R.J. Gilkes. 1992. Properties of soil kaolinites from south-western Australia. *Journal of Soil Science* 43: 645–667.
Stone, W.E.E., and J. Sanz. 1980. Distribution of ions in the octahedral sheet of micas, p. 317–329. In: J.W. Stucki, and W.L. Banwart (eds.), *Advanced Chemical Methods for Soil and Clay Minerals Research*. D Reidel, Dordrecht.
Tazaki, K. 1979. Micromorphology of halloysite produced by weathering of plagioclase in volcanic ash. *Developments in Sedimentology* 27: 415–422.
Ufer, K., and M.D. Raven. 2017. Application of the Rietveldt method in the Reynolds Cup contest. *Clays and Clay Minerals* 65: 286–297.
Vadrine, J.C. 1980. General theory and experimental aspects of electron spin resonance, p. 331–389. In: J.W. Stucki, and W.L. Banwart (eds.), *Advanced Chemical Methods for Soil and Clay Minerals Research*. D Reidel, Dordrecht.
Viscarra Rossel, R.A., T. Behrens, E. Ben-Dor, D.J. Brown, J.A.M. Demattê, K.D. Shepherd, Z. Shi, B. Stenberg, A. Stevens, V. Adamchuk, H. Aïchi, B.G. Barthès, H.M. Bartholomeus, A.D. Bayer, M. Bernoux, K. Böttcher, L. Brodský, C.W. Du, A. Chappell, Y. Fouad, V. Genot, C. Gomez, S. Grunwald, A. Gubler, C. Guerrero, C.B. Hedley, M. Knadel, H.J.M. Morrás, M. Nocita, L. Ramirez-Lopez, P. Roudier, E.M.R. Campos, P. Sanborn, V.M. Sellitto, K.A. Sudduth, B.G. Rawlins, C. Walter, L.A. Winowiecki, S.Y. Hong, and W. Ji. 2016. A global spectral library to characterize the world's soil. *Earth-Science Reviews* 155: 198–230.
Viscarra Rossel, R.A., S.R. Cattle, A. Ortega, and Y. Fouad. 2009. In situ measurements of soil colour, mineral composition and clay content by vis-NIR spectroscopy. *Geoderma* 150: 253–266.
Viscarra Rossel, R.A., D.J.J. Walvoort, A.B. McBratney, L.J. Janik, and J.O. Skjemstad. 2006. Visible, near infrared, mid infrared or combined diffuse reflectance spectroscopy for simultaneous assessment of various soil properties. *Geoderma* 131: 59–75.
Wada, K. 1989. Allophane and imogolite, p. 1051–1087. In: J.B. Dixon, and S.B. Weed (eds.), *Minerals in Soil Environments*, 2nd edn. Soil Science Society of America, Madison, Wisconsin.

13 Surfaces, Surface Reactions and Particle Size Effects

In science there are no 'depths'; there is surface everywhere.

Rudolf Carnap (1891–1970)

Many reactions of soils occur at the surfaces of their particles, and, as a result, there are close relationships between measures of surface area and certain soil properties. In particular cation exchange capacities (CECs) have been found to correlate strongly with specific surface areas (SSAs) – or surface area per gram (or kg) – particularly when these are measured by the retention of polar molecules, such as ethylene glycol and ethylene glycol monoethyl ether (EGME) (Cihacek and Bremner, 1979; Tiller and Smith, 1990; Churchman and Burke, 1991). SSAs determined from the retention of these polar molecules were also found to correlate well with various measures of water retention (Banin and Amiel, 1970; Tiller and Smith, 1990; Newman, 1983; Churchman and Burke, 1991) and with the water-content-related soil physical properties of liquid limit and plastic limit (Churchman and Burke, 1991). Indeed, many researchers (Orchiston, 1953; Newman, 1983; Grismer, 1987; Churchman and Burke, 1991) have proposed that SSAs of soils are most appropriately determined using the retention of water as the polar liquid which is integral to the soil environment. The close correlation between CEC, as a measure of surface charge, and SSA, as measured by the adsorption of polar molecules, which carry at least partial charges, is therefore not surprising.

13.1 SOIL CLAYS AND SURFACE AREAS

Early studies of the surface areas of soils employed nitrogen gas adsorption with analysis by the BET equation (Brunauer et al., 1938). However, following a pioneering paper by Dyal and Hendricks (1950), who employed the sorption of ethylene glycol for this purpose, this polar liquid, or, more commonly, its more volatile derivative EGME, has often been used to measure surfaces in soils. In order to convert uptake of the polar liquids to specific surface areas, early work (Carter et al., 1965; Heilman et al., 1965) used conversion factors based on theoretical specific surface areas of smectite clays calculated from their molecular structure. This approach is subject to various provisos, including that of the exact configuration of the polar molecules on the surfaces, both interlayer (Nguyen et al., 1987) and external; whether a monolayer has been achieved in the adsorption process (Tiller and Smith, 1990; Churchman et al., 1991); and the effect of exchangeable cations causing a clustering of polar liquids, including EGME and water (Quirk, 1955). More important, as predicted by Martin (1955), a conversion factor for smectites cannot be expected to apply to other clay minerals, both within and out of soils. These would adsorb polar liquids to both a different extent and also probably in different configurations than smectites. The truth of Martin's (1955) prediction has been borne out by studies of EGME retention by different clay minerals (Tiller and Smith, 1990) and also by soils containing different clay minerals (Tiller and Smith, 1990; Churchman and Burke, 1991). In summary, the relative areas of soils and soil clays, as well as 'pure' clay minerals are best given as values for their adsorption of polar liquids, which will be different for, say, EGME than for water and also, less appropriately, of N_2 gas. If specific surface areas are to be calculated, these will be different according to the adsorbent used and according to the mineralogical composition of the soils.

The availability of a large number of samples (1318) of a variety of soils throughout New Zealand that had been analysed for mineralogy, as well as a range of chemical and physical properties enabled the correlation against CEC of these soils of one of these latter properties, the 'moisture factor'

TABLE 13.1
Linear Regressions of Moisture Factor (MF) and Cation Exchange Capacity (CEC) (in cmol(+) kg^{-1}) for a Wide Range of New Zealand Soils (where Organic C ≤2%)

Soils[a]	Number of Samples	Linear Regression (CEC = m.MF = constant)	r^2
All	1318	225.MF-249	0.38***
Kandic	135	202.MF-196	0.34***
Kandic (halloysite)	39	145.MF-453	0.55***
Kandic (kaolinite)	70	454.MF-196	0.68***
Amorphic	122	145.MF-141	0.55***
Ferritic	24	160.MF-160	0.59***
Vermiculitic	227	324.MF-320	0.59***
Illitic	667	547.MF-545	0.74***
Smectitic	143	544.MF-542	0.88***

Source: Data from Churchman, G.J., and C.M. Burke, 1991, *Journal of Soil Science* 42: 463–478.

[a] Classifications from Whitton, J.S., and C.W. Childs, 1989, The New Zealand experience in applying the Soil Taxonomy (1975): Key to mineralogy classes and a proposed revision of the key to mineralogy classes, Division of Land and Soil Sciences Technical Record LH1, DSIR, Lower Hutt, New Zealand.

*** Significant at 0.1% level.

(Churchman and Burke, 1991). This measures the long-term retention of water at ambient relative humidity (around 60%; Churchman et al., 1991), as a measure of water adsorption. The moisture factor correlated very closely with water sorption from the BET equation (regression factor $r^2 = 0.93$) on a range of New Zealand (sub-)soils (Churchman et al., 1991). Table 13.1, from Churchman and Burke (1991), shows that the overall relationship of moisture factor to CEC for all soils was relatively poor ($r^2 = 0.38$). However, when soils were classified mineralogically, at the family level of the Soil Taxonomy (Whitton and Childs, 1989), the moisture factor was significantly linearly correlated with CEC to a greater or lesser extent. r^2 was greater than 0.55 for all correlations except that of all so-called kandic minerals (kaolinite + halloysite), for which r^2 was 0.38, as for all soils. However, correlations were much better when soils with substantial kaolinite ($r^2 = 0.68$) and those with substantial halloysite ($r^2 = 0.55$) were separately regressed against CEC. The closest correlations of moisture factor with CEC were those with smectitic soils, for which $r^2 = 0.88$. Hence it appears that the surfaces of soils reflect the chemical surface properties of their dominant clay minerals.

13.2 EFFECT OF ASSOCIATIONS OF CLAYS ON SURFACE AREAS

13.2.1 Effect of Organic Matter on Surface Areas

Removal of organic matter (OM) increases the specific surface area (SSA) of soils (e.g. Theng et al., 1999; Kahle et al., 2002; Mikutta et al., 2005; Ketrot et al., 2013; Singh et al., 2016). Generally, this has been attributed to the effect that OM has on blocking adsorption sites, including those on mineral surfaces and pores. Caution is advised when comparing SSA values across different studies because values obtained using different sorbents, whether nitrogen gas N_2, EGME (Tiller and Smith, 1990; Churchman and Burke, 1991), water (Churchman and Burke, 1991) or *para*-nitrophenol (*p*NP) – introduced to enable measuring surface areas of soils containing much organic matter (Theng et al., 1999) – may be expected to be different from one another for a given sample. This is because they each measure the interaction of these quite dissimilar compounds with soil surfaces.

13.2.2 Effect of Oxides, Oxyhydroxides and Hydroxides on Surface Areas

The contribution that Al hydroxides and Fe oxyhydroxides and oxides (together 'oxides') make to the surface area of soils reflects their crystallinity and particle size, which itself relates to crystallinity. Soil surface areas were all reduced when Fe oxides (and any Al hydroxide) were removed from soils with different dominant phyllosilicates (Singh et al., 2016), indicating that metal oxides, oxyhydroxides and hydroxides made a positive contribution to their surface area. This was also reflected in a reduction of their capacities to adsorb dissolved organic carbon (DOC). When non-soil phyllosilicates were coated by different synthetic Fe oxides, the SSA of kaolinite was greatly increased with a goethite coating, but coating with goethite had no effect on the SSAs of an illite or a smectite (Saidy et al., 2013). A coating of ferrihydrite increased the SSA of an illite, while a coating of hematite decreased this value below that of the uncoated illite. The uptake of DOC reflected the effect of changes in coating on SSA of these various synthetic mixtures (Saidy et al., 2013).

13.3 CHARGES ON SOILS AND SOIL CLAYS

The charges on soil particles are of two main types. They may derive from isomorphous substitutions and defects in the layers of aluminosilicate clay minerals (i.e. permanent charge) or else from interactions of H^+ and OH- in surrounding and bathing aqueous solutions with surface groups, such as Al-OH and Fe-OH on layer silicates and metal oxides, and with those such as COOH on organic matter (i.e. variable charge). Probably all soil clays and hence all soils have variable charge. The extent of total charge that is variable governs – or should govern – the method(s) used to measure charge on soils.

Generally, the charge on soils is measured as CEC because most soil clays are net negatively charged, as is organic matter. However, some, notably iron oxides, may be net positively charged (e.g. Table 13.3). In this case, the charge is measured as anion exchange capacity (AEC).

If a soil is dominated by permanently charged minerals, its CEC is commonly measured by the extent of exchange with ammonium NH^+ ions, as 1M NH_4Cl or NH_4COOCH_3, buffered at pH 7, unless it is calcareous when the pH is buffered at 8.5 (Rengasamy and Churchman, 1999). If a soil is dominated by variably charged components, CEC (or AEC) is most appropriately measured at the pH of the soil, using so-called compulsive exchange (Gillman and Sumpter, 1986) with barium chloride and ammonium chloride or else exchange with 0.01 M silver thiourea (Rengasamy and Churchman, 1999). CECs of soils range up to ~100 cmol(+) kg^{-1}, reflecting the types and amounts of soil clays and other components, especially organic matter, that are present. Most are much lower in value than 100 cmol(+) kg^{-1}, with typical values for soils in New Zealand estimated as between 5 and 30 cmol(+) kg^{-1} by McLaren and Cameron (1996), a range which is likely to apply to most, except smectite-rich, soils worldwide (e.g. Churchman, 2010).

13.4 EFFECT OF ASSOCIATIONS UPON CHARGES ON SOILS AND SOIL CLAYS

Organic matter generally contributes >25% of the CEC of surface layers of soils (Baldock and Broos, 2012); it contributes virtually 100% of the CEC to organic soils, but this is not relevant when assessing the effect of OM on the CEC of soil minerals. The CEC (measured at pH 5.8) of organic matter per se was found to range from 283 to 653 cmol(+) kg^{-1} (Oorts et al., 2003). Soil organic matter (SOM) was calculated to contribute 40%–50% of the CEC of British agricultural soils at pH 6–8 (Loveland and Webb, 2003). Naturally, SOM contributes mostly to the CEC of surface horizons. This can be seen in analyses of two sets of soils, each with a different dominant clay mineralogy. In Table 13.2, the CECs of horizon samples of soils dominated by kaolinite–smectite clays are shown as measured before and after peroxidation to remove organic matter. In Table 13.3, the net charges (negative or positive) are compared between samples at different depths in profiles dominated by kaolinite and/or iron oxides and/or gibbsite. All have a low pH (<5.5).

TABLE 13.2
Effect on CEC of Peroxidation of Some Australian Soils Dominated by Kaolinite-Smectite

Soil name[a]	H53		SA/71/P7
Location	Cornwall, Tasmania		Kongorong, S. Aust.
Classification[b]	Grey-brown podzolic		Yellow podzolic
Depth (cm)	0–8	8–4	10–20
CEC before peroxidation[c]	40	40	40
CEC after peroxidation[c]	22	24	19
CEC due to SOM (%)	45	40	53

Source: Data from Churchman, G.J., et al., 1994, *Soil Research* 32: 805–822.
[a] From database, CSIRO Land and Water, Glen Osmond, South Australia.
[b] Great Group classification (from Stace, H.C.T., et al., 1969, *A Handbook of Australian Soils*, Rellim, Glenside, South Australia).
[c] cmol(+) kg^{-1}

TABLE 13.3
Net Surface Charge[a] of Samples at Different Depths in Soils from North Queensland, Australia

Soil Name[b]	Depth (cm)	Organic C (%)	pH (in Water)	Main Clay Minerals[c]	Net Charge[d] IS = 0.002	Net Charge[d] IS = 1
Krasnozem	0–10	6.3	5.2	Fe ~ G > K	−1.2	−5.0
(Rhodic ferralsol)	30–60	1.9	5.4	Fe ~ G > K	0	0
	210–240	0.2	4.8	K ~ Fe > G	+2.4	+5.0
	450–480	0.2	4.8	K > Fe > G	+0.6	+1.4
Xanthozem	0–10	11.3	4.2	K	−2.0	−6.7
(Xanthic ferralsol)	30–40	2.2	4.7	K	−1.3	−3.0
	90–120	0.2	4.7	K	−0.5	−2.5
Red podzolic	0–10	8.7	5.0	K	−4.2	−9.5
(Orthic ferralsol)	20–30	3.1	5.2	K	−1.0	−4.3
	90–120	0.2	5.2	K	−1.3	−3.1
Yellow earth	0–10	1.9	5.2	K	−1.6	−2.9
(Ferric acrisol)	60–75	0.2	5.3	K	−0.2	−0.5
	120–135	0.1	5.3	K	−0.6	−1.1

Source: Data from Gillman, G.P., and L.C. Bell, 1976, *Soil Research* 14: 351–360.
[a] Determined at ionic strengths (IS) of 0.002 and 1.0.
[b] Classification by Great Soil Group (from Stace, H.C.T., et al., 1968, *A Handbook of Australian Soils*, Rellim, Glenside, South Australia); and World Soil Map, in brackets (from FAO-UNESCO, 1974, *Soil Map of the World, 1: 5,000,000*. Vol. 1: *Legend*, UNESCO, Paris).
[c] Fe, iron oxides (mainly goethite); G, gibbsite; K, kaolinite.
[d] cmol(+) kg^{-1}.

Table 13.2 shows that as much as 53% of the CEC in these (top)soils was attributable to SOM.

Table 13.3 shows that, in each of the profiles, the uppermost sample (0–10 cm) was the most negatively charged, i.e. had the highest CEC. These samples had substantially more organic C than samples from all other depths in each soil, indicating that organic matter contributes significant negative charge, i.e. CEC, to the topsoil samples. Generally, the negative charge, i.e. CEC, decreased with depth, but in one soil (the krasnozem) the charge reversed to become positive with depth, so that AEC > CEC in these samples. This indicates that the point of zero charge (PZC) of the dominant components of this soil was pH ~5.4, as net charge was zero at this particular pH and positive at the lower pHs of the samples at greater depth. The net negative charge of all of the samples in the other soils suggests that the PZC of the dominant kaolinite in these soils was pH <4.7.

Nonetheless, the CEC of soils are not simply proportional additions of those of minerals and those of organic matter; Loveland and Webb (2003) found that removal of SOM actually increased CEC in a soil containing montmorillonite, implying that SOM blocked exchange sites on the clay mineral. Although unblocking of exchange sites when SOM is removed may occur on other soils, Kahle et al. (2002) found that removal of organic C diminished the CEC in some illitic soils, with the loss of exchange sites on the SOM able to overcome any blocking of exchange sites on the minerals. CECs were measured in a range of New Zealand soils with different mineralogies and were plotted against typical values for the mineral components from the literature and a value of 221 $cmol_c$ kg^{-1} was indicated for OM (Theng et al., 1999). There was excellent correspondence between measured and calculated values. This suggests that the mineral and organic components of the soils behaved essentially independently towards the exchanging agent, which was NH_4^+ ions. The general rule is that, provided the texture and mineralogy are reasonably similar throughout the depth of a soil, surface horizons, with the most organic matter, tend to have a higher CEC than lower horizons, due to the influence of the SOM, as in Table 13.3. However, it is also noteworthy that the CEC of SOM is highly dependent on pH. An early comprehensive study of the separate effects of OM and clay was determined for 60 soils from Wisconsin over the pH range of 2.5 to 8.0 (Helling et al., 1964). Multiple regression analyses found that the average CEC of the organic matter varied linearly from 36 to 213 cmol(+) kg^{-1} as pH was raised from 2.5 to 8.0, while that of the clay also varied linearly, but less steeply, from 38 to 64 $cmol_c$ kg^{-1} over the same range of pHs. The mean relative contribution of OM to total soil CEC varied from 19% at pH 2.5 to 45% at pH 8.0 for soils with a mean organic matter content of 3.28% and a mean clay content of 13.3%. The wide range of values found in the literature for the CEC of organic matter (e.g. Helling et al., 1964; Theng et al., 1999; Oorts et al., 2003) once more emphasises the heterogeneity of SOM. Even so, regardless of the particular association of OM with minerals, e.g. whether surface-adsorbed or occluded, it can make a significant contribution to the CEC of the minerals.

Fe oxides and Al hydroxides alike have pH-variable charge. With PZCs ranging between 7 and 9 for all types of Fe oxides (Schwertmann and Taylor, 1989) and those of Al hydroxides also high – probably close to pH 9 (Hsu, 1989) – Fe oxides and Al hydroxides alike contribute negatively to the CEC of soils in which they occur when soil pH is acid and also often mildly alkaline. Instead, Fe oxides and Al hydroxides confer an AEC to some soils (e.g. the krasnozem in Table 13.3). Because of their specific, inner-sphere adsorption of many anions, e.g. HPO_4^{2-} and silicate ions, both Fe oxides and Al hydroxides with adsorbed compounds may sometimes contribute positively to soil CEC (Schwertmann and Taylor, 1989). Allophane and imogolite also have pH-variable charge, with their PZCs (strictly, PZNC, or point of zero net charge) dependent upon their Al:Si ratio, being 8.4 for imogolite (Al:Si = 2), 7.9 for 2:1 allophane (both invariant in NaCl from 0.1 and 0.01M), from 5.8 to 6.7 (varies in NaCl from 0.1 and 0.01M) for 1.6:1 allophane and 4.1 to 5.4 for 1.2:1 allophane (over same range on NaCl concentrations) (Harsh, 2012). These PZC values are for synthetic minerals, and the close associations these types of minerals form with organic matter in soils (Parfitt, 2009; Calabi-Floody et al., 2011; Huang et al., 2016) means that their effective PZC values will be driven to lower pH values as a result, for, as Helling et al. (1964) found, SOM remains net negatively charged over the pH range of almost all soils (2.5 to 8.0).

13.5 EFFECTS OF PARTICLE SIZE

The vast majority of studies of soil clays have concentrated on the <2 μm fraction, i.e. the 'clay fraction'. Occasionally, studies have also been made of the <0.2 μm fraction, or 'fine clay fraction'. However, it is well recognised that the finest particles of clays, the nanoparticles, generally <100 nm, are the most reactive for movement in the environment, often with attached contaminants and/or nutrients (Seta and Karathanasis, 1997; Noack et al., 2000; Waychunas et al., 2005; Tsao et al., 2013), as well as organic matter (Monreal et al., 2010; Calabi-Floody et al., 2011). This is not surprising, since the specific surface (m^2 g^{-1}) or surface area/volume of a 10 nm (spherical) particle is 100 times larger than that of a 1 μm particle of the same shape. It is for this reason that allophane and ferrihydrite, which occur as nanoparticles, confer the greatest part of the surface area to a soil they occupy (see Figure 1.6). Furthermore, particles at the nanoscale are likely to have irregular surfaces and different molecular structures from their larger counterparts and these may confer increased reactivity, which enhances uptake of contaminants, nutrients and organic matter (Waychunas et al., 2005; Monreal et al., 2010). As well, small particle size alone is likely to lead to higher hydraulic conductivities than clays with the same mineralogical composition but generally coarser particles. This effect is illustrated for a smectite (with some interstratification with illite) made up of very small particles, hence an exceptionally (external) high surface area (Churchman et al., 2002). The effect may be a positive attribute for a clay for dam construction or for a liner for landfill or nuclear waste material, but a negative one for the transport of water and nutrients, and root penetration in an agricultural soil.

There have been several studies recently that have examined the composition of soils in relation to particle size fractions within the <2 μm clay fraction (Tsao et al., 2013; Zhang et al., 2016, 2017). Generally, soils have been chemically pretreated to remove organic matter and metal oxides, centrifuged at various speeds to obtain fractions at, e.g. 450 nm and 100 nm, then by ultrafiltration to obtain the smallest fraction (1–100 nm for Tsao et al., 2013; 25–100 nm for Zhang et al., 2016, 2017). Several soil types according to the soil taxonomy classification (Soil Survey Staff, 1999) were fractionated this way and their mineralogies examined by conventional XRD and also by synchrotron XRD, to give enhanced peaks.

In an Oxisol (Tsao et al., 2013), the yield of kaolinite increased with decreasing particle size, while that of illite tended to decrease, and these phyllosilicates occurred alongside hematite and goethite in the 1–100 nm fraction. The trend was for increasing disorder to occur with decreasing particle size in two Ultisols at a subtropical and also a tropical site (Zhang et al., 2017). The 25–100 nm fractions were predominantly kaolinite, but of low crystallinity, probably due to some interstratification with 2:1 minerals and high structural Fe, and some hydroxy-interlayered (HI) mineral also occurred in the lower temperature, lower rainfall subtropical site. These authors suggested that there was a transformation occurring from illite to HI mineral to kaolinite. Zhang et al. (2016) examined two Alfisols in the same way and discovered that, while the coarser clay fractions contained illite, vermiculite, smectite, kaolinite and 'interstratified kaolinite', the 25–100 nm nanoparticles comprised illite and kaolinite, together with interstratified kaolinite. In this case, the authors suggested that there was a transformation occurring from the 2:1 minerals (illite, vermiculite, smectite) to interstratified kaolinite and kaolinite. In all of the studies by Tsao, Zhang and co-workers, the finest (nano)particles are taken to represent the latest stages in the transformation of 2:1 to 1:1 minerals. This represents a desilication process. In an Andisol, Calabi-Floody et al. (2011) found that the allophane dominating the clay fraction of the soil contained an enrichment of organic matter in the nanoclay particles, which were obtained without chemical pretreatment (using ultrasound), and much of this OM was protected from subsequent oxidation by peroxide. Small (nano)aggregates of allophane appear to protect the OM, as was found by Huang et al. (2016) for DNA sorbed onto synthetic allophane.

Especially where soils were chemically treated and then successively stirred prior to centrifugation, as well as being forced through ultrafilters, the resulting nanoparticles are hardly likely to represent the fine material that is responsible for transport of contaminants, nutrients and organic matter in the environment. This material will arise from the gentler breakdown of clays *in situ* by water, wind or agricultural cultivation. In one representation of the effect of breakdown without any mechanical force being applied, Churchman and Weissmann (1995) examined the effect of dialysis of a soil (a Mollisol) which had been saturated with sodium, then leached with water, thereby applying a 'chemical hammer' (Clapp and Emerson, 1965). The fine fractions (<~1 μm) moved against gravity (by Brownian motion) within the dialysis bag and were sampled at successively deeper intervals from the dispersion front. It was found that the Mollisol, containing kaolinite–smectite, broke down into successively more smectitic interstratifications as the particle size (determined by photon correlation spectroscopy) decreased from an average of 428 nm, through 383 nm, 314 nm and 206 nm to one of 148 nm. Even so, the particles obtained this way hardly represent most practical situations, except perhaps for breakdown by rainfall in a highly saline landscape. Instead, studies of the leaching of fine particles of minerals, including fine soil clays (Noack et al., 2000) and of soil clays exclusively (Seta and Karathanasis, 1997) through soil columns are likely to mirror most real situations where clay colloids travel within the environment. The results of these two studies agree that it is the size of the colloids – or nanoclays – which largely determine their mobility through soils. Particles need to be small enough to enter and then exit pores in the soils. Noack et al.'s (2000) study was concerned only with illites as the mobile materials, while Seta and Karathanasis (1997) studied mobile phases with a variety of clay mineralogies, and concluded that their relative mobilities, sizes, total exchangeable bases and pH (other factors affecting mobility) being similar were smectitic > mixed (combinations of HI minerals, mica, smectite and kaolinite) > kaolinitic. Among soil clays, small smectitic particles may be responsible for the easiest transport of contaminants, nutrients and organic matter.

13.6 INTEGRATION: IMPORTANCE OF PHENOMENA; LIMITATIONS OF MEASUREMENTS

Soil chemistry is largely surface chemistry and large surface areas are largely a consequence of fine particle sizes. The smaller the particles, the more extensive and hence adsorptive are their surfaces. Thus, measurements of surface area, generally expressed as specific surface area, or surface area per unit weight, e.g. m^2g^{-1}, are likely to be important in comparing the surface reactivities of different soils. Among all soil particles, by far the largest contributions to surface reactivities are those of the soil clays. Furthermore, within the clay (<2 μm) fractions of soils, the smallest particles contribute much more to surface adsorption larger particles. For instance, those with a 10 nm size (as equivalent spherical diameter, or esd) have 100 times the surface area of those with a 1 μm esd. In addition, mineralogy and also therefore particle shape can change with particle size. The smallest particles within the clay fraction (nanoparticles) are likely to represent its most disordered and also the most recently formed or transformed minerals. They also represent the most common carriers of contaminants, nutrients and organic matter into the wider environment.

In spite of its importance, there is probably no absolute measure of SSA of soil mineral particles. Measurements of SSA, generally made by the extent of adsorption of gases or volatile liquids, differ according to the adsorbate used. Water is the most realistic adsorbate for soils, which are invariably found in aqueous milieu. Different soil clays also differ in their reactivity towards, hence uptake, of each gas or liquid. For example, the uptake of polar liquids, including water, is enhanced by a high CEC for the mineral or soil. For this reason, uptake of water, for example, relates well to the CEC of the dominant mineral in a soil or set of soils. Therefore, the reactivity of their surfaces and that of soils can only be expressed as a weight of the particular gas or liquid adsorbed per unit weight of the mineral or soil.

Similarly, there is probably no absolute measure of particle size distribution of a soil (see also Section 11.1 herein). Particles are often bound together, with each other and also with other particles –

most notably metal oxides or organic matter. Particle size distributions and determinations inevitably follow disaggregation and/or dispersion procedures that are unrealistic for natural soil associations. These associations, in turn, affect measurements of the extent of surfaces, with soil organic matter apparently blocking adsorption sites for different adsorbates, while iron oxides tend to make positive contributions towards their uptake and hence measurements of surface area.

Charge is contributed to soils by its fine components – both clays and organic matter. It can greatly affect the reactivity of soils. Mostly the net charge is negative, giving a CEC. The CEC is measured according to the source of the charge, whether mainly from isomorphous substitutions within aluminosilicate layers, giving permanent charge or mainly from interactions between surface groups and H^+ and OH^- ions in solution, giving (pH-)variable charge. All clays have some variable charge but 2:1 aluminosilicates have a preponderance of permanent charge. Their CEC is conventionally determined at pH 7, or, if the soils are calcareous, at pH 8.5. Metal oxides and also organic matter are variably charged, and their CEC and that of soils with substantial proportions of them is determined at the pH of the soil. 1:1 aluminosilicates have both permanent and variable charge in similar proportions, so their CEC is best determined at soil pH. Some (acidic) soils with substantial contents of metal oxides have a net positive charge, particularly when organic matter is absent, so have an AEC. Organic matter can contribute substantial CEC to topsoils, especially at high pH.

REFERENCES

Baldock, J.A., and K. Broos. 2012. Soil organic matter, p. 11-1–11-52. In: P.M. Huang, Y. Li, and M.E. Sumner (eds.), *Handbook of Soil Sciences, Properties and Processes*, 2nd edn. CRC Press/Taylor & Francis Group, Boca Raton, Florida.

Banin, A., and A. Amiel. 1970. A correlative study of the chemical and physical properties of a group of natural soils of Israel. *Geoderma* 3: 185–198.

Brunauer, S., P.H. Emmett, and E. Teller. 1938. Adsorption of gases in multimolecular layers. *Journal of the American Chemical Society* 60: 309–319.

Calabi-Floody, M., J.S. Bendall, A.A. Jara, M.E. Welland, B.K.G. Theng, C. Rumpel, and M. de la Luz Mora. 2011. Nanoclays from an Andisol: Extraction, properties and carbon stabilization. *Geoderma* 161: 159–167.

Carter, D.L., M.D. Heilman, and C.L. Gonzalez. 1965. Ethylene glycol monoethyl ether for determining surface area of silicate minerals. *Soil Science* 100: 356–360.

Churchman, G.J. 2010. Is the geological concept of clay minerals appropriate for soil science? A literature-based and philosophical analysis. *Physics and Chemistry of the Earth, Parts A/B/C* 35: 927–940.

Churchman, G.J., M. Askary, P. Peter, M. Wright, M.D. Raven, and P.G. Self. 2002. Geotechnical properties indicating environmental uses for an unusual Australian bentonite. *Applied Clay Science* 20: 199–209.

Churchman, G.J., and C.M. Burke. 1991. Properties of subsoils in relation to various measures of surface area and moisture contents. *Journal of Soil Science* 42: 463–478.

Churchman, G.J., C.M. Burke, and R.L. Parfitt. 1991. Comparison of various methods for the determination of specific surfaces of subsoils. *Journal of Soil Science* 42: 449–461.

Churchman, G.J., P.G. Slade, P.G. Self, and L.J. Janik. 1994. Nature of interstratified kaolin-smectites in some Australian soils. *Soil Research* 32: 805–822.

Churchman, G.J., and D.A. Weissmann. 1995. Separation of sub-micron particles from soils and sediments without mechanical disturbance. *Clays and Clay Minerals* 43: 85–91.

Cihacek, L.J., and J.M. Bremner. 1979. A simplified ethylene glycol monoethyl ether procedure for assessment of soil surface area. *Soil Science Society of America Journal* 43: 821–822.

Clapp, C.E., and W.W. Emerson. 1965. The effect of periodate oxidation on the strength of soil crumbs. *Soil Science Society of America Proceedings* 29: 127–130.

Dyal, R.S., and S.B. Hendricks. 1950. Total surface of clays in polar liquids as a characteristic index. *Soil Science* 69: 503–509.

FAO-UNESCO. 1974. *Soil Map of the World, 1: 5,000,000.* Vol. 1: *Legend.* UNESCO, Paris.

Gillman, G.P., and L.C. Bell. 1976. Surface charge characteristics of six weathered soils from tropical North Queensland. *Soil Research* 14: 351–360.

Gillman, G.P., and E.A. Sumpter. 1986. Modification to the compulsive exchange method for measuring exchange characteristics of soils. *Soil Research* 24: 61–66.

Grismer, M.E. 1987. Water vapour adsorption and specific surface. *Soil Science* 144: 233–236.

Harsh, J. 2012. Poorly crystalline aluminosilicate clay minerals, p. 23-1–23-13. In: P.M. Huang, Y. Li, and M.E. Sumner (eds.), *Handbook of Soil Sciences: Properties and Processes*, 2nd edn. CRC Press, Boca Raton, Florida.

Heilman, M.D., D.L. Carter, and C.L. Gonzalez. 1965. The ethylene glycol monoethyl ether (EGME) technique for determining soil-surface area. *Soil Science* 100: 409–413.

Helling, C.S., G. Chesters, and R.B. Corey. 1964. Contribution of organic matter and clay to soil cation-exchange capacity as affected by the pH of the saturating solution. *Soil Science Society of America Journal* 28: 517–520.

Hsu, P.H. 1989. Aluminum oxides and oxyhydroxides, p. 331–378. In: J.B. Dixon, and S.B. Weed (eds.), *Minerals in Soil Environments*, 2nd edn. Soil Science Society of America, Madison, Wisconsin.

Huang, Y.-T., D.J. Lowe, G.J. Churchman, L.A. Schipper, R. Cursons, H. Zhang, T.-Y. Chen, and A. Cooper. 2016. DNA adsorption by nanocrystalline allophane spherules and nanoaggregates, and implications for carbon sequestration in Andisols. *Applied Clay Science* 120: 40–50.

Kahle, M., M. Kleber, and R. Jahn. 2002. Predicting carbon content in illitic clay fractions from surface area, cation exchange capacity and dithionite-extractable iron. *European Journal of Soil Science* 53: 639–644.

Ketrot, D., A. Suddhiprakarn, I. Kheoruenromne, and B. Singh. 2013. Interactive effects of iron oxides and organic matter on charge properties of red soils in Thailand. *Soil Research* 51: 222–231.

Loveland, P., and J. Webb. 2003. Is there a critical level of organic matter in the agricultural soils of temperate regions: A review. *Soil and Tillage Research* 70: 1–18.

Martin, R.T. 1955. Ethylene glycol retention by clays. *Soil Science Society of America Journal* 19: 160–164.

McLaren, R.G., and K.C. Cameron. 1996. *Soil Science: Sustainable Production and Environmental Protection.* Oxford University Press, Auckland.

Mikutta, R., M. Kleber, and R. Jahn. 2005. Poorly crystalline minerals protect organic carbon in clay subfractions from acid subsoil horizons. *Geoderma* 128: 106–115.

Monreal, C.M., Y. Sultan, and M. Schnitzer. 2010. Soil organic matter in nano-scale structures of a cultivated Black Chernozem. *Geoderma* 159: 237–242.

Newman, A.C.D. 1983. The specific surface of soils determined by water sorption. *Journal of Soil Science* 34: 23–32.

Nguyen, T.T., M. Raupach, and L.J. Janik. 1987. Fourier-transform infrared study of ethylene glycol monoethyl ether adsorbed on montmorillonite: Implications for surface area measurements of clays. *Clays and Clay Minerals* 35: 60–67.

Noack, A.G., C.D. Grant, and D.J. Chittleborough. 2000. Colloid movement through stable soils of low cation-exchange capacity. *Environmental Science and Technology* 34: 2490–2497.

Oorts, K., B. Vanlauwe, and R. Merckx. 2003. Cation exchange capacities of soil organic matter fractions in a Ferric Lixisol with different organic matter inputs. *Agriculture, Ecosystems and Environment* 100: 161–171.

Orchiston, H.D. 1953. Adsorption of water vapour. 1. Soils at 25°C. *Soil Science* 76: 453–465.

Parfitt, R.L. 2009. Allophane and imogolite: Role in soil biogeochemical processes. *Clay Minerals* 44: 135–155.

Quirk, J.P. 1955. Significance of surface areas calculated from water vapour isotherms by use of the B.E.T. equation. *Soil Science* 80: 423–430.

Rengasamy, P., and G.J. Churchman. 1999. Cation exchange capacity, exchangeable cations and sodicity, p. 147–157. In: K.I. Peverill, L.A. Sparrow, and D.J. Reuter (eds.), *Soil Analysis: An Interpretation Manual.* CSIRO, Collingwood, Australia.

Saidy, A.R., R.J. Smernik, J.A. Baldock, K. Kaiser, and J. Sanderman. 2013. The sorption of organic carbon onto differing clay minerals in the presence and absence of hydrous iron oxide. *Geoderma* 209–210: 15–21.

Schwertmann, U., and R.M. Taylor. 1989. Iron oxides, p. 379–438. In: J.B. Dixon, and S.B. Weed (eds.), *Minerals in Soil Environments*, 2nd edn. Soil Science Society of America, Madison, Wisconsin.

Seta, A.K., and A.D. Karathanasis. 1997. Stability and transportability of water-dispersible soil colloids. *Soil Science Society of America Journal* 61: 604–611.

Singh, M., B. Sarkar, B. Biswas, J. Churchman, and N.S. Bolan. 2016. Adsorption-desorption behavior of dissolved organic carbon by soil clay fractions of varying mineralogy. *Geoderma* 280: 47–56.

Soil Survey Staff. 1999. *Soil Taxonomy: A Basic System of Soil Classification for Making and Interpreting Soil Surveys*, 2nd edn. USDA, Natural Resources Conservation Service, Washington DC.

Stace, H.C.T., G.D. Hubble, R. Brewer, K.H. Northcote, J.R. Sleeman, M.J. Mulcahy, and E.G. Hallsworth. 1968. *A Handbook of Australian Soils*. Rellim, Glenside, South Australia.

Theng, B.K.G., G.G. Ristori, C.A. Santi, and H.J. Percival. 1999. An improved method for determining the specific surface areas of topsoils with varied organic matter content, texture and clay mineral composition. *European Journal of Soil Science* 50: 309–316.

Tiller, K.G., and L.H. Smith. 1990. Limitations of EGME retention to estimate the surface area of soils. *Soil Research* 28: 1–26.

Tsao, T., Y. Chen, H. Sheu , Y. Tzou, Y. Chou, and M. Wang. 2013. Separation and identification of soil nanoparticles by conventional and synchrotron X-ray diffraction. *Applied Clay Science* 85: 1–7.

Waychunas, G.A., C.S. Kim, and J.F. Banfield. 2005. Nanoparticulate iron oxide minerals in soils and sediments: Unique properties and contaminant scavenging mechanisms. *Journal of Nanoparticle Research* 7: 409–433.

Whitton, J.S., and C.W. Childs. 1989. The New Zealand experience in applying the Soil Taxonomy (1975): Key to mineralogy classes and a proposed revision of the key to mineralogy classes. Division of Land and Soil Sciences Technical Record LH1. DSIR, Lower Hutt, New Zealand.

Zhang, Z.Y., L. Huang, F. Liu, M.K. Wang, Q.L. Fu, and J. Zhu. 2016. Characteristics of clay minerals in soil particles of two Alfisols in China. *Applied Clay Science* 120: 51–60.

Zhang, Z.Y., L. Huang, F. Liu, M.K. Wang, Q.L. Fu, and J. Zhu. 2017. The properties of clay minerals in soil particles of two Ultisols, China. *Clays and Clay Minerals* 65: 273–285.

14 Role of Soil Clays in Agriculture, the Environment and Society

> Whoever could make two Ears of Corn, or two Blades of Grass to grow upon a Spot of Ground where only one grew before; would deserve better of Mankind, and do more essential Service to his Country, than the whole Race of Politicians put together.
>
> **Jonathan Swift (1667–1745)**

The useful properties of clays in soils can be categorised as those of their bulk or those of their surfaces. A property relates to its bulk because a particular soil can supply the property in an amount that is related to the mass of the soil. Surfaces may also supply properties (e.g. plant nutrients) but are particularly useful for enabling associations with other solids and hence developing soil structure through the formation of aggregates and hence pores.

14.1 PLANT NUTRITION

Clay minerals can play a role in the supply of some of the major chemical elements needed for plant growth, i.e. K, Ca, Mg, Si, P, N and C. They are the major source of potassium, for which they act as a reservoir for plants. Unlike other essential elements for plants, potassium does not occur in the soil in complexes with organic components. It occurs principally in micas (or illites, which are low charge micas). The only other common source of K in soils is K-feldspar, but the rate of release of K from these feldspars, if present, involves their dissolution so is slower than that from micas because release of K from micas occurs via ion exchange. Velde and Barré (2010) wrote that "most terrestrial plants are potassic" (p. 172). Furthermore, potassium is frequently a limiting factor in plant growth. Unlike other cations held on clay minerals for supplying plants, potassium is unique in that the quantities of K that are available for release are far greater than the quantities of K which are considered to be held on sites that contribute to the cation exchange capacity (CEC) of a soil. Typically, the CEC is determined by the sum of the cations released upon contacting the soil with a strong solution of ammonium NH_4^+ ions. Ammonium ions are useful for this purpose because ammonium does not occur as a cation in natural, i.e. non-agricultural, occurrences of soils and clays. Potassium required in plant nutrition is usually obtained by the depletion of interlayer K^+ ions from illites or micas, whether as separate phases or in mixtures with other layers. Recent work (Adamo et al., 2016) indicates that potassium can be lost from illite layers during plant growth but restored at the end of the growing season.

Nonetheless, it should be noted that K in micas has sometimes been considered to be 'fixed' owing to the close fit of potassium ions into the aluminosilicate layer. Even so, this is a relative expression, for many studies have shown that plants have the capacity to break down micas, and especially biotite, in their pursuit for potassium as an essential nutrient. As Loveland et al. (1999) stated, it was an early promise of clay mineralogy that it would explain the supply of potassium for plants. That this promise was not fulfilled may in large part be due to the difficulty of quantifying potassium-containing phases present. This difficulty needs to be re-examined in the light of the availability of XRD peak decomposition software (see Annex) with the ability to quantitatively separate micaceous phases into their constituent phases, both fully and partially K-depleted (Adamo et al., 2016).

Of additional use are the statistical methods linking XRD spectra to soil properties that have shown a close relationship to phyllosilicate K but not to K-feldspars, which are usually more prevalent in soils (see Chapter 12, Section 12.7).

Illitic layers, and even those which have been partially or completely converted to expanded layers of vermiculite and smectite, possess the useful property for plant nutrition of being able to take up added potassium ions and to return K^+ to interlayers that have been depleted by transformation processes (e.g. Marchuk et al., 2016). It has even been proposed (by Roth et al., 1969) that the transformation of primary micas into illites may be reversible, with both protons and electrons being lost from the aluminosilicate layers during oxidation but returned there upon reduction (Fanning et al., 1989).

It is possible that ammonium ions can proxy for potassium in the high charge 2:1 soil clay minerals (recent work by B. Velde, 2018). It is known that ammonium displaces cations in soil clay materials and recent X-ray diffraction (XRD) work indicates that it can be stored in interlayer spaces as is potassium.

With regard to another essential plant macronutrient, phosphorus, the capacity of a soil for P uptake is usually dependent on its content of Fe and Al oxides, and particularly those of Fe (Norrish and Rosser, 1983). Allophane, where present, also has a strong attraction for P. However, both allophane and the oxides may constitute a problem for plant nutrition of P insofar as its release may be difficult due to the strong links that P forms with these minerals.

Of the other major nutrients, sulphur, like P also presents as an anion and is also held strongly by the metal oxides and by allophane. Nitrogen, when present as the anion nitrate NO_3^-, is similarly attracted to the surfaces of these same minerals. However, when it is present as ammonium ions NH_4^+, it can be taken up as an adsorbed species in the interlayers of 2:1 phyllosilicates, ranging from unexpanded micas to fully expanded vermiculites and smectites.

Other, minor, nutrients are held on soils by mechanisms which depend on whether they are positively or negatively charged. Their chemical form and also charge may depend upon soil pH. Thus, cobalt is held by manganese oxides (Adams et al., 1969).

14.2 SOIL STRUCTURE AND WATER HOLDING AND SUPPLY

It has been seen (Chapter 10) that clays are important components for the formation and stabilisation of microaggregates and hence for keeping soils in place against the possibly erosive forces of water and wind. However, from the point of view of the plants, it is in their role in the formation and stabilisation of pores, via microaggregates – and also macroaggregates – that clays play a particularly important role in soils (see Figure 11.1, where pores free of fine clay appear in the virgin soil but not in the cultivated soils). Pores help to retain water in soils and also provide spaces for the growth of plant roots.

It has been observed herein that (aluminosilicate) soil clays seldom if ever occur as distinct entities but are almost invariably found in associations, with metal oxides, organic matter and other like particles. In other words, they generally occur in aggregates. Aggregates and particularly microaggregates explain many advantageous aspects of soils.

There is a long tradition of explanation of associations of the different components of soils into aggregates, including both macro- and microaggregates (Six et al., 2004). There is widespread acceptance now of the validity of the hierarchical model proposed by Tisdall and Oades (1982), at least for soils formed in temperate climate zones. Generally, these comprise mainly 2:1 minerals. In the hierarchical model, aggregates occur in several different broad size steps, spanning from 2 mm to 2 μm, i.e. 2–0.2 mm, 200–20 μm, 20–2 μm and 2–0.2 μm, and the different sizes of aggregates are bound by different agents with different strengths of bonding at each level. Microaggregates (generally sizes below 200 μm) are bound by the most persistent agents and hold together for longer times than macroaggregates. They can persist for hundreds of years.

Pores, responsible for the transport, uptake and release of water for plant uptake, are formed both within and between components of the hierarchy of aggregates in soils dominated by 2:1 minerals.

In general, pores may be characterised, after Hillel (1998) as (a) micropores, (b) capillary pores and (c) macropores, for the purpose of explaining the hydraulic properties of soils (Ghezzehi, 2012). The micropores, up to only a few micrometres in diameter, hold water tightly by adsorptive forces. It is the capillary pores (or 'mesopores'), several micrometres to ~1 mm in diameter (Regelink et al., 2015), that hold water subject to capillary forces, and it is this water which represents the bulk of plant-available water. Macropores provide for the easy passage of water, air, nutrients, roots and soil microorganisms. Soils which display a hierarchical structure of aggregates as a result of the action of organic materials, whether dead, processed or alive, also have a hierarchy of pores. This is the situation for soils dominated by 2:1 minerals (Oades and Waters, 1991; Denef and Six, 2005). By contrast, soils dominated by 1:1 minerals comprise aggregates which are formed by abiotic means (Oades and Waters, 1991; Denef and Six, 2005). Lacking a hierarchy of aggregates, these break down just to particles <20 μm and thus have very few, if any stable capillary pores for the uptake and supply of water.

The reason for the different pattern to the provision of pores between 2:1 minerals and 1:1 minerals very likely reflects the mechanism by which particles of each of the two different mineral types associate with one another. This reflects the nature of their surface charge. Particles of 2:1 minerals, which carry a substantial negative charge, associate with other particles of the same minerals mainly through organic linkages. With organic matter also having a net negative charge, minerals and organic matter are often linked together by polyvalent cations, of which Ca^{2+} is commonly encountered in soils, especially at higher pHs. Hence there is a network of associations of minerals and organic matter that itself is bound into larger particles by a succession of largely organic binding forces made up of processed organic matter, and, for larger entities, by living fungal hyphae and roots. Organo-mineral linkages at all levels lead to pores with a wide range of sizes. On the other hand, 1:1 minerals, with a much smaller overall charge and with centres of both negative and positive charges in the acidic conditions in which they are usually found, form links by ionic associations of the negative and positive charges, giving a card house structure. The card houses made up of 1:1 minerals like kaolinite can be joined to other similar particles through oxides, mainly those of iron, which are ubiquitous in soils containing kaolinite (or halloysite). Studying 18 kaolinitic tropical topsoils, mostly Oxisols, Barthès et al. (2008) found that the stability of aggregates at all size levels related strongly to their contents of Al-substituted hematite and goethite. The linkages are thus mainly abiotic in nature. Organic matter in such soils is bound more weakly to the associations of the 1:1 minerals with iron oxides and does not provide the main linkages between these associations. The arrangement is more brittle than that for the 2:1 minerals involving organic matter, so that the only stable associations are those between the mineral particles, leading to mainly only small interparticle pores.

Agents responsible for the stabilisation of aggregates range from living roots and fungal hyphae for macroaggregates to well-processed organic compounds such as root exudates and polysaccharides along with metal oxides and aluminosilicates for microaggregates. The most poorly crystalline Fe oxides such as ferrihydrite act as strong stabilising agents for microaggregates in soils where organic matter contents are low, most notably in subsoils (Churchman and Tate, 1987; Duiker et al., 2003). Al oxides may play a stronger or similarly important role as Fe oxides in stabilising soils, including some from tropical zones (Schaefer et al., 2004; Barthès et al., 2008). The most effective form of Al may derive from Al-substituted Fe oxides (Barthès et al., 2008).

Statistical and electron optical evidence, as well as the disaggregating effects of (relatively) selective dissolution treatments for Fe oxides suggest that these participate in interparticle binding (Goldberg, 1989; Schwertmann and Taylor, 1989). The common treatment for dissolution of Fe oxides, sodium citrate–dithionite–bicarbonate (Mehra and Jackson, 1958), also dissolves Al hydroxides, so that its disaggregation effect could point to the role of Al and/or that of Fe in binding. Nonetheless, the strength of metal oxide–phyllosilicate interactions was found to vary according to the dominant phyllosilicate mineral present. The interaction between metal oxides and clays appeared to provide the main cohesive forces in dominantly kaolinitic soils. Transmission electron microscopy showed deposits of Fe oxides on the surfaces of kaolinite platelets in a number of Australian soils (Fordham and Norrish, 1979). In dominantly smectitic soils, however, oxide–phyllosilicate interactions provided less effective cohesion than those arising from cation bridging. Given that the points of zero charge (PZCs) of Fe oxides are pH ca. 7–9 (McBride, 1989), they are positively charged at pHs below this value, and it is likely that Fe oxides (and also Al hydroxides, with a PZC of pH ca. 9–10) are attached to phyllosilicate surfaces in acid soils, where kaolinite (PZC ~3) commonly occurs (Kämpf et al., 2012). Accordingly, Fe oxides appear to play a large role in microaggregation in Oxisols (Beinroth et al., 2012; Kämpf et al., 2012), so that microaggregates in some Oxisols are characterised as having a "coffee powder structure" (Kämpf et al., 2012, p. 22–28). As ferrihydrites, they also enhance interparticle binding in Andisols (McDaniel et al., 2012) and their widespread occurrence in soils suggest that they bring about microaggregation in many soils, especially including those with high contents of Fe, e.g. in the eluviated spodic horizon of Spodosols (Schaetzl and Harris, 2012). In all cases, it is the poorly crystalline species, ferrihydrite for Fe and allophane for Al, that are most effective in forming aggregating associations in all types of soils (Arduino et al., 1989).

Even so, metal oxides are not essential for stabilising microaggregates in soils. Where they are in relatively short supply, as in many soils at near-neutral pH in temperate zones with dominantly 2:1 Si:Al aluminosilicates, it appears that the most reactive aluminosilicates, i.e. those with higher (specific) surface areas, stabilise microaggregates (e.g. Heister et al., 2011; Pronk et al., 2012; Fernandéz-Ugalde et al., 2013). Barré et al. (2014) noted that much remained to be known about the influence of minerals on organic matter stabilisation in soils. However, these authors concluded (p. 382) that "the most efficient phyllosilicates for SOC protection would be the phyllosilicates that preferentially form and stabilise clay-size aggregates".

14.3 FORMATION AND STABILISATION OF PORES

Some effects of clay mineralogy upon pore sizes and their distribution are shown by Churchman and Payne (1983). Soils dried by a method (critical point drying) that may lead to a collapse of larger pores upon insertion of mercury for analysis of pores by mercury porisimetry nonetheless showed that fine pores in the $10–10^4$ nm equivalent cylindrical diameter differed according to the mineralogy of the soils, examined as subsoils to minimise the effect of organic matter. A halloysitic soil had a high concentration of pores in the 10–100 nm range. These were largely retained upon oven drying. A smectitic soil had pores mainly in the $10–10^3$ nm range that were largely lost on oven drying. A soil with dominantly allophanic clays had quite a low volume of small pores, with its pore size distribution dominated by large pores $>10^3$ nm, consistent with this particular soil's pumiceous nature. Vesicles in the pumice may provide large pores. Using a combination of nitrogen gas sorption (for smaller pores) and mercury porisimetry, Sills et al. (1974) found that a number of soils containing kaolinite and/or montmorillonite showed similar distributions of pore sizes, with a peak near 3 nm. However, one soil, kaolinitic, with a high concentration of sesquioxides of Fe, contained larger pores near 6 nm, tentatively attributed to the Fe oxides. It therefore appears that the clay mineralogy can affect the nature of fine pores.

However, pores in soils, and especially topsoils, are also a function of the aggregates formed by the clay minerals with other soil components, principally organic matter. When aggregates are common, there tend to be more pores. Figure 14.1 shows this for the case of pores formed as a result of different farming practices on the same soil, as shown at the microaggregate level in Chapter 11, Figure 11.1. The soil at the level of larger aggregates (and primary particles) is shown in Figure 14.2. Pores with diameters of 10–100 μm are common in the virgin soil and, to a lesser extent, in the adjacent farmed soil, but not in the soils under no till and conventional cultivation, following long-term cultivation. The virgin soil has a greater volume of pores in the 0.1–10 μm range than all other soils. Figure 14.2 shows that the virgin soil has a strong tendency to form into aggregates of 50–100 μm in size. There is some tendency to do the same in the adjacent, lightly farmed soil, but there are

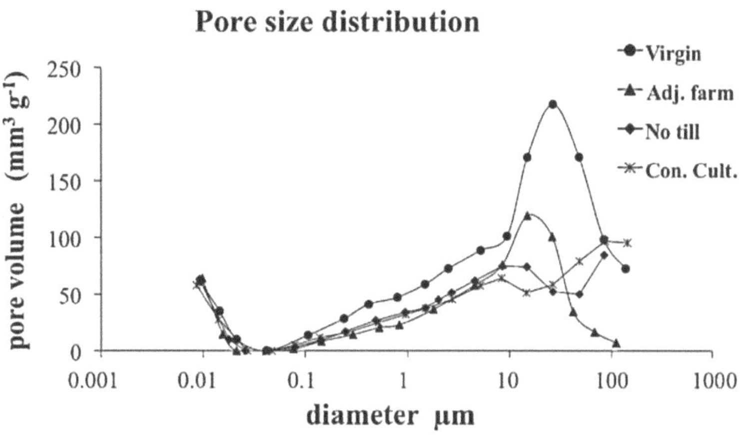

FIGURE 14.1 Pore size distribution from mercury porisimetry for surface layers of an Australian Alfisol resulting from different agricultural treatments, as in Figure 11.1 (see Chapter 11) and Figure 14.2. (From Churchman, G.J., et al., 2010, *Soil and Tillage Research* 109: 23–35. With permission from Elsevier.)

few aggregates in this size range in the two soils that had been subject to conventional cultivation for a long time. These features of the micrographs are consistent with the differences in the volume of pores in the 10–100 μm range in Figure 14.1. The differences in the volume of pores in the 0.1–10 μm range between the virgin soil and the soil from the other three treatments is consistent, in turn, with the observation of many clear spaces in the areas between microaggregates that constitute the

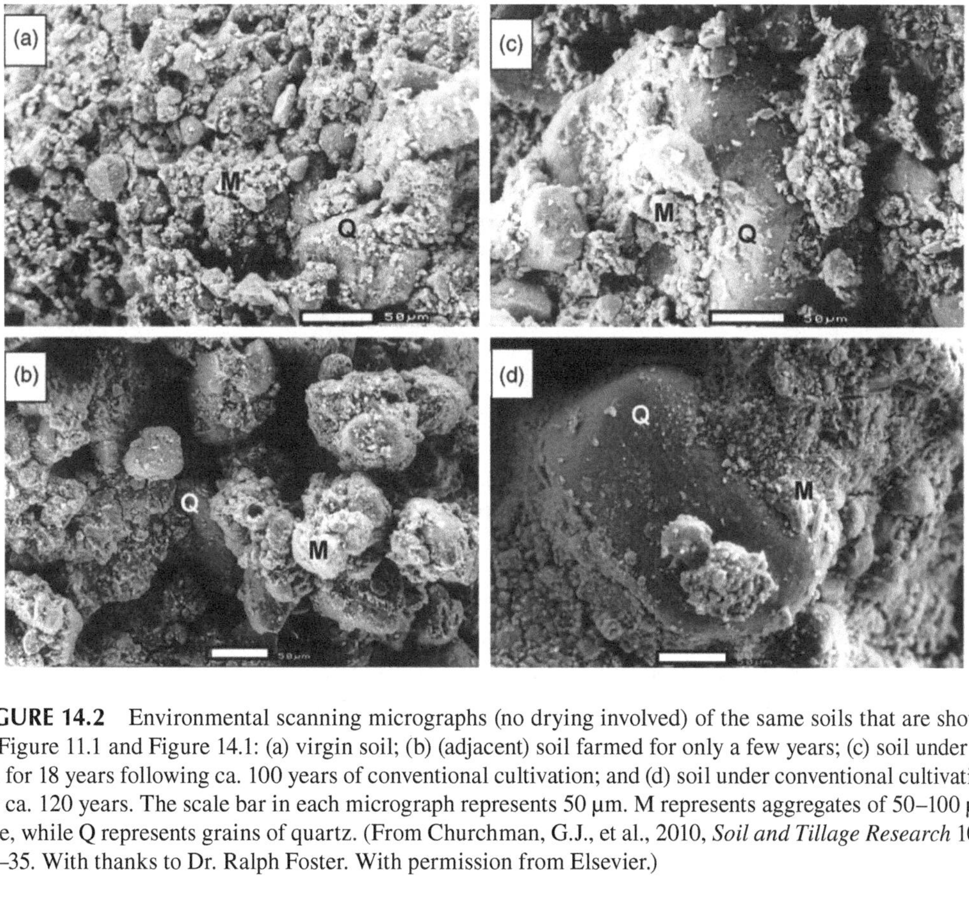

FIGURE 14.2 Environmental scanning micrographs (no drying involved) of the same soils that are shown in Figure 11.1 and Figure 14.1: (a) virgin soil; (b) (adjacent) soil farmed for only a few years; (c) soil under no till for 18 years following ca. 100 years of conventional cultivation; and (d) soil under conventional cultivation for ca. 120 years. The scale bar in each micrograph represents 50 μm. M represents aggregates of 50–100 μm size, while Q represents grains of quartz. (From Churchman, G.J., et al., 2010, *Soil and Tillage Research* 109: 23–35. With thanks to Dr. Ralph Foster. With permission from Elsevier.)

pores in the virgin soil, whereas the pores in the soil from the other three treatments are blocked with small particles (see Figure 11.1 in Chapter 11).

It appears that small pores are the most important for the accumulation and protection of organic matter (OM). In particular, Juarez et al. (2013) found that pores >13 μm showed no effects on mineralisation in soil of added organic matter, although different soils had clear differences in their physical structure. The use of ultra-small angle X-ray scattering enabled a study of the organic matter within very small pores in microaggregates of 53–250 μm in diameter (McCarthy et al., 2008). It was found that most of the pores holding OM were in the 10–10^3 nm size range, with little OM in pores larger than this size, i.e. >1 μm. Large bacteria could inhabit these larger pores, but it is noted that soil microbes degrade complex organic molecules largely via extracellular enzymes and it may not be useful, in terms of energy expended for that gained, for microbes to express enzymes in order to degrade OM from within submicron-sized pores (McCarthy et al., 2008). It is only necessary for the throats of pores to be blocked by OM for protection to take place in pores. This work confirms the finding by Mayer and Xing (2001) that occlusion of OM within microaggregates and, more particularly those comprising pores, is likely to be more effective for the protection of OM than are dispersed adsorbed coatings of OM on mineral grains. The minerals, metal oxides as well as phyllosilicates, likely play their most important role in the preservation of organic matter and hence carbon sequestration (see Section 14.6) through their role in forming pores via microaggregates.

At the finest, molecular scale, the use of synchrotron-based near-edge X-ray spectromicroscopy (NEXAFS) has revealed that organic matter in contact with mineral surfaces is highly heterogeneous, whereas the organic carbon forms at the macro, visible scale were remarkably similar for acid and calcareous soils from around the world and even for soils at different stages of degradation (Lehmann et al., 2008). Generally, at the molecular scale, different forms of organic matter (with different functional groups) bind to different mineral surfaces. Short-range order minerals were found to bind more organic matter than other minerals.

With an emphasis on microbes, Young and Crawford (2004) described soil as a complex adaptive system. While this description may be easy to visualise in the case of living entities such as microorganisms, it may also be that minerals play an active role in the life of soils at the molecular scale; they may be more than just templates for adsorption or components of walls for occlusion and for pore formation. For example, bacteria have been shown to reduce structural Fe(III) on the surfaces of smectite clay minerals (Kostka et al., 2002), albeit that smectite from a non-soil source was studied. Also using 'pure' clay, Alimova et al. (2009) showed evidence suggesting that biofilms in contact with smectite actively create organo-clays within the clay mineral and they proposed that the resultant organo-clay could provide a possible storage medium for carbon within a microbial colony. Furthermore, the composition of the microbial community has been shown to change with different minerals added to a soil (Carson et al., 2007, 2009). Studying an unaltered soil, Kotani-Tanoi et al. (2007) found through DNA analyses of single soil particles that the bacterial communities located on soil particles of the same (mineralogical) type were more similar in composition than communities located on particles of the other types. Hence, it appears that minerals can play a direct role in microbial ecology rather than by simply providing a matrix for microbial activity. The mineral matrix may not be inert.

14.4 AGGREGATION, PORES AND SOIL QUALITY

For at least 20 years, the concept of 'soil quality' has been proposed as an indicator of "the capacity of a soil to sustain biological productivity, maintain environmental quality and promote plant and animal health" (Wienhold et al., 2005, p. 349). However, there is no consensus about the soil properties that constitute soil quality. Generally, it is considered to be indicated by a variety of soil properties or characteristics such as pH, organic matter, microbial biomass C, respiration and/or enzyme activities (e.g. Bastida et al., 2008). There is also a lack of consensus on the methods to be used for its quantification (e.g., Bastida et al., 2008). Soil quality as a concept has also met strong criticism from some, such as Sojka et al. (2003, p. 1), who labelled it "elusive and value-laden".

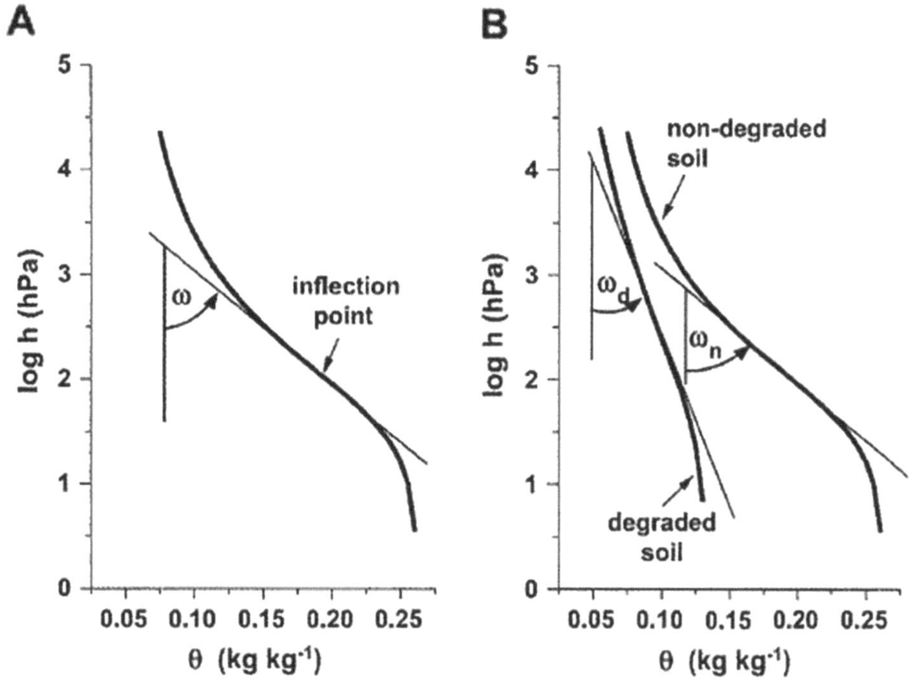

FIGURE 14.3 (a) Example of a soil water retention curve showing the inflection point and the slope, tan ω, of the tangent to the curve at the inflection point. (b) Water retention curves of the same sandy clay loam soil at two different bulk densities. Soil physical degradation occurs when the soil is compacted, and this reduces the slope of the retention curve at the inflection point. h is water suction and θ is the gravimetric water content. (From Dexter, A.R., 2004, *Geoderma* 120: 201–214. With permission from Elsevier.)

Among the various soil quality indices, one proposed by Dexter (2004) that is based on the water characteristic curve (Figure 14.3) gives a measure which is related to the microstructure of a soil that is sensitive to degradation by, for example, extensive tillage (Dexter and Czyż, 2007). Using a modern representation of the water retention equation, Dexter et al. (2008) have been able to separately distinguish the pore spaces due to structure and to texture. Regardless of the method of its interpretation, the water retention curve offers the prospect of defining the extent and nature of aggregation, leading to stable pores, in soils. The strength of the (physical) soil quality index proposed by Dexter is that it is able to separate structural porosity from textural porosity (see also Dexter and Richard, 2009). The former reflects soil structure and varies with management of a soil, while the latter is an invariant measure, reflecting soil texture.

A high contribution to structural porosity means a soil has good physical quality. Good soil physical quality may be expected to enhance the biological and chemical qualities, including the organic matter content, of a soil.

Since the water retention curve reflects the porosity of soils, it may be that other measures of porosity, e.g. mercury porosimetry, could be used to reflect soil physical quality. In any case, the involvement of clays in the formation and stabilisation of pores will only add to the physical quality of soils.

14.5 BULK SOIL PHYSICAL PROPERTIES

Because they are hydrophilic and take up water on their surfaces, all common clays in soils can swell, and, on losing water, can shrink. However, smectites and, to a lesser extent, vermiculites are liable to swell and shrink extensively, while other clays experience swelling and shrinking to only limited extents. Smectites are capable of swelling to an infinite extent if they contain substantial Na or Li ions in their interlayer sites (Norrish, 1954; Borchardt, 1989). Beyond a certain point of swelling, the interactions between adjacent layers become insignificant, and smectite layers act as dispersed individual particles. By contrast, Ca-saturated smectites show limited swelling. They can swell linearly by 100% of their oven-dry state: from a c-spacing of 1.0 nm to one of 2.0 nm (Borchardt, 1989).

Clays swell on wetting and shrink on drying to an extent that is dependent upon their charge and also upon the nature of the exchangeable cation associated with the clay. Their ability to swell and their plasticity have the same cause. The effect of exchangeable sodium on swelling capacity is particularly remarkable, especially for smectite clays. Its effect may be contrasted with that of calcium. Both sodium and calcium commonly occur as ions in soils. When a soil is dry, sodium and calcium ions are both located quite close to the clay surfaces, and adjacent surfaces are similarly close to each other in both cases. However, when the soil is wet, a sodium ion comes to be surrounded by a large number of associated water molecules, especially in comparison to a calcium ion in the calcium clays.

As a result of both this effect and also its single charge, a sodium ion, cannot approach the negatively charged aluminosilicate layers as closely as can a calcium ion, which also has a greater attractiveness for the clay layer from its double charge. Consequently, sodium ions do not bridge adjacent clay layers together as well as calcium ions. They do not neutralise the charge on the clay layers as effectively as do the less hydrated, and hence smaller and more highly charged hydrated calcium ions. This means that, in the sodium-saturated clays, the inherent tendency of similarly (negatively) charged adjacent aluminosilicate layers to repel each other is less strongly opposed by the exchangeable, interlayer cations. Indeed, in calcium smectites, the attractive forces due to the cations actually exceed the repulsive forces due to the adjacent layers when the layers are <2 nm apart. Calcium smectites are capable of swelling to about 2 times their dry volume (see earlier). However, the net attraction between adjacent layers prevents expansion beyond this point. In stark contrast, the repulsive forces exceed the attractive forces at all distances of layer separation in sodium smectites. As a result, adjacent layers can continue to expand relative to one another until they are so far apart that they effectively exist in suspensions as single particles. They may therefore swell, by imbibing water to an infinite extent, although their volume expands by about 5 times before they cease to exist as coherent solids and go into suspension, i.e. they exceed their liquid limit, in soil engineering terms. The clay then becomes dispersed.

However, both swelling and also the associated dispersion only occur in salt-free water of interparticle or pore solutions. When electrolytes are introduced into solutions, hydrated exchangeable cations, including sodium, are themselves forced closer into clay minerals surfaces where they tend to neutralise layer charge more effectively and thereby suppress interparticle repulsion. The common use of gypsum to stabilise soils relies partly on an electrolyte effect, which occurs rapidly, and also on the replacement of sodium ions by calcium ions in the clay mineral interlayers, which occurs over the long term (Churchman, 2002a).

The influence of clay mineralogy on physical properties of soils relating to tillage has been indicated by the different rheological behaviours of some soils (Markgraf et al., 2006). High clay contents lead to a sliding shear behaviour, in contrast to silty soils, which show a more turbulent shear. Furthermore, clayey soils with swellable smectitic clay minerals have a more distinctive sliding character than those with non-expandable kaolinites.

Swelling of soils can decrease their permeability, as measured by hydraulic conductivity (HC), provided the electrical conductivity (EC) of the soil solution is low (Quirk and Schofield, 1955; Sumner, 1993). This is exacerbated if their component particles are especially small (see Chapter 13, Section 13.5). The extensive swelling (and shrinkage) of smectitic soils has wider consequences for land use. Buildings and other structures built on soils containing smectites (so-called reactive soils) are subject to damage as a result and have been the main cause of structural damage in the United States, being more destructive than earthquakes and floods (Reid-Soukup and Ulery, 2002).

FIGURE 14.4 **(See colour insert.)** Examples of cracks developed in buildings in Australia as a result of the expansion and contraction of reactive soils. (Left) From https://www.abis.com.au/cracking-movement; (Right) From https://www.domain.com.au/news/is-your-house-cracking-up-what-to-do-about-cracks-in-the-home-20160420-goamq7/ (both accessed August 15, 2018).

Examples from Australia of the damage they can cause are given in Figure 14.4. The expansion and contraction of reactive – generally smectitic – soils that is the cause of the cracking arises from changes in water content that are likely to be seasonal but may also be exacerbated by the drying action of roots of trees near the buildings.

In another case, optical fibre cable was to be laid Australia-wide over thousands of kilometres. The cable was to be buried, which protects it from various types of damage. However, soils which shrink and swell (e.g. Figure 14.5a) could lead to damage of vital and expensive telecommunication links and, through a shearing action, could even break the cable and stop transmission (Fitzpatrick, 2013). The solution to this problem was to either avoid shrink–swell soils by diverting the cables around them or, if this was impractical, to use heavy-duty cables to traverse these soils.

Soils containing smectites are also responsible for landslides, soil creep and erosion. These are exacerbated when there is considerable sodium on the exchange sites of the clay. However, landslides can also arise with soils containing clays besides smectites. In Hong Kong, destructive landslides have occurred on slopes which are dominantly kaolinitic or halloysitic (Kirk et al., 1997; Campbell et al., 1998; Churchman et al., 2010b). In New Zealand, landslides have occurred in highly halloysitic altered pyroclastic tephra (Smalley et al., 1980; Moon, 2016; Kluger et al., 2017). Particle shape appears to play a role, with halloysites having a spheroidal shape being particularly prone to landsliding (Smalley et al., 1980; Kluger et al., 2017) (see Figure 14.6). There appear to be few, if any, interparticle links between the spheroids, and their shapes suggest that they 'roll' upon one another like ball bearings. Allophane has also been seen to trigger engineering failure in soils, whether on slopes or in dams to restrain water (Wesley, 1973, 1977, 2001). Other factors such as the rate of supply of water, including rainfall intensity, and the degree of slope play a role in the initiation of landslides. Highly smectitic soils may also lead to their slumping into depressions in the landscape. These are known as 'gilgai' and often also include 'slickenslides' or smooth edges to cracks that are formed (see Figure 14.5).

All clays can experience dispersion when sodium occupies a substantial proportion of their exchangeable sites (Churchman et al., 1993). This phenomenon, known as 'sodicity', can have deleterious effects on soils as a result of the dispersion of clay particles that takes place in low EC solutions. The dispersed clay can then block pores, impairing plant growth and enhancing erosion (Quirk and Schofield, 1955; Sumner, 1993; Rengasamy, 2002; So, 2002). Dispersion also occurs when considerable potassium occupies exchangeable sites, but to a lesser extent than with sodium (Sumner, 1993; Rengasamy and Marchuk, 2011). Amelioration is generally carried out by the addition of gypsum, which both diminishes dispersion in the short term and remediates in the longer term through replacement of Ca for Na (or K) on exchange sites (Churchman, 2002a).

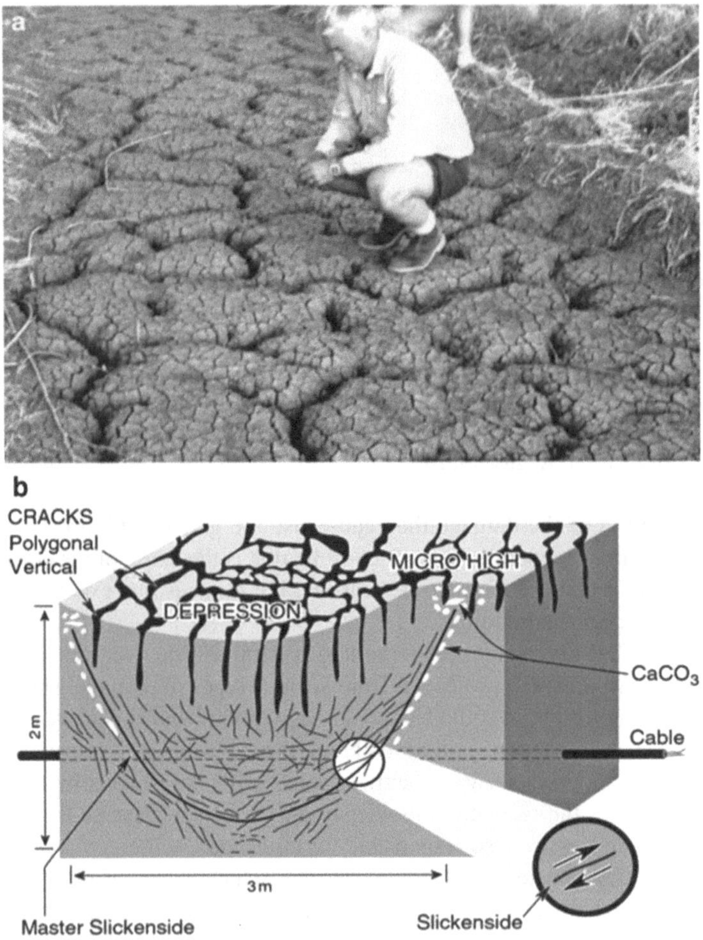

FIGURE 14.5 (a) Cracks greater than 40 mm wide to a depth of 1 m in a deep clay soil, which shrinks and swells during seasonal wetting and drying cycles (Queensland, Australia). (b) Schematic section through a swelling clay soil or Vertosol (Isbell, 1996) showing micro relief ('gilgai'), cracks zones with slickensides ('shearing zone'), where optical fibre cable distortion occurs due to soil movement ('shearing action'). (From Fitzpatrick, R.W., 2013, Demands on soil classification and soil survey strategies: Special-purpose soil classification systems for local practical use, p. 52–83, in S.A. Shahid, F.K. Taha, and M. Abdelfattah (eds.), *Developments in Soil Classification, Land Use Planning and Policy Implications: Innovative Thinking of Soil Inventory for Land Use Planning and Management of Land Resources*, Springer, Dordrecht. With permission from Springer Nature.)

On the positive side, soils rich in clays and especially smectite, and particularly those with considerable exchangeable Na, have a low permeability upon swelling and, provided they are kept moist or wet, can provide barriers to water in farm dams and to wastes in landfills. If sourced locally, especially highly smectitic soils may provide a low-cost alternative to commercial bentonite, albeit usually with a higher permeability. The same principles apply as for commercial bentonites in barriers (Gates et al., 2009). Even so, the prolonged association of the soil clays in dams with water can lead to the partial dissolution of the clay minerals and even the generation of new minerals or different forms of the original minerals. This latter process, a metamorphosis, appears to have occurred in an earth dam in Hawaii which had been filled with water for 112 years, when fine 'spherical' halloysite appeared within the dam materials; originally these comprised mainly halloysite in a variety of particle shapes – tubular, blocky, spheroidal and 'onion-like' (Shaller et al., 2016). Halloysites

FIGURE 14.6 Scanning electron micrograph of spheroids of halloysite from tephra in New Zealand thought to be responsible for landsliding. (From Kluger, M.O., et al., 2017, *Geology* 45: 131–134. With permission from Geological Society of America.)

often form initially as small sub-spherical particles (e.g. Papoulis et al., 2004; Cunningham et al., 2016). In the dam, the newly formed small particles of near-spherical shapes (Figure 14.7) probably helped – along with the loss by reduction of Fe oxides – to weaken the structure. Once again, there were few, if any, interparticle links, and the mass of the water could cause the particles to roll over one another, leading to the eventual catastrophic failure of the dam in this case.

FIGURE 14.7 Scanning electron micrograph of sub-spherical particles of halloysite presumed to have formed in a dam comprising larger particles of halloysite of various shapes in contact with a large body of water for 112 years. (From Shaller, P., et al., 2016, *Clay Minerals* 51: 499–515. With permission from Mineralogical Society of Great Britain and Ireland.)

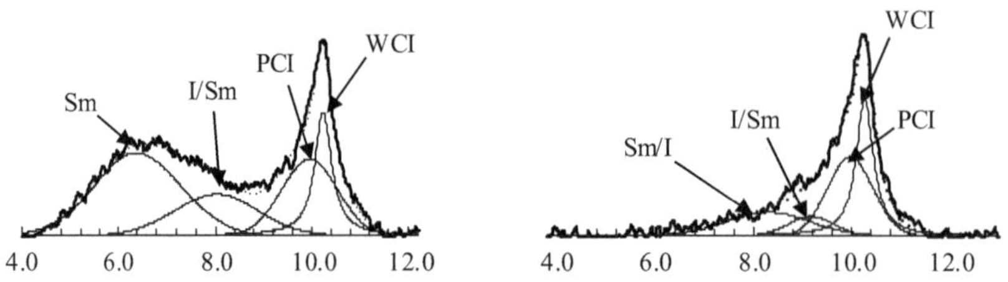

FIGURE 14.8 X-ray diffraction trace over the 001 peak zone for illite and smectite of the clay fraction of a Red Sodosol (Isbell, 1996) soil from Australia before (left) and after (right) additions of winery wastewater. The heavier line represents experimental data and the lighter line represents the best-fit-computed data. (Reprinted, with amendment, from Marchuk, S., J. Churchman, and P. Rengasamy, 2016, *Soil Research* 54: 857–868. With permission from CSIRO Publishing.)

Potassium, as K^+, is common, often in high concentrations (hundreds and thousands of $mg\,L^{-1}$) in a wide variety of industrial wastewaters (Marchuk et al., 2016). When these wastewaters are used to irrigate crops, as is often the case, K^+ is taken up by expandable clays leading to its fixation and the 'illitisation' (i.e. conversion to illite) of smectitic and vermiculitic layers (Figure 14.8). These mineralogical changes in the soils can affect their physico-chemical properties, such as hydraulic conductivity (Rengasamy and Marchuk, 2011). They can also provide a renewed source of potassium for plant nutrition (see Section 14.1).

Soils, and particularly sandy soils, can become hydrophobic, i.e. non-wetting, after long dry periods (McKissock et al., 1998; Cann, 2000; Churchman, 2002c; Churchman et al., 2014) or as a result of wildfires (DeBano et al., 1970). The problem is widespread, occurring in, for example, Florida (Jamison, 1946), California (DeBano et al., 1970) and Australia (Ward and Oades, 1993; McKissock et al., 1998; Cann, 2000). In Australia, at least, it is thought to be caused by deposits of waxy organic compounds on the sands (Franco et al., 1995; McKissock et al., 2002). The effect of this phenomenon is crops and pastures that are patchy and produce poor yields (e.g. Figure 14.9).

The solution to the problem of non-wetting soils is relatively simple, with farmers discovering over 40 years ago that addition of clays to the surface of the hydrophobic sandy soils enabled water

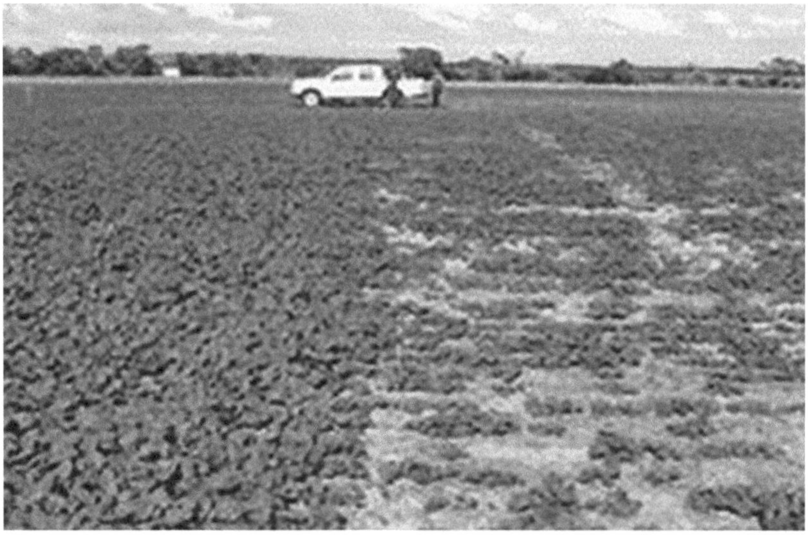

FIGURE 14.9 **(See colour insert.)** A non-wetting soil in Western Australia with patchy plant growth (right) and improved growth through the addition of clay (at 300 t ha^{-1}) (left). (With thanks to David Hall, Western Australia Department of Primary Industries and Regional Development.)

penetration and raised plant yields (Cann, 2000; Churchman et al., 2014). Up to 160,000 ha of land in South and Western Australia has been improved for cropping and pastures over 40 years or more. Since the sandy soils in Australia are generally 'duplex' in nature, with sandy topsoils over clayey subsoils (see Figure 14.10a), much of their amelioration is carried out by 'delving', or the mechanical inversion of subsoil material (Betti et al., 2015; see also Figure 14.10b), so that it is brought to or near the surface, thereby introducing clayey material into the sandy topsoils (see Figure 14.10c).

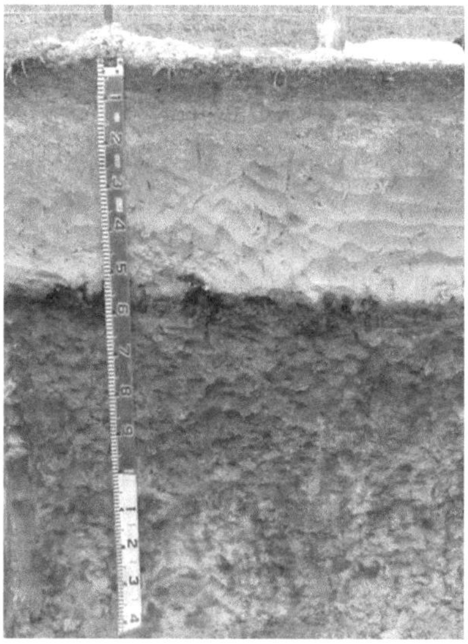

(Untreated soil)

OC (%)	Σ Exch cats(~CEC) (cmol(+)/kg)
3.6	6.5
1.5	2.6
0.7	0.9
0.3	0.4
0.1	0.3
0.8	2.3
0.5	10.7
0.2	9.4

(a)

(b) **(c)**

FIGURE 14.10 (See colour insert.) (a) Profile of a duplex (sand over clay) soil, common in southern and western Australia. Organic carbon (OC) and exchangeable cation (Σ Exch cats) values are low in all but the top of the A horizon, indicating low organic C and low clay except in the thin topsoil, while Σ Exch cats values are higher in the B horizon, indicating higher clay content. (b) Clay from B horizons is moved into sandy A horizons by delving, giving the result in (c). The depth scale is in centimetres. (Photographs reproduced from Churchman, G.J., et al., 2014, Clay addition and redistribution to enhance carbon sequestration in soils, p. 327–335, in A.E. Hartemink and K. McSweeney (eds.), *Soil Carbon*, Progress in Soil Science, Springer, Switzerland. With permission from Springer Nature.)

14.6 CARBON SEQUESTRATION

Sequestration of carbon occurs when carbon sources (principally CO_2 and CH_4) are taken up by soils and retained there for long periods (many years). Uptake of CO_2 is generally via plants, which remove CO_2 through photosynthesis and supply soils with the products of their growth and eventual death. It has sometimes – but not always – been shown that the carbon content of a soil is related, in the first instance, to its clay content (Spain, 1990; Ladd et al., 1996) (see also Chapter 10, Section 10.3). Ladd et al. (1996) found that substituting CEC for clay content only marginally improved the relationship with carbon content. Spain (1990) found a similarly good relationship of soil carbon content to clay content in most soils studied. However, this relationship was not shown by soils from basalt, in which iron minerals were supposed to interact with organic matter. Saggar et al. (1996) found that the amount of (radioactive) carbon retained by soils in an incubation experiment did not relate well to clay content but showed a good relationship with the surface area of the soils, as determined by the adsorption of *para*-nitrophenol. Several studies (e.g. Kaiser and Zech, 1998; Richards et al., 2009; Saidy et al., 2013; Regelink et al., 2015; Singh et al., 2016) have identified that oxides of iron and also aluminium relate well to carbon content and propose that these are the principal soil component responsible for uptake of carbon in soils that contain the oxides in reasonable concentrations. Allophane may be particularly useful for the sequestration of carbon, given studies (Calabri-Floody et al., 2011; Huang et al., 2016) which showed nanoaggregation of organic carbon apparently occurring in nanoaggregates and hence protected from degradation. A comparison of the rates of incorporation and turnover of carbon from earlier atmospheric bomb testing (C^{14} was measured and incorporated into a model) is given in Figure 14.11, from Parfitt (2009), and shows that the residence time of C in soils containing allophane, as measured by ^{14}C, is much greater than that of a nearby soil lacking substantial allophane.

Whether and how sequestration of carbon can be achieved on unamended soils is a subject for much debate (Stockmann et al., 2013). Even so, it is highly likely that addition of clays to soils should increase the capacity of soils to uptake and hold organic matter. When the clay content

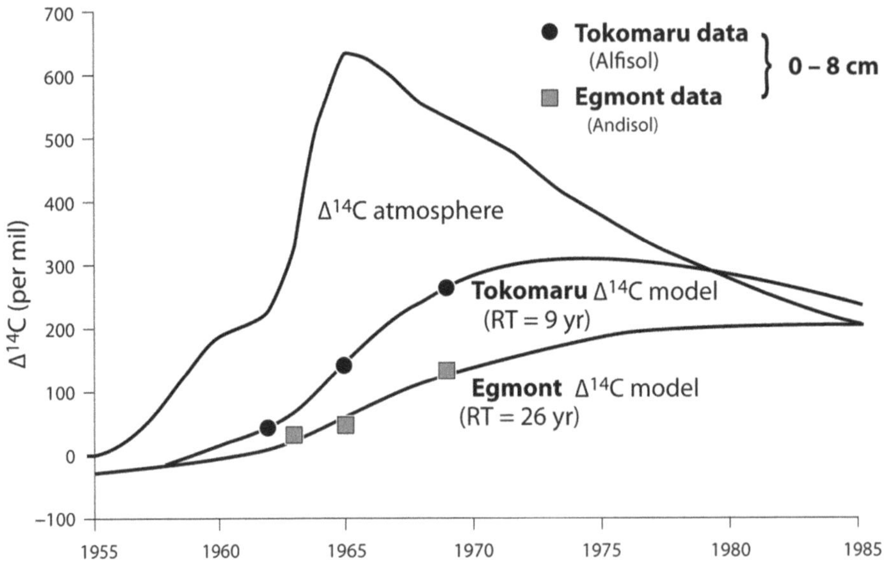

FIGURE 14.11 Model showing slower rates of incorporation and turnover of organic C using bomb-derived ^{14}C in an allophanic soil (Egmont series) compared with those of a soil without appreciable allophane (Tokomaru series) under similar climate and land-use in New Zealand. RT, residence time of carbon. (Modified diagram from Parfitt, R.L., 2009, *Clay Minerals* 44: 135–155. With permission from Mineralogical Society of Great Britain and Ireland.)

of sandy topsoils was increased through the mechanical process of delving (see Section 14.5 and Figure 14.10b), the organic carbon contents of the soils were also increased (Schapel et al., 2018). They are a function of the increased plant growth in the soils, leading to increased production of roots, increased microbial activity and hence increased carbon in the soils. It is likely that the clays enabled new aggregates to form, most probably because of the action of roots from the newly established plants. These form macroaggregates in the first instance and, within them, microaggregates, with the capacity to store the new organic carbon over the long term. Figure 14.12 shows how the addition of clay to a sandy soil, not otherwise affected by non-wetting, has led to spectacularly increased growth of forage sorghum. The yield increases in this trial are given in Churchman et al. (2014, Figure 34.3). The effect of the added clay on plant growth was much greater than that of compost. Furthermore, the effect of the clay is permanent. The clay that was added in the trial pictured in Figure 14.12 is a bentonite, but clay from soil sources would act similarly, as shown by Schapel et al. (2018, 2019). With many millions of hectares of sandy soils occurring worldwide, this technology offers the prospect of simultaneously increasing plant yield, thus helping to feed the world, and storing more carbon, hence alleviating global climate change. Of course, the limitations to its success depend upon easy (local) sources of clays, especially soil clays, and adequate supplies of water, including from rainfall.

The mechanism and amount of carbon stabilised against mineralisation and loss is dependent on soil type. A comparison of topsoils and subsoils in both temperate and tropical climate zones led to the conclusion that in acid subsoils under temperate and also tropical forests, the amount of oxidation-resistant organic matter related strongly to the content of poorly crystalline mineral phases (Rumpel and Kögel-Knabner, 2011). Subsoils in podzols were stabilised by Al-organic matter complexes. While pedogenic oxides are likely to stabilise organic matter in acidic subsoils, their capacity for carbon stabilisation is strongly reduced in near-neutral and calcareous soils. In these soils, links between organic matter and minerals are likely to occur via Ca^{2+} ions (Rumpel and Kögel-Knabner, 2011). Indeed, it has been found that, where oxide contents are low, uptake of organic matter nonetheless occurs via attraction to phyllosilicate clays. Uptake may be via polyvalent cations linking the two negatively charged moieties (clays and organic matter) and/or by covalent bonding through exposed surface and edge hydroxyl groups (Si-OH and Al-OH) on the phyllosilicates (Barré et al., 2014).

Each soil has a limited capacity for the uptake of carbon (Hassink, 1997; Stewart et al., 2008). Soils can become saturated with carbon (Stewart et al., 2008). In view of the great propensity of

FIGURE 14.12 **(See colour insert.)** The effect of adding bentonite clay to sandy soils in tropical Thailand upon the growth of forage sorghum. Crop under normal farming practice in foregrounds, crop on clay-amended soil in backgrounds. (Left) With average rainfall. (Right) In a dry year. (From Saleth, R.M., et al., 2009, Economic gains of improving soil fertility and water holding capacity with clay application: The impact of soil remediation research in Northeast Thailand, IWMI Research Report 130. International Water Management Institute, Colombo, Sri Lanka. With permission from International Water Management Institute.)

organic matter and minerals to bind together in soils, it is clear that the amount and types of minerals in soils have a large bearing on their capacity for taking up and also for sequestering carbon in the long term. It is most likely that the formation of microaggregates plays a leading part in retaining organic carbon in soils against its degradation and removal by microbes and by oxidation (e.g., Angst et al., 2017). It appears that there are selective spots for the uptake and eventual sequestration of carbon in soils (Vogel et al., 2014). These favoured spots have been described as "organo-mineral clusters with rough surfaces" (Vogel et al., 2014, p. 5). These spots are likely to be etch pits, micropores and cracks on mineral surfaces. The larger part of mineral surfaces is not involved in the uptake and sequestration of carbon. It thus appears that it is the nature of the minerals and their propensity to form associations with both organic matter and other mineral particles to provide such features as micropores that can encourage the uptake and eventual sequestration of carbon by soils.

14.7 POLLUTION AND ITS REMEDIATION

Clays, both phyllosilicates and oxides, have an affinity for heavy metals. Cationic heavy metals are held more strongly than alkali and alkaline earth cations on phyllosilicates. Heavy metal uptake involves more than simple ion exchange (Churchman et al., 2006; Yuan et al., 2013). Surface complexation and surface precipitation may also be involved. Anionic forms may be taken up by oxides. Iron oxides, either as separate phases or in association with phyllosilicate minerals, have a particularly strong affinity for the sorption of trace metals in soils, although their effectiveness varies with soil pH (Sipos et al., 2018). Some complex organic molecules may also be taken up by clay minerals. Organic pesticides such as paraquat and diquat are cationic and are strongly attracted to phyllosilicates.

Modification of expandable clay minerals through the uptake of quaternary ammonium compounds (QACs) (surfactants) into their interlayers renders the minerals hydrophobic and also organophilic, so that they can take up non-ionic organic pollutants such as hydrocarbons (e.g. Churchman et al., 2006). Figure 14.13 shows that not only pure clay minerals but also soil clays can be modified for this purpose, and they also are effective at taking up non-ionic compounds. Generally, surfactant-modified clays can also simultaneously adsorb heavy metals (e.g. Sarkar et al., 2012).

The use of polycations rather than monomer QACs can enable the uptake of both anions and non-ionic compounds by clays that have been rendered both hydrophobic and organophilic and also cationic. Figure 14.14 shows the effect of adding the polycationic poly-diallyl dimethylammonium chloride (poly-DADMAC) to a bentonite. As the content of the polycation in the complex increased (at the expense of Na^+ and Ca^{++} from the interlayers of the clay mineral), toluene uptake occurred and the charge (given by zeta potential) became less negative and, eventually, positive. Poly-DADMAC–modified clays have been used for the removal of both an anionic phenol and a non-ionic phenol from water (Ganigar et al., 2010). While the work for Figure 14.14 and Ganigar et al. (2010) used 'pure' clays, work by B. Pittan (2005) and T. Wright (2007) has shown that modification of some Australian soils has rendered them positive in charge and therefore likely to also have organophilic properties (following Figure 14.13). The modification of the soils enabled their uptake of phosphate (Pittan, 2005; Wright, 2007) and also arsenate (Pittan, 2005), nitrate (Pittan, 2005; Wright, 2007) and dissolved organic carbon (DOC) (Wright, 2007). Soils containing both expandable (smectite) and non-expandable (illite and vermiculite) minerals were successfully treated with poly-DADMAC in both studies. Leaching experiments, and also repacked trays and field plots subjected to a rainfall simulator, all showed reductions in P, DOC and N in leached or run-off water (Wright, 2007). Potentially the treatment could be used to remove anionic (and also non-ionic) pollutants from run-off in farms, especially those containing grazing animals, for the protection from eutrophication of water bodies receiving the run-off. However, our experience with long-term treatment of soil clays *in situ* with poly-DADMAC to remove excess phosphate in run-off from dairy farms by adsorption suggests that the complexes formed between the soil clays and the polycation, and perhaps also QACs, may be compromised by microbial breakdown of the QACs (G.J. Churchman, unpublished results).

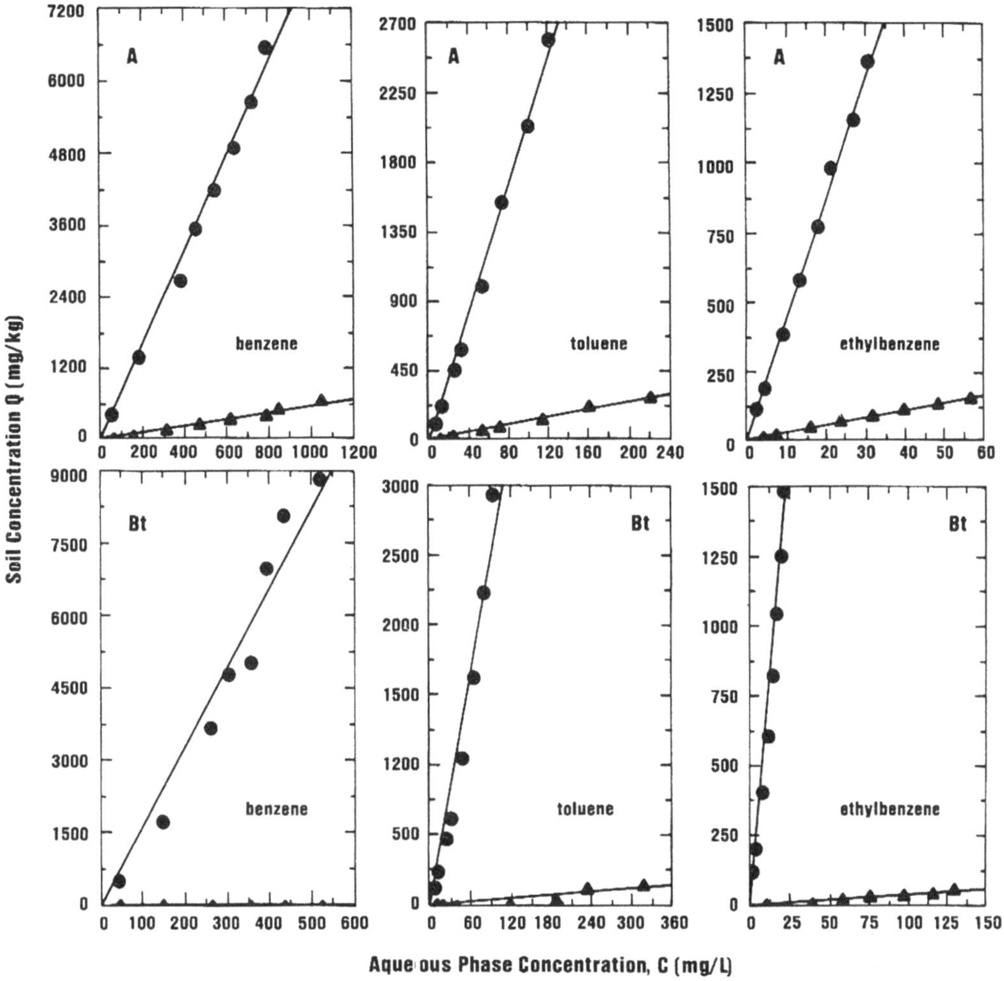

FIGURE 14.13 Sorption of benzene, toluene and ethylbenzene on untreated (triangles) and hexadecyltrimethylammonium (HDTMA)-treated (circles) St. Clair A horizon (top plots) and Oshtemo B horizon (lower plots) soils, each containing substantial illite, vermiculite and kaolinite. (From Lee, J.F., J.R. Crum, and S.A. Boyd, 1989, *Environmental Science and Technology* 23: 1365–1372. With permission from American Chemical Society.)

Hydrophobic organic compounds HOCs are hardly adsorbed by soil minerals, including metal oxides, except in the complete absence of water (Yang and Xing, 2012), which probably never occurs in soils. Instead, HOCs are taken up by SOM associated with the minerals, by partition rather than adsorption. Some polar pollutants, such as nitroaromatic compounds (NACs), can be adsorbed onto the siloxane (Si-O) surfaces of clay minerals such as kaolinite, illite and montmorillonite, but water is a strong competitor for these compounds (Yang and Xing, 2012). The ubiquitous occurrence of water in soil systems precludes the adsorption of polar compounds like NACs in the soil system. This is particularly the case when the clays are exchanged with strongly hydrating cations such as Na^+, Ca^{2+}, Mg^{2+}, Mg^{2+} and Al^{3+}, as is usually the situation in soils (Yang and Xing, 2012).

Pure clays per se (Warr et al., 2009, 2016), QAC-modified clays (Kemnetz and Cody, 1995; Churchman, 2002c), and acid-activated clays remaining following the bleaching of cooking oils (Churchman and Anderson, 2001; Churchman, 2002c) have shown the capability of remediating spills of oil, which commonly occur at sea. Pure clays are considered to act as templates for the biodegradation of oil by bacteria, especially when fertilisers are added (Warr et al., 2009, 2016).

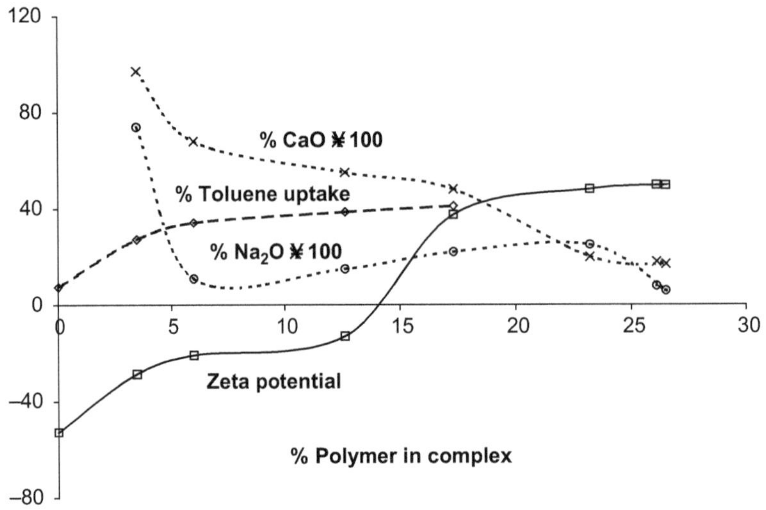

FIGURE 14.14 Zeta potential, percentage removal of toluene from a solution and percentages of both calcium and sodium (expressed as oxides and multiplied by 100), plotted simultaneously against changes in the actual percentage of poly-DADMAC (PD) of the polymer in the complex of the polycation with Wyoming bentonite. (From Churchman, G.J., 2002b, *Applied Clay Science* 21: 177–189. With permission from Elsevier.)

The acid-activated clays are considered to have become hydrophobic, hence oleophilic, so attract oil. Their subsequent degradation could also be carried out by bacteria. Soils, hence clay-bearing, could conceivably be used for this purpose, particularly when highly organic, as they would supply both attractants for oil and bacteria for its degradation.

On the other side of the ledger for soil clays and pollution, soil clays may themselves be pollutants, especially when transported by wind (e.g. Syers et al., 1969; Collyer et al., 1984) and may carry contaminants when they are moved through profiles by water into groundwater, streams and lakes.

14.8 MEDICINE

It has long been known that clays can have medicinal uses (Robertson, 1986). Generally, clays from monomineralic deposits have been used in medicine. Montmorillonite, and probably also sepiolite or palygorskite, was used in Ancient Rome for covering wounds and drying the skin, and kaolinite was taken internally for stomach disorders (Robertson, 1986). Soil clays could conceivably find similar uses, depending on their purity. Sudanese villagers living along the banks of the River Nile have traditionally used clay taken from soil along the river to flocculate the turbid water in the river and hence to clear it of viruses, parasites (the source of schistosomiasis, or bilharzia, which infects about 200 million worldwide; Olsen, 1987) and bacteria (Churchman et al., 2006). The clay has been identified as a smectite and its origin in riverbanks suggests it is a soil clay.

Ingestion of soils, known as ‘geophagy’, has been a common practice throughout human history (and pre-history). In Ancient Greece, Hippocrates (ca. 460–377 BC) and Aristotle (384–322 BC) each noted the occurrence of geophagy, for satiating hunger and therapeutic or religious purposes (Abrahams, 2010). Later, in Rome, Pliny the Elder writes in 77 AD of the use of a specific earth, “Lemnian”, or *Terra sigillata,* from the island of Lemnos, which was “taken as a draught for troubles of the spleen and kidneys and for excessive menstruation; and likewise as a remedy for poisons and snake bites – hence it is in common use for all antidotes” (Abrahams, 2010, p. 371). Soranus, a 2nd century Greek gynaecologist, obstetrician and paediatrician, noted that pregnant women had a craving for soil, among other foods, to overcome a condition called ‘pica’, and this observation is confirmed by Aetius of Amida (now located in Turkey) in an obstetrics textbook in the 6th century AD.

Geophagy was observed by early explorers of the New World, including Alexander von Humboldt and Davis Livingstone (Abrahams, 2010). In more recent times, Vermeer and Frate (1979) found the practice to be common in the African-American population in an area of the State of Mississippi. They observed that illuvial B horizon soils were taken for the practice from soils in roadcuts. Hook (1978) found that pregnancy increased the craving for geophagy among women in New York State. While the practice has declined in the West, with other remedies for pica being available, it is still found in tropical countries, including throughout Africa and in Indonesia (Abrahams, 2010).

There may be a number of drawbacks to the practice of geophagy, especially when topsoils are ingested (Carretero et al., 2006, 2013; Abrahams, 2010). In a medical textbook by Celsus in Ancient Rome, an association was noted between geophagy and anaemia, due to the ingested soils absorbing iron in the human gut, carrying parasites that led to the loss of blood (and iron) and the replacement of iron-rich food by the soils (Abrahams, 2010). Other maladies associated with geophagy include excessive tooth wear, intestinal blockage, deficiencies of nutrients (including potassium, zinc and iron), parasite infestations and toxicity through exposure to potentially harmful elements (such as fluorine, lead, potassium [in excess] and radionuclides) (Abrahams, 2010).

Generally speaking, while clays offer a number of benefits in medicine (for example, as gastro-intestinal protectors, antacids and antidiarrhoeics; Carretero et al., 2006, 2013), and also in cosmetics, it is safer to use pure clays from deposits or from some subsoils rather than topsoils for these purposes. However, clays are also used in pelotherapy to treat dermatological diseases and alleviate pains from rheumatological inflammations, among other external ailments, as well as in spa and beauty treatments. Organic matter, together with minerals, are sometimes thought desirable for the 'peloids' applied to the skin in these treatments, so some soils may be suitable.

There has been a great upsurge recently in research on the clay mineral halloysite (Abdullayev and Lvov, 2016; Churchman et al., 2016; Fizir et al., 2018). This mineral, especially in its common tubular form, has been shown to be useful as both a template for other reactions and a carrier for the transport and subsequent slow release of agents ranging from pesticides in agriculture to drugs in medicine. The latter have included the transport within the body of drugs for the treatment of cancer. This derives in large part from the observation that the interior lumen of tubular halloysites are positively charged at pHs <8. They can therefore take up anionic drugs for their subsequent slow release. The idea has been put forward (Roberts, 2015) that such drugs could be transported to the sites of cancerous tumours by magnetising their carriers (in general) and using MRI scanners to propel the magnetised carriers to these sites. This latter step is likely to be unnecessary if halloysite from soils – but not necessarily monomineralic halloysite from deposits – is used to carry the drugs because soil halloysites are commonly associated with iron oxides in high concentrations (see also Chapter 8, Section 8.7; and Churchman, 2010) so should be magnetic. It may be argued that halloysites in soils can occur in much smaller tubular particles than geological halloysites (see e.g. Churchman and Theng, 1984). Nonetheless, these particles will also exhibit a positive charge on their inner surfaces within the digestive system and, while its release may be more rapid than from a halloysite with a longer tube, a shorter release time may even be favourable. Of course, the halloysites would first need to be sterilised. More research is called for on this potential use of halloysite from soil.

14.9 FORENSICS

Forensic soil science is concerned with samples of soil that have been disturbed, augmented or moved by human activity. Its objective is to compare defining characteristics of these samples with those of natural samples, with the aim of locating the scene of a crime and linking the scene to its perpetrator(s) (Fitzpatrick, 2009; Fitzpatrick et al., 2009). In general, there are five main reasons why soils are useful in criminal investigations: soils are highly individualistic; soils have a high probability of transfer and retention; soil is nearly invisible; soil can quickly be collected, separated and concentrated; and soil materials are easily characterised (Fitzpatrick, 2009; Fitzpatrick et al., 2009).

One of the common characteristics that is identified is that of mineralogy, both of clays and coarser fractions. According to Fitzpatrick et al. (2009, p. 106), "X-ray powder diffraction methods are arguably the most significant for both qualitative and quantitative analyses of solid materials in forensic soil science". Fitzpatrick (2009) and Fitzpatrick et al. (2009, 2017) detail a number of criminal cases in which the identification of minerals by XRD has been crucial to the solution of the crimes. They include the location of two missing bodies through identification of soils, and thence their origin, on the shovel used to dispose of them (Fitzpatrick, 2009). A common situation is the linking of the minerals in the soil at the scene of a crime to those of the soil scraped off the suspect's footwear (or car tyres). Another is the finding of recently transported foreign materials in a soil sample.

14.10 ARCHAEOLOGY AND ENVIRONMENTAL HISTORY

Clays have played a vital role in the daily life of people since the Neolithic era in the Middle East. Roughly at the time when people there began to select grain-bearing plants and farm them, they discovered the use of clays to make ceramic objects. This occurred in the period 7500 to 7000 BC (Mellaart, 1975; de Kéroualin, 2003). Their development made farming much more efficient and useful for the consumption of the foodstuffs produced.

First, food heated by cooking becomes roughly 40% more nutritious to humans. Thus the early farmers could benefit from the growth of their crops in quality as well as quantity. Early ceramics were thus often cooking pots.

Second, ceramic vessels were used for storage of grains and other foodstuffs and thus prolonged their useful season, being protected from predator animals in the vicinity.

Third, of course, ceramics used as eating vessels made eating cooked substances much easier.

Thus ceramics made of clays were a fundamental part in the development of farming communities and made it possible, and in fact necessary, to establish permanent communities such as villages. Sedentarisation was fundamental to the development of human society. Clay-based ceramics were then fundamental to the development of society as we know it today and have been used throughout the ages since the Neolithic era.

The sources of clays used in ceramics were roughly those of surface deposits, soils and sediments, hence weathered. The use of hydrothermally altered rocks (kaolinite-bearing for example) is only several centuries old. The variability of the geologic substrata producing soil clays is then fundamental in the production of ceramics, and identification of these materials is used by archaeologists to determine the provenance of the ceramic products (Gliozzo et al., 2018, for example).

Components of ceramics are basically clays that give the solids a plasticity when wet, and grits (sand and other non-plastic mineral grain materials) are added that modify the plasticity in order to make products of different sizes and wall thickness. The bigger the object, the more grits necessary to form thicker walls in the object. The grits added to the clays were often from slightly different sources than those of the clay deposits. However, potters often attempted to use naturally mixed clay-sand materials in deposits that were convenient for the type of product they were working on. These different methods of modifying the plasticity of the initial clay-rich materials of a ceramic are often important in the archaeological investigations of materials in a given community (see Velde and Druc, 1999). Local products show minerals in the grits common to the geology of the site while imported ceramics show minerals not found locally. Thus commerce of ceramics can be detected by archaeologists using the components of the ceramics found on different community sites.

In more recent periods (several centuries), commerce can be seen in the very special types of clays used (kaolinite) to produce porcelain or other very special products that were transported thousands of miles.

Pottery found in the South Pacific has been used as evidence for the migration of people there from Taiwan (Irwin, 2017). The manufacture of ceramics began in Taiwan around 3000 BC and produced red-slipped pottery. Over the next 1500 years the descendants of these ceramic-making people (known as the 'Lapita people', after the style of pottery) migrated to the western part of the South Pacific (northeast of Papua New Guinea) where these now Austronesian people mixed with the indigenous inhabitants and produced their distinctive pottery.

The relationship of clays and ceramics is thus related to the type of object produced and as such can be used to describe the commercial activity of a community (Gliozzo et al., 2018).

Clays have been a fundamental aspect of societal development for humans from the Neolithic period.

Clays can be sourced that may contain DNA representative of past natural environments. This possibility is most likely when the clays adsorb organic matter, hence DNA, particularly strongly, and poorly crystalline (nano-)clays like allophane and ferrihydrite are a case in point (see Chapter 10). Huang et al. (2016) have developed a method to isolate DNA from allophanic soils at depth and have amplified DNA extracted by the polymerase chain reaction (PCR) to show its origin in plants that differ from the European grasses growing currently on the soil's surface.

REFERENCES

Abdullayev, E., and Y. Lvov. 2016. Halloysite for controllable loading and release, p. 554–605. In: P. Yuan, A. Thill, and F. Bergaya (eds.), *Nanosized Tubular Clay Minerals. Halloysite and Imogolite*. Developments in Clay Science, vol. 7. Elsevier, Amsterdam.

Abrahams, P.W. 2010. "Earth eaters": Ancient and modern perspectives on human geophagy, p. 369–398. In: E.R. Landa, and C. Feller (eds.), *Soil and Culture*. Springer, Dordrecht.

Adams, S.N., J.L. Honeysett, K.G. Tiller, and K. Norrish. 1969. Factors controlling the increase of cobalt in plants following the addition of a cobalt fertilizer. *Soil Research* 7: 29–42.

Alimova, A., A. Katz, N. Steiner, E. Rudolph, H. Wei, J.C. Steiner, and P. Gottlieb. 2009. Bacteria-clay interaction: Structural changes in smectite induced during biofilm formation. *Clays and Clay Minerals* 57: 205–212.

Angst, G., K.E. Mueller, I. Kögel-Knabner, K.H. Freeman, and C.W. Mueller. 2017. Aggregation controls the stability of lignin and lipids in clay-sized particulate and mineral associated organic matter. *Biogeochemistry* 132: 307–324.

Arduino, E., E. Barberis, and V. Boero. 1989. Iron oxides and particle aggregation in B horizons of some Italian soils. *Geoderma* 45: 319–329.

Barré, P., O. Fernandez-Ugalde, I. Virto, B. Velde, and C. Chenu. 2014. Impact of phyllosilicate mineralogy on organic carbon stabilization in soils: incomplete knowledge and exciting prospects. *Geoderma* 235–236: 382–395.

Barthès, B.G., E. Kouakoua, M.-C. Larré-Larrouy, T.M. Razafimbelo, E.F. de Luca, A. Azontonde, C.S.V.J. Neves, P.L. de Freitas, and C.L. Feller. 2008. Texture and sesquioxide effects on water-stable aggregates and organic matter in some tropical soils. *Geoderma* 143: 14–25.

Bastida, F., A. Zsolnay, T. Hernández, and C. García. 2008. Past, present and future of soil quality indices: A biological perspective. *Geoderma* 147: 159–171.

Beinroth, F.H., H. Eswaran, G. Uehara, C.W. Smith, and P.F. Reich. 2012 Oxisols, p. 33-177–33-190. In: P.M. Huang, Y. Li, and M.E. Sumner (eds.), *Handbook of Soil Sciences: Properties and Processes*, 2nd edn. CRC Press/Taylor & Francis Group, Boca Raton, Florida.

Betti, G., C. Grant, G. Churchman, and R. Murray. 2015. Increased profile wettability in texture-contrast soils from clay delving: case studies in South Australia. *Soil Research* 53: 125–136.

Borchardt, G. 1989. Smectites, p. 675–727. In: J.B. Dixon, and S.B. Weed (eds.), *Minerals in Soil Environments*, 2nd edn. Soil Science Society of America, Madison, Wisconsin.

Calabi-Floody, M., J.S. Bendall, A.A. Jara, M.E. Welland, B.K.G. Theng, C. Rumpel, and M. de la Luz Mora. 2011. Nanoclays from an Andisol: Extraction, properties and carbon stabilization. *Geoderma* 161: 159–167.

Campbell, S.D.G., N.P. Koor, C.A.M. Franks, and W.L. Shum. 1998. Geological assessment of slopes in areas close to major landslides on Hong Kong Island, p. 121–128. In: K.S. Li, J.N. Kay, and K.K.S. Ho (eds.), *Proceedings of the Annual Seminar on Slope Engineering in Hong Kong*. Balkema, Rotterdam.

Cann, M.A. 2000. Clay spreading on water repellent sands in the south east of South Australia – Promoting sustainable agriculture. *Journal of Hydrology* 231–232: 333–341.

Carretero, M.I., C.S.F. Gomes, and F. Tateo. 2006. Clays and human health, p. 717–752. In: F. Bergaya, B.K.G. Theng, and G. Lagaly (eds.), *Handbook of Clay Science*. Developments in Clay Science, vol. 1. Elsevier, Amsterdam.

Carretero, M.I., C.S.F. Gomes, and F. Tateo. 2013. Clays, drugs, and human health, p. 711–764. In: F. Bergaya, and G. Lagaly (eds.), *Handbook of Clay* Science, 2nd edn. Developments in Clay Science, vol. 5. Elsevier, Amsterdam.

Carson, J.K., L. Campbell, D. Rooney, N. Clipson, and D.B. Gleeson. 2009. Minerals in soil select distinct bacterial communities in their microhabitats. *FEMS Microbiology Ecology* 67: 381–388.

Carson, J.K., D. Rooney, D.B. Gleeson, and N. Clipson. 2007. Altering the mineral composition of soil causes a shift in microbial community structure. *FEMS Microbiology Ecology* 61: 414–423.

Churchman, G.J. 2002a. Sodic soils, reclamation of, p. 1224–1228. In: R. Lal (ed.), *Encyclopedia of Soil Science*. Marcel Dekker, New York.

Churchman, G.J. 2002b. Formation of complexes between bentonite and different cationic polyelectrolytes and their use as sorbents for non-ionic and anionic pollutants. *Applied Clay Science* 21: 177–189.

Churchman, G.J. 2002c. The role of clay in the restoration of perturbed ecosystems. *Developments in Soil Science* 28A: 333–350.

Churchman, G.J. 2010. Is the geological concept of clay minerals appropriate for soil science? A literature-based and philosophical analysis. *Physics and Chemistry of the Earth, Parts A/B/C* 35: 927–940.

Churchman, G.J., and J.S. Anderson. 2001. Use of soil waste material. PCT Patent 80019036.

Churchman, G.J., R.C. Foster, L.P. D'Acqui, L.J. Janik, J.O. Skjemstad, R.H. Merry, and D.A. Weissmann. 2010a. Effect of land-use history on the potential for carbon sequestration in an Alfisol. *Soil and Tillage Research* 109: 23–35.

Churchman, G.J., W.P. Gates, B.K.G. Theng, and G. Yuan. 2006. Clays and clay minerals for pollution control, p. 625–675. In: F. Bergaya, B.K.G. Theng, and G. Lagaly (eds.), *Handbook of Clay Science*. Developments in Clay Science, vol. 1. Elsevier, Amsterdam.

Churchman, G.J., A. Noble, G. Bailey, D. Chittleborough, and R. Harper. 2014. Clay addition and redistribution to enhance carbon sequestration in soils, p. 327–335. In: A.E. Hartemink, and K. McSweeney (eds.), *Soil Carbon*. Progress in Soil Science. Springer, Switzerland.

Churchman, G.J., P. Pasbakhsh, and S. Hillier. 2016. The rise and rise of halloysite. *Clay Minerals* 51: 303–308.

Churchman, G.J., and D. Payne. 1983. Mercury intrusion porisimetry of some New Zealand soils in relation to clay mineralogy and texture. *Journal of Soil Science* 34: 437–451.

Churchman, G.J., I.R. Pontifex, and S.G. McClure. 2010b. Factors affecting the formation and characteristics of halloysites or kaolinites in granitic and tuffaceous saprolites in Hong Kong. *Clays and Clay Minerals* 58: 220–237.

Churchman, G.J., J.O. Skjemstad, and J.M. Oades. 1993. Influence of clay minerals and organic matter on effects of sodicity on soils. *Soil Research* 31: 779–800.

Churchman, G.J., and K.R. Tate. 1987. Stability of aggregates of different size grades in allophanic soils from volcanic ash in New Zealand. *Journal of Soil Science* 38: 19–27.

Churchman, G.J., and B.K.G. Theng. 1984. Interactions of halloysites with amides: Mineralogical factors affecting complex formation. *Clay Minerals* 19: 161–175.

Collyer, F.X., B.G. Barnes, G.J. Churchman, T.S. Clarkson, and J.T. Steiner. 1984. A trans-Tasman dust transport event. *Weather and Climate* 4: 42–46.

Cunningham, M.J., D.J. Lowe, J.B. Wyatt, V.G. Moon, and G.J. Churchman. 2016. Discovery of halloysite books in altered silicic Quaternary tephras, northern New Zealand. *Clay Minerals* 51: 351–372.

DeBano, L.F., L.D. Mann, and D.A. Hamilton. 1970. Translocation of hydrophobic substances into soil by burning litter. *Soil Science Society of America Proceedings* 34: 130–133.

De Keroualin, K.M. 2003. *Genèse et diffusion de l'agriculture en Europe*. Editions Errance, Arles.

Denef, K., and J. Six. 2005. Clay mineralogy determines the importance of biological versus abiotic processes for macroaggregate formation and stabilisation. *European Journal of Soil Science* 56: 469–479.

Dexter, A.R. 2004. Soil physical quality. Part I. Theory, effects of soil texture, density and organic matter, and effects on root growth. *Geoderma* 120: 201–214.

Dexter, A.R., and E.A. Czyż. 2007. Application of S-theory in the study of soil physical degradation and its consequences. *Land Degradation and Development* 18: 369–381.

Dexter, A.R., E.A. Czyż, G. Richard, and A. Reszkowska. 2008. A user-friendly water retention function that takes account of the textural and structural pore spaces in soil. *Geoderma* 143: 243–253.

Dexter, A.R., and G. Richard. 2009. Tillage of soils in relation to their bi-modal pore size distributions. *Soil and Tillage Research* 103: 113–118.

Duiker, S.W., F.E. Rhoton, J. Torrent, N.E. Smeck, and R. Lal. 2003. Iron (hydr)oxide crystallinity effects on soil aggregation. *Soil Science Society of America Journal* 67: 606–611.

Fanning, D.S., V.Z. Keramidas, and M.A. El-Desoky. 1989. Micas, p. 551–634. In: J.B. Dixon, and S.B. Weed (eds.), *Minerals in Soil Environments*, 2nd edn. Soil Science Society of America, Madison, Wisconsin.

Fernández-Ugalde, O., P. Barré, F. Hubert, I. Virto, C. Girardin, E. Ferrage, L. Caner, and C. Chenu. 2013. Clay mineralogy differs qualitatively in aggregate-size classes: Clay-mineral-based evidence for aggregate hierarchy in temperate soils. *European Journal of Soil Science* 64: 410–422.

Fitzpatrick, R.W. 2009. Soil: Forensic analysis, p. 2377–2388. In: A. Jamieson, and A. Moenssens (eds.), *Wiley Encyclopedia of Forensic Science*. John Wiley & Sons, Chichester, United Kingdom.

Fitzpatrick, R.W. 2013. Demands on soil classification and soil survey strategies: Special-purpose soil classification systems for local practical use, p. 52–83. In: S.A. Shahid, F.K. Taha, and M. Abdelfattah (eds.), *Developments in Soil Classification, Land Use Planning and Policy Implications: Innovative Thinking of Soil Inventory for Land Use Planning and Management of Land Resources*. Springer, Dordrecht.

Fitzpatrick, R.W., M.D. Raven, and S.T. Forrester. 2009. A systematic approach to soil forensics: Criminal case studies involving transference from crime scene to forensic evidence, p. 105–127. In: K. Ritz, L. Dawson, and D. Miller (eds.), *Criminal and Environmental Soil Forensics*. Springer, Berlin.

Fitzpatrick, R.W., M.D. Raven, and P.G. Self. 2017. The role of pedology and mineralogy in providing evidence for 5 crime investigations involving a wide range of earth materials. *Episodes* 40: 148–156.

Fizir, M., P. Dramou, N.S. Dahiru, E. Ruya, T. Huang, and H. He. 2018. Halloysite nanotubes in analytical sciences and drug delivery: A review. *Microchimica Acta* 185: 389.

Fordham, A.W., and K. Norrish. 1979. Electron microprobe and electron microscope studies of soil clay particles. *Soil Research* 17: 283–306.

Franco, C.M.M., M.E. Tate, and J.M. Oades. 1995. Studies on non-wetting sands. I. The role of intrinsic particulate matter in the development of water-repellence in non-wetting sands. *Soil Research* 33: 253–263.

Ganigar, R., G. Rytwo, Y. Gonen, A. Radian, and Y.G. Mishael. 2010. Polymer-clay nanocomposites for the removal of trichlorophenol and trinitrophenol from water. *Applied Clay Science* 49: 311–316.

Gates, W.P., A. Bouazza, and G.J. Churchman. 2009. Bentonite clay keeps pollutants at bay. *Elements* 5: 105–110.

Ghezzehi, T.A. 2012. *Soil Structure*, p. 2-1–2-17. In P.M. Huang, Y. Li, and M.E. Sumner (eds.), *Handbook of Soil Sciences: Properties and Processes*, 2nd edn. CRC Press/Taylor & Francis Group, Boca Raton, Florida.

Gliozzo, E., M.Turchiano, P. Fantozzi, and A. Romano. 2018. Geosources for ceramic production and communication pathways: The exchange network and scale of chemical representative differences. *Applied Clay Sciences* 161: 242–255.

Goldberg, S. 1989. Interaction of aluminum and iron oxides and clay minerals and their effect on soil physical properties: A review. *Communications in Soil Science and Plant Analysis* 20: 1181–1207.

Hassink, J. 1997. The capacity of soils to preserve organic C and N by their association with clay and silt particles. *Plant and Soil* 191: 77–87.

Heister, K., C. Höschen, G.J. Pronk, C.W. Mueller, and I. Kögel-Knabner. 2011. NanoSIMS as a tool for characterizing soil model compounds and organomineral associations in artificial soils. *Journal of Soils and Sediments* 12: 35–47.

Hillel, D. 1998. *Environmental Soil Physics*. Academic Press, San Diego.

Hook, E.B. 1978. Dietary cravings and aversions during pregnancy. *The American Journal of Clinical Nutrition* 31: 1355–1362.

Huang, Y.-T., D.J. Lowe, G.J. Churchman, L.A. Schipper, R. Cursons, H. Zhang, T.-Y. Chen, and A. Cooper. 2016. DNA adsorption by nanocrystalline allophane spherules and nanoaggregates, and implications for carbon sequestration in Andisols. *Applied Clay Science* 120: 40–50.

Huang, Y.-T., D.J. Lowe, H. Zhang, R. Cursons, J.M. Young, G.J. Churchman, L.A. Schipper, N.J. Rawlence, J.R. Wood, and A. Cooper. 2016. A new method to extract and purify DNA from allophanic soils and paleosols, and potential for paleoenvironmental reconstruction and other applications. *Geoderma* 274: 114–125.

Irwin, G. 2017. Pacific migrations – Into Remote Oceania: Lapita people. *Te Ara – The Encyclopedia of New Zealand*, http://www.TeAra.govt.nz/en/photograph/1766/lapita-pottery (accessed August 30, 2018).

Isbell, R.F. 1996. *The Australian Soil Classification*. CSIRO Publishing, Melbourne.

Jamison, V.C. 1946. The penetration of irrigation and rain water into sandy soils of central Florida. *Soil Science Society of America Proceedings* 11: 103–109.

Juarez, S., N. Nunan, A.-C. Duday, V. Pouteau, and C. Chenu. 2013. Soil carbon mineralisation responses to alterations of microbial diversity and soil structure. *Biology and Fertility of Soils* 49: 939–948.

Kaiser, K., and W. Zech. 1998. Soil dissolved organic matter sorption as influenced by organic and sesquioxide coatings and sorbed sulfate. *Soil Science Society of America Journal* 62: 129–136.

Kämpf, N., A.C. Scheinost, and D.G. Schulze. 2012. Oxide minerals in soils, p. 22-1–22-34. In: P.M. Huang, Y. Li, and M.E. Sumner (eds.), *Handbook of Soil Sciences: Properties and Processes*, 2nd edn. CRC Press/Taylor & Francis Group, Boca Raton, Florida.

Kemnetz, S., and C.A. Cody. 1995. Oil spill flocculating agent and method of remediating oil spills. US Patent 5558777.

Kirk, P.A., S.D.G. Campbell, C.J.N. Fletcher, and R.J. Merriman. 1997. The significance of primary volcanic fabrics and clay distribution in landslides in Hong Kong. *Journal of the Geological Society* 154: 1009–1019.

Kluger, M.O., V.G. Moon, S. Kreiter, D.J. Lowe, G.J. Churchman, D.A. Hepp, D. Seibel, M.E. Jorat, and T. Mörz. 2017. A new attachment-detachment model for explaining flow sliding in clay-rich tephras. *Geology* 45: 131–134.

Kostka, J.E., D.D. Dalton, H. Skelton, S. Dollhopf, and J.W. Stucki. 2002. Growth of iron (III)-reducing bacteria on clay minerals as the sole electron acceptor and comparison of growth yields on a variety of oxidized iron forms. *Applied and Environmental Microbiology* 68: 6256–6262.

Kotani-Tanoi, T., M. Nishiyama, S. Otsuka, and K. Senoo. 2007. Single particle analysis reveals that bacterial community structures are semi-specific to the type of soil particle. *Soil Science and Plant Nutrition* 53: 740–743.

Ladd, J.N., M. Van Gestel, L. Jocteur Monrozier, and M. Amato. 1996. Distribution of organic 14C and 15N in particle-size fractions of soils incubated with 14C, 15N-labelled glucose/NH4, and legume and wheat straw residues. *Soil Biology and Biochemistry* 28: 893–905.

Lee, J.F., J.R. Crum, and S.A. Boyd. 1989. Enhanced retention of organic contaminants by soils exchanged with organic cations. *Environmental Science and Technology* 23: 1365–1372.

Lehmann, J., D. Solomon, J. Kinyangi, L. Dathe, S. Wirick, and C. Jacobsen. 2008. Spatial complexity of soil organic matter forms at nanometre scales. *Nature Geoscience* 1: 238–242.

Loveland, P.J., I.G. Wood, and A.H. Weir. 1999. Clay mineralogy at Rothamsted: 1934–1988. *Clay Minerals* 34: 165–183.

Marchuk, S., J. Churchman, and P. Rengasamy. 2016. Possible effects of irrigation with wastewater on the clay mineralogy of some Australian clayey soils: Laboratory study. *Soil Research* 54: 857–868.

Markgraf, W., R. Horn, and S. Peth. 2006. An approach to rheometry in soil mechanics –Structural changes in bentonite, clayey and silty soils. *Soil and Tillage Research* 91: 1–14.

Mayer, L.M., and B. Xing. 2001. Organic matter-surface area relationships in acid soils. *Soil Science Society of America Journal* 65: 250–258.

McBride, M.B. 1989. Reactions controlling heavy metal solubility in soils, p. 1–56. In: B.A. Stewart (ed.), *Advances in Soil Science*. Springer, New York.

McCarthy, J.F., J. Ilavsky, J.D. Jastrow, L.M. Mayer, E. Perfect, and J. Zhuang. 2008. Protection of organic matter in soil microaggregates via restructuring of aggregate porosity and filling of pores with accumulating organic matter. *Geochimica et Cosmochimica Acta* 72: 4725–4744.

McDaniel, P.A., D.J. Lowe, O. Arnalds, and C.-L. Ping. 2012. Andisols, p. 33-29–33-48. In: P.M. Huang, Y. Li, and M.E. Sumner, (eds.), *Handbook of Soil Science*, 2nd edn. CRC Press/Taylor & Francis, Boca Raton, Florida.

McKissock, I., R.J. Gilkes, R.J. Harper, and D.J. Carter. 1998. Relationships of water repellency to soil properties for different spatial scales of study. *Soil Research* 36: 495–507.

McKissock, I., R.J. Gilkes, and E.L. Walker. 2002. The reduction of water repellency by added clay is influenced by clay and soil properties. *Applied Clay Science* 20: 225–241.

Mehra, O.P., and M.L. Jackson. 1958. Iron oxide removal from soils and clays by a dithionite-citrate system buffered with sodium bicarbonate. *Clays and Clay Minerals* 7: 317–327.

Mellaart, J. 1975. *The Neolithic of the Near East*. Scribner, New York.

Moon, V.G. 2016. Halloysite behaving badly: Geomechanics and slope behaviour of halloysite-rich soils. *Clay Minerals* 51: 517–528.

Norrish, K. 1954. The swelling of montmorillonite. *Discussions of the Faraday Society* 18: 120–134.

Norrish, K., and H. Rosser. 1983. Mineral phosphate, p. 335–361. In: *Soils: An Australian Viewpoint*. CSIRO/Academic Press, Melbourne.

Oades, J.M., and A.G. Waters. 1991. Aggregate hierarchy in soils. *Soil Research* 29: 815–828.

Olsen, A. 1987. Low technology water purification by bentonite clay and *Moringa oleifera* seed flocculation as performed by Sudanese villages: Effects on *Schistosoma mansoni cercarae*. *Water Research* 21: 517–522.

Paola, A., B. Pierre, C. Vincenza, and V. Bruce. 2016. Short term clay mineral release and re-capture of potassium in a Zea mays field experiment. *Geoderma* 264: 54–60.

Papoulis, D., P. Tsolis-Katagas, and C. Katagas. 2004. Progressive stages in the formation of kaolin minerals of different morphologies in the weathering of plagioclase. *Clays and Clay Minerals* 52: 275–286.

Parfitt, R.L. 2009. Allophane and imogolite: Role in soil biogeochemical processes. *Clay Minerals* 44: 135–155.

Pittan, B. 2005. Einfluss des Polykation Poly-DADMAC auf die Verminderung von Auswaschungsverlusten und Nachweis Dessen Wirkung auf Nährstoffaufnahme und Nährstoffverfügbarkeit bei Weizen Mittels Iotopenverdünnung, Diplomarbeit. Christian-Albrechts Universität zu Kiel.

Pronk, G.J., K. Heister, G.-C. Ding, K. Smalla, and I. Kögel-Knabner. 2012. Development of biogeochemical interfaces in an artificial soil incubation experiment; aggregation and formation of organo-mineral associations. *Geoderma* 189–190: 585–594.

Quirk, J.P., and R.K. Schofield. 1955. The effect of electrolyte concentration on soil permeability. *Journal of Soil Science* 6: 163–178.

Regelink, I.C., C.R. Stoof, S. Rousseva, L. Weng, G.J. Lair, P. Kram, N.P. Nikolaidis, M. Kercheva, S. Banwart, and R.N.J. Comans. 2015. Linkages between aggregate formation, porosity and soil chemical properties. *Geoderma* 247–248: 24–37.

Reid-Soukup, D.A., and A.L. Ulery. 2002. Smectites, p. 467–499. In: J.B. Dixon, and D.G. Schulze (eds.), *Soil Mineralogy with Environmental Applications*. Soil Science Society of America, Madison, Wisconsin.

Rengasamy, P. 2002. Sodic soils, p. 1210–1212. In: R. Lal (ed.), *Encyclopedia of Soil Science*. Marcel Dekker, New York.

Rengasamy, P., and A. Marchuk. 2011. Cation ratio of soil structural stability (CROSS). *Soil Research* 49: 280–285.

Richards, A.E., R.C. Dalal, and S. Schmidt. 2009. Carbon storage in a Ferrasol under subtropical rainforest, tea plantations, and pasture is linked to soil aggregation. *Soil Research* 47: 341–350.

Roberts, M. 2015. MRI scans can deliver cancer therapy. www.bbc.co.uk/news/health-33957105 (accessed August 3, 2017).

Robertson, R.H.S. 1986. *Fuller's Earth: A History*. Volturna Press, Hythe, Kent, United Kingdom.

Roth, C.B., M.L. Jackson, and J.K. Syers. 1969. Deferration effect on structural ferrous-ferric iron ratio and CEC of vermiculites and soils. *Clays and Clay Minerals* 17: 253–264.

Rumpel, C., and I. Kögel-Knabner. 2011. Deep soil organic matter – A key but poorly understood component of terrestrial C cycle. *Plant and Soil* 338: 143–158.

Saggar, S., A. Parshotam, G.P. Sparling, C.W. Feltham, and P.B.S. Hart. 1996. 14C-labelled ryegrass turnover and residence times in soils varying in clay content and mineralogy. *Soil Biology and Biochemistry* 28: 1677–1686.

Saidy, A.R., R.J. Smernik, J.A. Baldock, K. Kaiser, and J. Sanderman. 2013. The sorption of organic carbon onto differing clay minerals in the presence and absence of hydrous iron oxide. *Geoderma* 209–210: 15–21.

Saleth, R.M., A. Inocencio, A. Noble, and S. Ruaysoongnern. 2009. Economic gains of improving soil fertility and water holding capacity with clay application: The impact of soil remediation research in Northeast Thailand. IWMI Research Report 130. International Water Management Institute, Colombo, Sri Lanka.

Sarkar, B., Y. Xi, M. Megharaj , G.S.R. Krishnamurti , M. Bowman , H. Rose, and R. Naidu. 2012. Bioreactive organoclay: A new technology for environmental remediation. *Critical Reviews in Environmental Science and Technology* 42: 435–488.

Schaefer, C.E.G.R., R.J. Gilkes, and R.B.A. Fernandes. 2004. EDS/SEM study on microaggregates of Brazilian Latosols, in relation to P adsorption and clay fraction attributes. *Geoderma* 123: 69–81.

Schaetzl, R., and W. Harris. 2012. Spodosols, p. 33-113–33-127. In: P.M. Huang, Y. Li, and M.E. Sumner (eds.), *Handbook of Soil Sciences: Properties and Processes*, 2nd edn. CRC Press/Taylor & Francis Group, Boca Raton, Florida.

Schapel, A., P. Marschner, and J. Churchman. 2018. Clay amount and distribution influence organic carbon content in sand with subsoil clay addition. *Soil & Tillage Research* 184: 253–260.

Schapel, A., P. Marschner, and J. Churchman. 2019. Influence of clay clod size and number for organic carbon distribution in sandy soil with clay addition. *Geoderma* 335: 123–132.

Schwertmann, U., and R.M. Taylor. 1989. Iron oxides, p. 379–438. In: J.B. Dixon, and S.B. Weed (eds.), *Minerals in Soil Environments*, 2nd edn. Soil Science Society of America, Madison, Wisconsin.

Shaller, P., D. Sykora, M. Doroudian, and G.J. Churchman. 2016. Rapid *in situ* conversion of late-stage volcanic materials to halloysite implicated in catastrophic dam failure, Hawaii. *Clay Minerals* 51: 499–515.

Sills, I.D., L.A.G. Aylmore, and J.P. Quirk. 1974. Relationship between pore size distributions and physical properties of clay soils. *Soil Research* 12: 107–117.

Singh, M., B. Sarkar, B. Biswas, J. Churchman, and N.S. Bolan. 2016. Adsorption-desorption behavior of dissolved organic carbon by soil clay fractions of varying mineralogy. *Geoderma* 280: 47–56.

Sipos, P., V.K. Kis, R. Balázs, A. Tóth, I. Kovács, and T. Németh. 2018. Contribution of individual pure or mixed-phase mineral particles to metal sorption in soils. *Geoderma* 324: 1–8.

Six, J., H. Bossuyt, S. Degryze, and K. Denef. 2004. A history of research on the link between (micro) aggregates, soil biota, and soil organic matter dynamics. *Soil and Tillage Research* 79: 7–31.

Smalley, I.J., C.W. Ross, and J.S. Whitton. 1980. Clays from New Zealand support the inactive particle theory of soil sensitivity. *Nature* 288: 576–577.

So, H.B. 2002. Slaking, dispersion, and crust formation, p. 1206–1209. In: R. Lal (ed.), *Encyclopedia of Soil Science.* Marcel Dekker, New York.

Sojka, R.E., D.R. Upchurch, and N.E. Borlaug. 2003. Quality soil management or soil quality management: Performance versus semantics. *Advances in Agronomy* 79: 1–68.

Spain, A.V. 1990. Influence of environmental conditions and some soil chemical properties on the carbon and nitrogen contents of some tropical Australian rainforest soils. *Soil Research* 28: 825–839.

Stewart, C.E., A.F. Plante, K. Paustian, R.T. Conant, and J. Six. 2008. Soil carbon saturation: Linking concept and measurable carbon pools. *Soil Science Society of America Journal* 72: 379–392.

Stockmann, U., M.A. Adams, J.W. Crawford, D.J. Field, N. Henakaarchchi, M. Jenkins, B. Minasny, A.B. McBratney, V.D. de Courcelles, K. Singh, I. Wheeler, L. Abbott, D.A. Angers, J. Baldock, M. Bird, P.C. Brookes, C. Chenu, J.D. Jastrow, and M. Zimmermann. 2013. The knowns, known unknowns and unknowns of sequestration of soil organic carbon. *Agriculture, Ecosystems & Environment* 164: 80–99.

Sumner, M.E. 1993. Sodic soils: New perspectives. *Soil Research* 31: 683–750.

Syers, J.K., M.L. Jackson, V.E. Berkheiser, R.N. Clayton, and R.W. Rex. 1969. Eolian sediment influence on pedogenesis during the Quaternary. *Soil Science* 107: 421–427.

Tisdall, J.M., and J.M. Oades. 1982. Organic matter and water-stable aggregates in soils. *Journal of Soil Science* 33: 141–163.

Velde, B., and P. Barré. 2010. *Soils, Plants and Clay Minerals: Mineral and Biologic Interactions.* Springer, Heidelberg.

Velde, B., and I.C. Druc. 1999. *Archaeological Ceramic Materials.* Springer-Verlag, London.

Vermeer, D.E., and D.A. Frate. 1979. Geophagia in rural Misssissippi: Environmental and cultural contexts and nutritional implications. *The American Journal of Clinical Nutrition* 32: 2129–2135.

Vogel, C., C.W. Mueller, C. Höschen, F. Buegger, K. Heister, S. Schulz, M. Schloter, and I. Kögel-Knabner. 2014. Submicron structures provide preferential spots for carbon and nitrogen sequestration in soils. *Nature Communications* 5, article 2947: 1–7.

Ward, P.R., and J.M. Oades. 1993. Effect of clay mineralogy and exchangeable cations on water-repellency in clay amended sandy soils. *Soil Research* 31: 351–364.

Warr, L.N., A. Friese, F. Schwarz, F. Schauer, R.J. Portier, L.M. Basirico, and G.M. Olson. 2016. Experimental study of clay-hydrocarbon interactions relevant to the biodegradation of the Deepwater Horizon oil from the Gulf of Mexico. *Chemosphere* 162: 208–221.

Warr, L.N., J.N. Perdrial, M.-C. Lett, A. Heinrich-Salmeron, and M. Khodja. 2009. Clay mineral-enhanced bioremediation of marine oil pollution. *Applied Clay Science* 46: 337–345.

Wesley, L.D. 1973. Some basic engineering properties of halloysite and allophane clays in Java, Indonesia. *Géotechnique* 23: 471–494.

Wesley, L.D. 1977. Shear strength properties of halloysite and allophane clays in Java, Indonesia. *Géotechnique* 27: 125–136.

Wesley, L.D. 2001. Consolidation behaviour of allophane clays. *Géotechnique* 51: 901–904.

Wienhold, B.J., G.E. Varvel, and J.W. Doran. 2005. Soil quality, p. 349–353. In: D. Hillel (ed.), *Encyclopaedia of Soils in the Environment.* Elsevier, Amsterdam.

Wright, T. 2007. Attenuating nitrogen, phosphorus and dissolved organic carbon movement in runoff: Poly-DADMAC application to soil. BSc (Hons) thesis, University of Adelaide.

Yang, K., and B. Xing. 2012. Soil physicochemical and biological interfacial; processes governing the fate of anthropogenic organic pollutants, p. 9-1–9-44. In: P.M. Huang, Y. Li, and M.E. Sumner (eds.), *Handbook of Soil Sciences, Resource Managements and Environmental Impacts*, 2nd ed. CRC Press/ Taylor & Francis Group, Boca Raton, Florida.

Young, I.M., and J.W. Crawford. 2004. Interactions and self-organization in the soil-microbe complex. *Science* 304: 1634–1637.

Yuan, G.D., B.K.G. Theng, G.J. Churchman, and W.P. Gates. 2013. Clays and clay minerals for pollution control, p. 587–644. In: F. Bergaya, and G. Lagaly (eds.), *Handbook of Clay* Science, 2nd edn. Developments in Clay Science, vol. 5. Elsevier, Amsterdam.

15 Summary

Begin at the beginning, the King said, gravely, and go on till you come to the end, then stop.

Lewis Carroll (1832–1898)

15.1 SOILS (FROM CHAPTER 1)

Soils began in geological time with the advent of vascular plants on the Earth's surface, but some clay minerals preceded soils. Weathering results in the breakdown of minerals from rocks when they are out of equilibrium at earth surface conditions. Silicates are the most abundant minerals in soils. They break down in water to an extent and ease that reflects the degree to which they are out of equilibrium with the earth environment, so those originally formed from magmatic rocks break down first and those from sediments only slightly or not at all. Their disintegration reflects the strength of their silicate bond so that those with cations other than Si or Al are most susceptible to disintegration, and quartz, with only Si-O bonds, is especially stable. It may be that there is no soil without clay.

15.2 CLAYS (FROM CHAPTER 2)

Clays in soils are best defined as "secondary inorganic phases of clay (<2 μm) size in soils, regardless of their crystalline or nanocrystalline order, or their degree of disorder". Thus, they include a variety of phyllosilicates as well as oxides, oxyhydroxides or hydroxides (collectively 'oxides'). They contribute almost all of the surface area, and, along with organic matter, all of the charge to soils. Clays do not occur as such in undisturbed soils, but instead are associated into larger units. Furthermore, soil clays may not be well represented by clays from deposits having the same names.

Phyllosilicates are made up of tetrahedral sheets, comprising Si-O covalent bonds, and octahedral sheets, generally based on Al-O and Al-OH bonds, but often also containing other cations, either divalent or trivalent, in place of Al. Al may also replace some of the Si in the tetrahedral sheet. Sheets are associated in either of two arrangements: (1) 1:1 Si:Al with one tetrahedral T and one octahedral O sheet linked together to give TO, or (2) 2:1 Si:Al with one octahedral sheet sandwiched between two tetrahedral sheets to give TOT. When divalent cations dominate in the octahedral sheet, it is a trioctahedral arrangement (three cations for each $O_{10}(OH)_2$ unit), and when trivalent cations dominate, it is a dioctahedral arrangement (two cations per $O_{10}(OH)_2$). The layer made up of these sheets has a charge given by the extent of replacement of Al in the octahedra by other, usually divalent, cations, leading to a net negative charge and/or by the extent of replacement of tetrahedral Si by Al, similarly conferring a negative charge on the layers. TO structures show little or no replacement of Si and/or Al in the structure.

The charge balance in 2:1 phyllosilicates can be affected by oxidation or reduction of octahedral cations, particularly iron. This can affect structural coherence and/or migration of cations out of the layers. Potassium ions fit very closely in the interlayers of 2:1 phyllosilicates. Cations from the bathing solution can be taken up by 2:1 minerals; most are taken up into the interlayer region. If the exchanging cation is hydrated, the expanded phases, vermiculite or smectite, are formed upon complete replacement of the original cations, typically K^+ in a mica or illite, or an Al-hydroxy ion in a chlorite. If the product is a smectite, it will expand further upon addition of a polar organic molecule, which then occupies the interlayer region.

Cation exchange is governed by the law of mass action. Cation exchange takes place within the interlayers of expanded 2:1 minerals and at the edges of all clays. The amount of edge exchange depends on particle size and is affected by pH.

15.3 FORMATION OF CLAYS IN SOILS (FROM CHAPTERS 3–7)

The fundamental principle of soil formation, due to Dokuchaev, has soils related to five factors of formation: time, climate, organisms, relief and parent material. The earliest weathering changes are probably inorganic, involving oxidation of Fe and Mn and loss of alkali and alkaline earth elements and silicon. Ferrihydrite is often formed early, and Al is taken up by nanocrystalline minerals, mainly allophane. Organic matter is built up early in soil formation. Transformations of chlorite and mica, and especially trioctahedral types, including biotite, occur early, and smectites can form from this process. A range of mixed-layer minerals can also form. Soils may include inputs from loess and also fine dust.

Dissolution in slightly acid rainwater – the product of the solution of CO_2 in water – is probably the most rapid process in the alteration of primary minerals. Feldspars can give rise to a wide variety of minerals following dissolution and recrystallisation. Micas and chlorites are essentially altered by changes within their interlayer regions. Little change produces an illite from micas and more change may lead to mixed-layer minerals such as illite-vermiculite and chlorite-smectite. Replacement of all potassium in interlayers of micas or illites leads to vermiculite and loss of some ions from vermiculite layers produces smectites. The results of the various alteration processes are likely to be heterogeneous.

New clay minerals, silicates and oxides, form by the destabilisation of similar materials in rocks that are exposed to surface chemical conditions of altering aqueous solutions that are slightly acidic and saturated with atmospheric oxygen. The processes that affect alteration are diffusion of hydrogen ions into the structures, or that of electrons, affecting oxidation. The interactions are slightly different depending upon the amount of water present and its residence time in the alteration system. The less time the solutions are in contact with the minerals, the more superficial the interactions.

When the soil-forming rocks are sedimentary, many of the raw materials for alteration may already be secondary minerals, having been formed from primary minerals by alteration in one or more previous weathering cycles. Transformation of the 2:1 and 2:1:1 (chlorite) minerals from these minerals in sediments and sedimentary rocks leads to expanded soil clay phases of these minerals. These changes are common in more or less flat-lying continental basins.

Amphiboles, pyroxenes and olivine tend to produce trioctahedral smectites in closed systems within disintegrating rocks but are replaced by dioctahedral smectites and other minerals, including kaolinite, halloysite, gibbsite and iron oxides, in open systems. Serpentines also yield trioctahedral smectites in closed systems, then vermiculite, chlorite-vermiculite and, ultimately, kaolinite and quartz, in open systems.

Alteration profiles are also heterogeneous. The lower portion (or 'alterite' zone) often overlies the saprolite (or saprock), which comprises disintegrated rock and includes secondary minerals. The upper portion of the profile (the 'soil' zone) reflects the influence of plants. Minerals tend to become oxidised in the uppermost horizon (A horizon), hence few trioctahedral minerals occur there.

Iron generally occurs as Fe(II) in primary minerals and can be oxidised to Fe(III) simply by drying, while Mn(II) is oxidised to Mn(IV). Iron oxides probably occur in all soils.

Some minerals, including carbonates and quartz, are highly insoluble and dissolve congruently. Silicates tend to dissolve incongruently, and the soluble components tend to recrystallise close to their parent primary minerals, often occurring as polymorphs of these minerals.

The rate at which fluids move through the altering minerals greatly influences reaction rate. This is affected by topography.

Primary minerals are subject to cracking as a result of thermal expansion and contraction. Soil is produced by the action of living organisms on roots, leaves and other biological substrates and on finely ground minerals from rocks. The interface between water and rock occurs deeper in the profile with time. Weathering begins in the saprock stage. In the alterite zone above this, the material becomes more homogeneous with time. The soil zone above this generally contains phyllosilicates and oxides. Some of this material is moved downwards and accumulates in the B horizon.

The dominant characteristics of soil arise from interactions of organic matter – primarily from plants and also microbes – with silicate and oxide minerals. Soil that is stabilised by plants becomes the primary substrate for further plant growth. Without plant life – in the past or present – there is no soil.

Alteration of minerals occurs as a result of the diffusion of hydrogen ions, causing exchange of cations and hydrolysis, and of electrons into the mineral structures, causing oxidation. The course of these reactions is affected by the amount of water present and its residence time in the alteration system. The older the profile, the deeper the alterite zone. Soils can result from either bottom-up or top-down alterations. In order for new clay minerals to be formed out of solutions of dissolved elements, the solution must travel sufficiently slowly for recrystallisation to occur. The important climate factor is the availability of water. Essentially water plus rock makes clay, and the more diluted the solutions, the more clay will occur.

The ionic exchange of hydrogen ions with K, Na, Ca and Mg can result in a gel with a strongly modified structure that eventually becomes a clay mineral. Silica-rich 2:1 minerals tend to form first, followed by 1:1 minerals, then Al or Fe oxides when Si disappears from the system. The formation of new minerals depletes the elements in solution, ensuring that more dissolution takes place.

Ease of removal of cations is related to the electrostatic valency of the cation, denoted by the ratio of its valency to its coordination number. Organic compounds, pH and temperature all affect the rate of dissolution. Dissolution often occurs within cracks and fissures in the rocks. These make up a wide range of 'plasmic microsystems'. There is a wide range of products, reflecting the availability of water within these, which range from closed cracks with little water to gravity-fed systems with rapidly flowing water. Many different products can be formed from the one type of primary mineral, reflecting many different environments of alteration, including the other minerals undergoing alteration. An early promise of the prediction of minerals formed in contact with solutions using equilibrium thermodynamics appears to founder on the many assumptions made, including that of purely inorganic systems, and on the experimental difficulties of determining the thermodynamic data for many minerals and also of obtaining meaningful soil solutions. In any case, given enough time, the kaolinite–gibbsite–iron oxide suite of minerals will dominate the products.

Minor modifications of 2:1 phyllosilicates occur by transformation, largely in the interlayer region, while the large part of the covalently bonded silicate structure remains intact. Initiation of transformation largely occurs through oxidation of Fe and/or Mn divalent ions that changes the charge balance on the 2:1 structure. Inhomogeneous minerals with apparent mixed layering can result. Contact with aqueous solutions affects the composition of the interlayer regions of 2:1 minerals. Protons from solutions play a key role, especially in the loss of potassium ions from interlayers. It appears that direct transformations might also occur sometimes from 2:1 to 1:1 minerals in the solid state.

Some soil minerals, notably illites, defy conventions for the naming of minerals that have been agreed upon for type minerals. This illustrates the particular peculiar properties of soil clays.

Acidification accompanies plant growth because roots take up an excess of cations over anions and also because of the presence of organic acids, hence protons play a key role in the release of K^+ from 2:1 minerals by transformation. It has been found that there is an apparent reversal of the transformation reaction as a result of plants uplifting K^+ from lower horizons to re-form illites in order that they can supply potassium to the plants. Uplift of Si can also occur following the decomposition of plants that contained phytoliths. There thus appears to be a symbiotic relationship between plants and 2:1 aluminosilicates.

Crystalline minerals have been seen to have formed and also altered as a result of microbial activity. Soil organic matter, like microbes, occurs in patches on the mineral surface. Most occur in micropores 10–10^3 μm in diameter, and particularly in their openings, as well as in biologically active areas like the rhizosphere. Clays may play an ecological function in the metabolism of microorganisms. In exchange, microorganisms can both reduce Fe(III) and oxidise Fe(II) in clay minerals – both 1:1 and 2:1 minerals – thereby aiding their alteration and transformation.

15.4 TYPES OF CLAYS AND THEIR ORIGINS (FROM CHAPTERS 8 AND 9)

Illites in soils mainly originate from the partial transformation of micas. They can also form out of solution by neogenesis. Vermiculites form by the complete transformations of micas and chlorites. Vermiculites are most often dioctahedral in surface soils but may be trioctahedral at depth, in the alterite zone. At pHs between 4.6 to 5.8 they often attract hydroxy-Al ions into their interlayers to form non-expanding pedogenic chlorite, among other names such as HI minerals. They may also be further transformed, usually at low pHs, to smectites. Incomplete transformation of illites is common, leading to mixed layers in crystals ('interstratifications') comprising a variety of combinations of layers of illite with those of vermiculite or smectite and often also HI minerals.

Smectites arise by neogenesis from solutions with high concentrations of Si and also basic cations, usually as a result of low rainfall and/or poor drainage, from a variety of rock types but especially from calcareous and volcanic parent materials. They also form by the transformation of micas and chlorites, usually by leaching at low pHs. Most soil smectites are dioctahedral. However, they tend to have more tetrahedral Al and octahedral Fe than smectites in bentonites. It has also been shown that the formation of a soil on a bentonite substrate led to most of the bentonite in the soil zone being altered to a smectite with a smaller grain size and indications of some transformation to more illitic phases.

Kaolin minerals may occur as interstratifications with smectites in intermediate sites on slopes or in intermediate horizons in profiles which have more smectitic minerals in lower horizons and more kaolinitic horizons at the surface. They could be formed by either addition, whereby hydroxy-alumina interlayers form and smectite sheets are cleaved to give 2:1 to 1:1 interstratifications, or by subtraction wherein some smectite sheets are dissolved with the same result. Interstratifications of kaolin and smectite have most likely been overlooked in many studies.

Halloysite-smectites can derive from a different mechanism to kaolinite-smectites. They can be a product of volcanic ash alteration with some restricted drainage. Kaolin minerals can also form interstratifications with vermiculite and mica/illite and can participate in interstratifications with more than one other crystal type, including illite and smectite together.

Kaolinite is the most common secondary phyllosilicate in soils and has a TO structure. In soils, kaolinites tend to be highly disordered, smaller in particle size, higher in surface area and charge and more irregular in shape than kaolinites in monomineralic geological deposits. Kaolinites are usually the products of the strong weathering of many different minerals. Both kaolinites and halloysites formed in soils are smaller and less well-ordered than those from mineral deposits, reflecting the more heterogeneous composition of the solutions from which they crystallised. Kaolinites are common in warm, humid climates in which there is substantial (but not strong) leaching.

Halloysite can form from a variety of rock types It has a kaolin layer structure (TO) that has or has had water in its interlayers. It occurs in a variety of particle shapes but is often tubular. Compared with kaolinites, halloysites tend to form at low pH and in an environment that is always wet (but not strongly leached). The octahedral layer, on the inside of tubes, has a positive charge at pH <8.5, while the tetrahedral layer, on the surface of tubes, has a negative charge. Together, they probably attract water molecules electrostatically into the interlayer region. The size of halloysite crystals has depended on the amount of impurities in the zone of formation. When impurities such as oxides of Fe and Mn are abundant in the zone of formation, halloysite crystals formed there comprise small particles (tubes). They are large (long tubes) in zones of formation without any impurities. It is likely that organic matter would also serve to constrain crystals (of halloysite, or other phyllosilicates) in the same way as Fe and Mn oxides, if present in the zone of formation.

Halloysite is often found at depth in weathering profiles, with kaolinite more common nearer the surface. On volcanic ash, drainage sequences show that halloysite forms under low throughflow of water, often from lower rainfall, while allophane results when throughflow is higher, associated often with higher rainfall.

Imogolite and the much more common nanocrystalline mineral allophane consist of isolated tetrahedra inside a coiled spherule of alumina. Allophane varies in composition from 1:2 to 1:1 Si:Al. It forms by strong leaching of volcanic glass and in the lower B horizons of podzolised soils. It is often found in the upper horizons of profiles from volcanic parent materials, with halloysite in lower horizons. Generally speaking, volcanic parent materials give rise to allophane under strong leaching, leading to low Si concentrations in solution, and when the substrate is thin, hence free drainage. They give rise instead to halloysite (or kaolinite on seasonal drying) rather than allophane under lower rainfall, so leaching is less, giving a build-up of Si in solution and on thick substrate, which impedes drainage. They may produce allophane in upper parts of weathering profiles (and catena) and halloysite in lower parts of these. The conformation of Al and Si in the structures of allophane and halloysite are different, so conversion from one to another takes place through dissolution and recrystallisation, rather than transformation in the solid phase.

Palygorskite and sepiolite often occur in the parent materials of soils in arid, calcareous environments and are largely found in lower horizons of the soils. Palygorskite, but not sepiolite, has been formed in some soils. They often appear in association with smectites.

Iron oxides (including oxyhydroxides) are almost ubiquitous in soils. They result from the oxidation of Fe (II) in primary minerals. Goethite is the most common iron oxide and is most often found in cool and temperate climates. Hematite is common in soils in warm and hot climates. Nanocrystalline ferrihydrite is widespread in soils except where hematite is present. It is formed when oxidation of Fe (II) occurs slowly and there are impurities in the forming solution. Lepidocrocite forms when the rate of oxidation is particularly slow, as in seasonally anaerobic soils. Maghemite is the product of heating other Fe oxides. Acid sulphate soils may contain schwertmannite and/or green rusts. Iron oxides are often closely associated with other minerals, especially aluminosilicates.

Gibbsite is the most common aluminium oxide, oxyhydroxide or hydroxide in soils. It forms when Si is in short supply, so under strong leaching. It may be inherited from parent saprolites. Al oxyhydroxides, like boehmite, and oxides occur under very strong weathering and, as corundum after heating, as in bush fires. Manganese oxides, such as birnessite, lithiophorite and todorokite, are widespread in soils but usually in small amounts. Si mobilised by leaching may precipitate deeper in profiles as hard pans, silcretes and as quartz or cristobalite. Biogenic Si occurs as opal, including plant opal (phytoliths). Titanium oxides mainly derive from parent materials, but pseudorutile, from the breakdown of ilmenite, can form rutile and anatase. Zircon, the most common Zr mineral, is usually inherited from parent materials.

Phosphates, from the weathering of forms of apatite or from reactions between aluminosilicates and phosphate fertilisers, are rare in soils. Sulphides, especially pyrite, result from the bacterial reduction of sulphate in seawater. They can re-oxidise to form sulphates, especially jarosite and schwertmannite, on drying. Pyrophyllite, talc and zeolites are rare in soils. Zeolites and also soluble minerals are found in saline soils, especially in arid areas.

In cold zones and in developing soils, such as are found after glaciers have retreated, the common change is the transformation of micas and chlorites to illite, vermiculite, smectite, and interstratifications of these minerals, also sometimes with Al-hydroxy interlayers. Regular interstratifications are confined to soils in cool mountainous areas. On volcanic parent materials, little formation of defined clay minerals occurs, especially when plant growth is poor as a result of the cold climate.

In warm zones, the two climatic factors of precipitation on the wettest month together with a leaching index have explained most mineral changes. Illite-smectites were found to be widespread in surface soils. These are randomly interstratified in all except soils in cold zones. In some areas in warm-cool climates, many clay minerals such as kaolinite and halloysite are inherited from their formation in an earlier, hotter climate. Volcanic parent materials give rise to a different suite of products, most generally allophane and/or halloysite.

All clay mineral types can form in hot zones, but kaolinite is virtually ubiquitous there. Halloysite can be found at depth, particularly in laterite profiles. Mica can be transformed to expanded 2:1 minerals with leaching. Smectites can also form in tropical soils, especially those on basalt, and kaolin-smectites often form in appropriate catena as intermediates between smectites and kaolinite. Parent materials are important in explaining the formation of clays in soils on volcanic materials, which mostly give rise to allophane and/or halloysite; and on serpentinite, which gives rise to chlorite, serpentine, smectite and vermiculite, together with interstratifications.

In general, an important process for clay formation is desilication, with Si:Al ratios changing from 2.0 (for mica/illite, chlorite, vermiculite and smectite and their interstratifications) through 1.5 for kaolin-smectites, 1.0 for kaolinite and halloysite, also 1:1 allophane, then 1:2 allophane and imogolite, with 0.5 as a ratio and finally to gibbsite and boehmite with a zero ratio due to the lack of silica. Volcanic parent materials mostly tend to be involved with alterations over the range encompassing kaolin minerals through to gibbsite. Not all alterations follow each step of this series. Resilication, especially from allophane to halloysite, can occur, and under low rainfall and/or poor drainage, smectite may form from a build-up of Si.

Illite, vermiculite and smectite (and sometimes also kaolin) can form mixed-layer interstratified phases with spacings reflecting the wide range encompassed by their constituent minerals. Computer software has been developed to enable decomposition of X-ray diffraction (XRD) traces in order to apportion minerals into their separate phases, including mixed-layer phases (see Annex).

15.5 ASSOCIATIONS OF CLAYS IN SOILS (FROM CHAPTER 10)

Because of their high reactivity, clay minerals can be associated with many other solids in soils. These include other aluminosilicates and oxides, and, most often, soil organic matter (SOM).

Organic matter in soils is a complex mixture of polymers and smaller, soluble molecules formed by the microbial depolymerisation of larger molecules. Microbes also exude a variety of substances which can bridge between minerals and SOM. Biological entities influence soil formation in a number of ways, most notably through their production of CO_2 to acidify rainwater.

Most of the organic matter in soils is closely associated with minerals. SOM can be associated with almost all mineral phases but shows a preference for poorly crystalline (nanocrystalline) oxides and silicates of Fe and Al. The bonding tends to occur via singly coordinated surface hydroxyl groups. At low pHs, inner-sphere bonds form between SOM and both oxides and nanocrystalline minerals, while both inner-sphere and outer-sphere complexes form between SOM and phyllosilicates. The latter commonly comprise bridges between negatively charged minerals and negatively charged SOM formed by polyvalent cations such as Ca^{2+}. pH governs the mechanism of stabilisation of SOC by minerals and ions.

Associations with organic matter can also take place through co-precipitation with Fe or Al.

Iron oxides and oxyhydroxides and aluminium hydroxide can form close associations in soils, both with themselves and also with phyllosilicates. Fe oxides and oxyhydroxides, in particular, bind together by crystal intergrowth, forming concretions, nodules or cements. Compounds of aluminium and iron are most often found to be closely associated with the surfaces of 1:1 clays rather than of 2:1 clays. Hydroxy-Al ions, often formed under mildly acidic conditions in soils, can be adsorbed by the external surfaces as well as the internal surfaces, as in the case of expandable 2:1 clays. These may then bind clay particles together.

There is a tendency for associations to form with clays because they have unsatisfied charges on the edges of phyllosilicates, but throughout all the surfaces for 'amorphous' (often nanocrystalline) minerals. These arise because covalent bonds are broken at the edges of mostly laminar phyllosilicate particles. In the amorphous materials, bonds, hence charges, tend to be spread over the whole particle.

TEM has shown that organic entities, both SOM and also bacteria and their exudates, such as polysaccharides, are often enclosed by minerals. Minerals of different types and organic materials are often formed together into microaggregates, which protect organic molecules from predation by microbes. They can be long lasting, with the SOM in some found to be several centuries old. Microaggregates can be regarded as the fundamental units of soils.

Fractionation by density has shown the densest material to have the oldest organic matter and also the highest content of Fe. Minerals can also have an anti-bacterial action from the toxicity of their component Al and Fe ions, regardless of whether they are in microaggregates. Minerals can have both quantitative and qualitative effects on OM in soils, and clay minerals may play an active part in the ecology of soils.

Minerals and organic matter and their associations evolved together in the formation of soils.

An Andisol showed regions with smooth surfaces characterised by well-crystallised minerals and containing little C, and contrasting regions with rough surfaces that comprised X-ray amorphous minerals and high concentrations of C. It had also been noted that organic matter added to an arable European soil (a Luvisol) was preferentially attracted to rough surfaces. The observation that it is preferentially held on these surfaces within nanocomposites formed with non-crystalline minerals may apply to Andisols and non-Andisols alike. The occurrence of both rough surfaces, with organic matter associated with minerals and smooth inorganic surfaces means that organic matter is largely associated with minerals, but minerals often occur devoid of any organic matter. Generally, most surfaces in soils comprise minerals alone.

SOM is held on mineral surfaces in patches rather than as an evenly spread layer. When it comes to their affinity for organic matter, not all minerals are equal.

The central role that clays in general play in the formation and stability of aggregates and particularly microaggregates means that clays are also major players in the formation and stability of pores, which are important for both soil and plant functioning. Soils in temperate zones, generally comprising mainly 2:1 minerals, contain a variety of sizes of pores, forming a hierarchy in sizes. They include micropores, capillary pores ('mesopores') and macropores. The capillary pores, from several micrometres to ~1 mm in diameter, are responsible for holding and releasing the water for plants. Soils dominated by 1:1 minerals, including many tropical soils, do not form aggregates in a hierarchy of sizes and instead all pores tend to be small (~20 μm).

The 2:1 minerals (phyllosilicates), through strong interaction by covalent chemical bonding at crystal edges and a less strong ionic bonding on particle surfaces in comparison to other clay types, leads to the formation of clay particle–organic networks that form organo-mineral complexes, which then can combine to form microaggregates with capillary microporosities. The initial smaller units are brought into contact during the drying cycle. In most soils, rainfall is not continuous but periodic, and the surface soils are saturated and subsequently dried, whereupon inter-unit pore space increases. The wetting cycle brings smaller units together through swelling of clays associated with water. This contact allows a new set of chemical bonding conditions between clays and organic materials, while plant growth allows physical binding to occur through entanglement of microaggregates and primary particles by root hairs and fungal hyphae. These units constitute macroaggregates.

The formation of new pore structures is not homogeneously distributed in the soils where organics are present to bind clay particles together, but it becomes more homogeneous as a function of depth. At greater depth there is a tendency to maintain larger aggregate units. In general, the formation of macroaggregates determines to a large extent the fertility of a soil for plants with significant root systems near the surface. Clay mineralogy can be important in the formation of macroaggregates in soils. This aggregation is the basis for the agricultural term 'tilth', which dates from the 12th century.

The differences between soils dominated by 2:1 minerals and those comprising mainly 1:1 minerals arise from the different methods of connection of their particles. Particles in the former are held together through organic matter, with which they link covalently or through polyvalent cations.

Particles in the latter soils are mostly held together by abiotic forces, often in a card-house type of structure. The interparticle links in soils dominated by 1:1 minerals are more brittle than those in soils which mainly comprise 2:1 minerals.

15.6 EXTRACTION OF CLAYS FROM SOIL, AND THEIR IDENTIFICATION AND QUANTIFICATION (FROM CHAPTERS 11 AND 12)

As we have seen, clays tend to occur in associations, and dissociation from their attachments is usually necessary to enable their characterisation, but the process can cause artefacts.

Extraction of clays involves both the disaggregation of soils and the dispersion (or peptisation) of the disaggregated particles. The common procedure for disaggregation comprises both oxidation of organic matter and reduction (with chelation) of mainly Fe oxides. Also, carbonate cements need to be removed from calcareous soils (by acidification) and salts from saline soils (by dilution). Dispersion is carried out by either saturating the soil with sodium or by the addition of (alkaline) sodium hexametaphosphate ('Calgon'). To obtain adequate dispersion of soils with variable charge it is necessary that the pH of the suspension be either well above, or below, the zero point of charge of the soil components. While there is no absolute clay content for a soil, the complete dispersion possible for each soil best enables its clay content to be meaningful as an index or a surrogate for properties, such as its capacity to retain and supply water and nutrients. The method chosen for complete dispersion may differ between soils.

It should be noted that oxidation with hydrogen peroxide, reduction with dithionite and citrate for chelation can lead to the loss of interlayer hydroxy ions from within clays and only physical treatments – shaking or ultrasonication – should be used to avoid this loss.

X-ray diffraction (XRD) is the main technique for the identification of soil clays. Clays are examined by XRD as both oriented and random samples. XRD is unsuitable for non-crystalline or nanocrystalline phases such as allophane and ferrihydrite. These are identified and measured by the yields of Si and Al (for allophane) and of Fe (for ferrihydrite) from the dissolution of the soil or soil clay fraction in ammonium oxalate. Iron oxides often show weak peaks in XRD traces, and total Fe oxides can be measured by their extraction by sodium dithionite as a reductant and sodium citrate as a chelating agent, often with sodium bicarbonate to buffer pH.

Infrared spectroscopy can most usefully be used to confirm the presence of kaolin minerals, while differential thermal analysis can be used to give a semi-quantitative analysis of the amount of kaolin mineral present in a sample. Compared with clays from deposits, many of which show distinctive particle shapes, soil clays often show indistinctive shapes in electron micrographs. Other techniques used to characterise soil clays and their properties include Mössbauer spectroscopy, nuclear magnetic resonance, electron paramagnetic resonance or electron spin resonance, neutron scattering and photoacoustic spectroscopy.

The poor crystallinity and variability of soil clays mean that all techniques fall short of providing precise and accurate quantitative analyses. In XRD, reflections of X-rays vary as a wave function with angle of diffraction and also vary with variations in the structure of the clays. Even with these factors taken into account, analyses can only be semi-quantitative, and, except in series where qualifying factors may be similar between samples, are often best given in ranges rather than as precise percentages. XRD patterns often show overlapping peaks, especially for the various 2:1 minerals. A computer program is available to decompose the patterns into individual contributing phases (see Annex). Some check on the values from XRD semi-quantitative analyses may be obtained by measuring the cation exchange capacities of soils or soil clays, and using an algorithm with assumed values for pure phases to provide a likely combination of phases giving rise to the overall value.

Infrared spectra in the mid- and visible-near ranges can be used for the identification and estimation of soil clay minerals. Furthermore, there is a widespread use of whole infrared spectra of whole soils as data sets to predict important chemical and physical soil properties following calibration with a training set of soils or a library of spectra. There is scope to apply this approach using whole X-ray diffraction spectra also as data sets.

15.7 SURFACES, SURFACE REACTIONS AND PARTICLE SIZE EFFECTS (FROM CHAPTER 13)

The relative areas of soils and soil clays, as well as 'pure' clay minerals are best given as values for their adsorption of polar liquids, which will be different for, say, ethylene glycol monoethyl ether (EGME) than for water and also, less appropriately, of N_2 gas. If specific surface areas (SSAs) are to be calculated, these will be different according to the adsorbent used and according to the mineralogical composition of the soils. There is probably no absolute measure of SSA of soil mineral particles.

Removal of SOM increases the (specific) surface area of soils probably because SOM may block some sites for uptake of the adsorbing species. If SOM per se has an SSA value, it is considerably less than that revealed by its removal. Metal oxides, oxyhydroxides and hydroxides make a positive contribution to their surface area. The contribution that Al hydroxides and Fe oxides make to the surface area of soils reflects their crystallinity and particle size, which itself relates to crystallinity.

The charges on soil particles may derive from isomorphous substitutions and defects in the layers of aluminosilicate clay minerals, giving permanent charge and/or variable charge from interactions of H^+ and OH- in surrounding and bathing aqueous solutions with surface groups such as Al-OH and Fe-OH on layer silicates and metal oxides, and with those such as COOH on organic matter. Probably all soil clays and hence all soils have variable charge. The extent of total charge that is variable governs – or should govern – the method(s) used to measure charge on soils.

Generally, the charge on soils is measured as cation exchange capacity (CEC) because most soil clays are net negatively charged, as is organic matter. However, some, notably iron oxides, may be net positively charged at (low) soil pHs. In this case, the charge is measured as anion exchange capacity (AEC).

Organic matter generally contributes >25% of the CEC of surface layers of soils. The general rule is that, provided the texture and mineralogy are reasonably similar throughout the depth of a soil, surface horizons, with the most organic matter, tend to have a higher CEC than lower horizons, due to the influence of the SOM. While the CEC is a combination of that of the minerals with that of SOM, the combination may not be by addition since SOM can block some sites on minerals for exchange. The CEC of SOM is highly dependent on pH.

The mineralogy of soils often changes with particle size within the clay (<2 µm) fraction. The few studies of this aspect have generally shown a tendency for kaolinite to increase at the expense of 2:1 minerals and for the crystallinity of minerals to decrease with decreasing size. The smallest particles within the clay fraction (nanoparticles) are likely to represent its most disordered and also the most recently formed or transformed minerals. Nonetheless, there is probably no measure of particle size distribution of a soil that is wholly meaningful for soil properties.

The finest (nano)particles are implicated in the transport of contaminants, nutrients and organic matter into the environment. The minerals that are most likely to be transported are fine materials found as leachates in columns of soils. Smectites appear to be the most easily transported. Small particle size alone is likely to lead to higher hydraulic conductivities than clays with the same mineralogical composition but generally coarser particles.

15.8 ROLE OF SOIL CLAYS IN APPLICATIONS IN AGRICULTURE, THE ENVIRONMENT AND SOCIETY (FROM CHAPTER 14)

Potassium needed for plant nutrition is usually obtained by depletion of interlayer K^+ from micas and illites. The use of new software for decomposing XRD peak profiles has meant that the explanation – and prediction – of K^+ supply to plants may now be feasible.

The capacity of soils for the uptake of P, and also S and N, usually depends on the contents of Fe and Al oxides and allophane. In the case of P especially, these minerals may constrain the anions so strongly that it is difficult for plants to release the nutrient elements.

The role of clays in the formation and stabilisation of microaggregates restrains fine materials against erosion and leads to pores for the supply of water to plants and as spaces for the growth of plant roots. It has been shown that the number of fine pores in a soil differs according to the dominant clay mineral. Minerals – metal oxides as well as phyllosilicates – likely play their main role in the protection of organic matter, hence carbon sequestration, through their role of forming pores via microaggregates. Occlusion of OM within microaggregates and, more particularly those comprising pores, is likely to be more effective for the protection of OM than are dispersed adsorbed coatings of OM on mineral grains. Aggregates and particularly microaggregates explain many advantageous aspects of soils.

Minerals may play an active role in the life of soils. Bacteria can reduce Fe(III) on mineral surfaces; biofilms on clays actively create organo-clays; and the composition of a microbial community was changed when different minerals were added to a soil. Further, DNA analyses showed that bacterial communities on soil particles of the same mineralogy were more similar in composition that those on particles of different mineralogies. Different forms of organic matter (with different functional groups) bind to different mineral surfaces. Short-range order minerals were found to bind more organic matter than other minerals. The mineral matrix may not be inert.

An index of soil (physical) quality that has been found to be useful in explaining effects of agriculture on soils is based on the water characteristic curve, hence pore size distribution of soils. Since clays affect the formation and stabilisation of fine pores, they can also affect soil (physical) quality, which may be the only objective measure of soil quality. The involvement of clays in the formation and stabilisation of pores will only add to the physical quality of soils.

Sandy soils can present a number of problems for agriculture. Among these may be non-wetting following dry weather. This can be alleviated by the addition of clay, either from external sources or from deeper in the B horizons of some profiles (by mechanical means). In any case, sandy soils are improved by the incorporation of clays, and their plant yields are increased as a result. Increases in plant growth lead to increases in organic carbon contents and ultimately should enhance the sequestration of carbon.

Clays in soils – both aluminosilicates and oxides – have the capacity to remove heavy metals from contaminated soils. Although generally hydrophilic, they can be modified with quaternary ammonium compounds to become hydrophobic and organophilic, hence can take up non-ionic organic pollutants such as hydrocarbons. Modification with polycations can impart a positive charge on clays and soils so that they can take up anionic pollutants while at the same time removing non-ionic organic contaminants from water. Since clays, and also organically modified clays or clays treated with organic materials have been shown to remediate oil spills at sea, soils containing clays and organic matter may also be useful for this purpose. Soil clays may themselves act as contaminants when transported by wind and may carry contaminants when leached from soil profiles by water.

All common clays in soils can swell with water and shrink on drying, but smectites swell and shrink extensively. This leads to smectitic clayey soils giving a strong sliding shear in rheology tests relating to tillage, compared with less marked sliding shear for kaolinitic soils and turbulent shear for silty soils. The swelling of smectites in soils can lead to damage to buildings and other structures, and also to landslides, soil creep and erosion. Landslides can also occur with other clays. Halloysites, in particular, can give rise to landslides, especially if composed of spheroidal particles. All clays can disperse when Na occupies exchange sites and amelioration is carried out by the addition of gypsum. Soils rich in clays, especially smectite, and particularly when Na-exchanged, have a low permeability on swelling and can provide barriers in farm dams and to waste in landfills. On the other hand, clays in contact with water in dams for a long period can change, with halloysites particularly having given rise to small spherical particles in this situation, leading to a weakening and breach of the dam structure.

Ancient practices have used clays for a number of medicinal purposes. Soil clays taken from the banks of the River Nile have long been used to flocculate, hence clarify, the river water for drinking and bathing. The clays have been shown to remove viruses, parasites and bacteria, and thereby avoid the spread of debilitating diseases.

Geophagy, or the eating of soils and clays, is an ancient practice that persists today in tropical regions, especially in Africa. It is most prevalent among pregnant women, who may develop a craving for soils and earth. It may have deleterious effects, including anaemia, and topsoils are especially dangerous to eat. Soil clays may be useful for external applications in pelotherapy.

As tubular minerals with positive charges on their inner surfaces at pH <8, the clay mineral halloysite can take up anionic drugs into its tubes, carry them through the body and release them slowly over time. They can be used in the delivery of cancer drugs. It has been suggested that magnetisation of carrier particles (in general) will enable their transport to a tumour or site of infection using MRI machines, available in all large hospitals. Halloysites occur in soils, especially in volcanic zones. They occur in association with Fe oxides in soil, so are already magnetised.

Soil clay minerals as well as coarser soil minerals may enable matching soils from crimes and suspects with those of natural samples, with the aim of linking the scene of the crime to its perpetrator(s).

Clays, including soil clays have been used as ceramic objects by people since Neolithic times and the discovery of these objects or their remains has assisted archaeologists in locating the origin of peoples worldwide. This is because the methods of manufacture of the objects and their uses are often characteristic of particular groups of people. Clays that have a high affinity for organic matter may be sourced for associated DNA, which can be extracted and processed to enable characterisation of plant species that formerly inhabited the area.

15.9 RÉSUMÉ

As shown in the preceding summary sections, this book attempts to illustrate, identify and explain the presence of the different minerals in soils. These include silicates and oxides and oxyhydroxides. Soils are defined here as the surface zone that involves plant growth. Plants are fundamental in the definition of the chemistry of soil clays. The importance of such information is in the observations that soil clays in this regime are the basis for agriculture and other uses for humans. Without this chemical regime of plant–mineral interactions, human occupation of the surface would not have occurred.

In the text, the origin of different clay minerals (silicates and oxyhydroxides and oxides) is indicated as a function of alteration of rocks and other materials at the surface under the influence of rainfall and the resulting interaction of hydrogen-bearing water solutions with rock minerals that become chemically unstable, dissolving partially and creating new hydrous minerals that are stable under such hydrous conditions (soil clay minerals). An important aspect of such interactions is the loss of certain chemical elements and the concentration of others in going from rocks to clay assemblages. The types of clays formed (silicate vs oxides) and their chemical properties are very important to the functioning of soil clay assemblages.

One aspect which is extremely important is the chemical activity of the surfaces of the mineral clay grains in the soils. These relations determine their interaction with organic matter present as small particles that are is fixed on their surfaces and the formation of clay mineral grain assemblages in soils (aggregates). Surface interactions are the basis of the formation of clay aggregates that have structures that retain water, and have active chemical surfaces that retain cations from the altering solutions that are the fundamentals of plant fertilisation such as potassium ions or ammonium ions. These active surfaces are of both mineral and organic particle origin.

An important aspect of such studies is the identification of the different types of clay particles present. which leads to an understanding of their physical and chemical properties and associations with organic matter.

The basic properties that interest farmers are those of water retention (aggregation of mineral and organic matter to form microstructures that hold capillary water), and their ability to retain and release fertilising cations over the period of growth and plant maturity. However, clays from soils have been important for other aspects of human societal development such in the use to make pottery to cook and store agricultural materials, and some use for medical purposes.

Understanding the properties of soil clays and their basis in the chemical and structural aspects of the minerals is a fundamental aspect of understanding the world we live in and have lived in for thousands of years.

BIBLIOGRAPHY

Bergaya, F., and G. Lagaly (eds.). 2013. *Handbook of Clay Science,* 2nd edn. Developments in Clay Science. Elsevier, Amsterdam.

Churchman, G.J., and D.J. Lowe. 2012. Alteration formation and occurrence of minerals in soils, p. 20.1–20.72. In: P.M. Huang, Y. Li, and M.E. Sumner (eds.), *Handbook of Soil Sciences: Properties and Processes*, 2nd edn. CRC Press, Boca Raton, Florida.

Dixon, J.B., and D.G. Schulze (eds.). 2002. *Soil Mineralogy with Environmental Applications.* Soil Science Society of America, Madison, Wisconsin.

Jenny, H., 1994. *Factors of Soil Formation: A System of Quantitative Pedology.* Courier Corporation, North Chelmsford, Massachusetts.

Kleber, M., K. Eusterhues, M. Keiluweit, C. Mikutta, R. Mikutta, and P.S. Nico. 2015. Mineral-organic associations: Formation, properties, and relevance in soil environments. *Advances in Agronomy* 130: 1–140.

Theng, B.K.G., 2012. *Formation and Properties of Clay-Polymer Complexes*, vol. 4. Elsevier, Amsterdam.

Velde, B., and A. Meunier, 2008. *The Origin of Clay Minerals in Soils and Weathered Rocks.* Springer-Verlag, Berlin.

Velde, B., and P. Barré. 2010. *Soils, Plants and Clay Minerals: Mineral and Biologic Interactions.* Springer-Verlag, Berlin.

Wilson, M.J., W.A. Deer, R.A. Howie, and J. Zussman. 2013. *Rock-Forming Minerals.* Volume 3C – *Sheet Silicates: Clay Minerals.* The Geological Society, London.

Annex: Simplified Methods for the Interpretation of X-Ray Diffraction Diagrams of Soil Clay Assemblages

A.1 INITIAL PHASES

Some studies have proposed to use the entire spectrum of X-ray diffraction (XRD) responses for a multiphase assemblage to directly compare the presence of soil clay minerals. This necessitates the use of maximum intensities for readily visible peaks to measure the changes in soil assemblages (Casetou-Gustafson et al., 2018) and to ignore the overlapping peaks that form the 'background'. However, the relative changes in mineral abundance are much more subtle in that the shapes of the peaks are different and they overlap a lot. It is therefore necessary to separate the different components of the combined mineral species and to compare them by their relative surface area. In order to do this one must use a decomposition method to model the peaks that are present and estimate their relative abundance. Such models can be found on operating modules of analysis machines using Raman excitation or others, and perhaps X-ray diffraction. Otherwise MATLAB® gives methods and means of decomposition. However, it is necessary to understand the numerical responses to the material analysed.

Therefore, we present a description of such methods in the following.

A.2 DECOMPOSITION METHODS

X-ray diffraction spectra of clay assemblages are the major means of identifying the mineral types present and, more important, in measuring, quantitatively, the changes from one environment to another of the different components of the clay assemblages. This tool, now routinely available via the Internet and by other means, is the key to any modern interpretation and assessment of clay mineral change and relative abundance. Identification is one thing, reasonably important in itself, but the change effected by different environments in clay abundance is the most important factor and tool that can be used to assess the use of different practices in using plants to effect changes in soils. Soil clays respond, rapidly, to changes in their local chemistry. This local chemistry is often, or can be, determined by the plants and bioagents in the soils. The changes in clays are an indicator of their potential fertility, or infertility, for the eventual growth of other plants. Some plants need certain elements that are found stocked in clays, other plants, other elements or perhaps none at all. The effects of plants on clays can be seen only through a quantitative and comparative analysis of X-ray diffraction spectra of the minerals present.

This has been demonstrated throughout the preceding chapters of this book, at least we hope so. Here we would like to present a short explanation of the principles of the interpretation of spectra using decomposition methods in order to obtain quantitative estimates of mineral type abundance. One must be careful in using such estimates in that the intensities of different minerals depends upon their composition, on their grain size and on the structure and interstratification of different layers, all of which change the peak surface for a given mass of material. Hence absolute abundance or mass is a difficult objective to attain. However, if one looks at a series of samples with approximately the same mineralogy, one can observe the changes of abundance in a relative manner and thus interpret the importance of the different chemical parameters affecting the clay assemblages. Relative change in mineral abundance is a very important estimate, especially in time series experiments, which are very common in agricultural practice. Comparing one sample against another is often more important than an absolute estimation of mineral abundance in a single sample.

The decomposition of complex spectra of electromagnetic signals is an old art. Problems arise when electromagnetic radiations interact with complex (often multiphase) materials. Here the individual bands of intensity, having each a specific intensity distribution, often overlap and present a vague distribution of intensities. Spectra of infrared interaction, Raman interaction and different X-ray energy interactions have been the subject of efforts at extracting the individual components from the complex, overlapping series of bands in a spectrum. The use of relatively proven methods for an interpretation of X-ray diffraction data is not new (Lanson, 1997). However, it is little employed at present in studies of clay minerals in soils. Nevertheless, a certain number of studies have used such methods to interpret the spectra of soil mineral samples.

The problems of interpretation can be seen very clearly in the example of Figure A.1.

In this background subtracted spectrum of a soil clay mineral assemblage one sees two sharp peaks which can be interpreted as diffraction peaks of specific minerals in the soil, illite and hydroxy-interlayer (HI) chlorite. The illite peak is in fact a major feature of the spectrum, but the HI mineral is in reality a minor component. However, the large majority of diffraction intensity is contributed to by other minerals; two types of illite-smectite mixed-layer minerals, S/I and I/S, indicating relative abundance of the illite component; and the poorly crystallised form of illite (PCI). An identification based solely upon identifiable sharp peaks is very misleading in interpreting X-ray diffraction spectra of soil clays. One would say illite plus HI or chlorite when most of the diffraction intensity is contributed by mixed-layer minerals of different sorts. This has been the problem with the interpretation of X-ray diffractograms for the last 50 years. It is almost impossible to estimate the relative abundance of the less well-characterisable minerals present. Clearly it is necessary to go into the problem on a different scale.

A.3 SOME FUNDAMENTAL CONCEPTS OF DECOMPOSITION TECHNIQUES

A.3.1 Background

X-ray diffraction diagrams of clays usually show a highly asymmetric background intensity which increases with decrease in diffraction angle (° 2 theta). This background must be subtracted in order to observe the diffraction maxima for the different components of the clay assemblage. Such subtraction methods are routine in most analysis programs. In Figure A.2 one can see the effect of background and its subtraction in going from A to B.

Given the need for decomposing the complex X-ray spectra of clay minerals, how can one go about the process without making too many errors? If one considers again Figure A.1, there are obviously two sharp maxima, but other peaks are present which overlap and give difficult interpretations upon visual inspection. In Figure A.1 in the lower figure it is clear that the large part of the diffraction intensity of the spectrum is made up of large peaks that often overlap, with the result of an apparent continuum of slightly varying diffraction intensity. How then can one go about identifying the different components of this diffraction intensity?

Initially there are several principles which must be recognised.

A.3.2 Principles for Decomposition

Principle 1. In the region of interest for clay minerals, 4 to 14° 2 theta for copper radiation, or 16–7 Å, *some of the peaks appear to be slightly asymmetrical*, with a slight broadening to the low angle, high d value (Å) side. This can be an effect of peak broadening due to physical effects of X-ray diffraction as is the case in Figure A.2. The asymmetry can be defined by a second peak of minor peak area, usually below 15% of the total peak surface area of the major reflection. The asymmetric

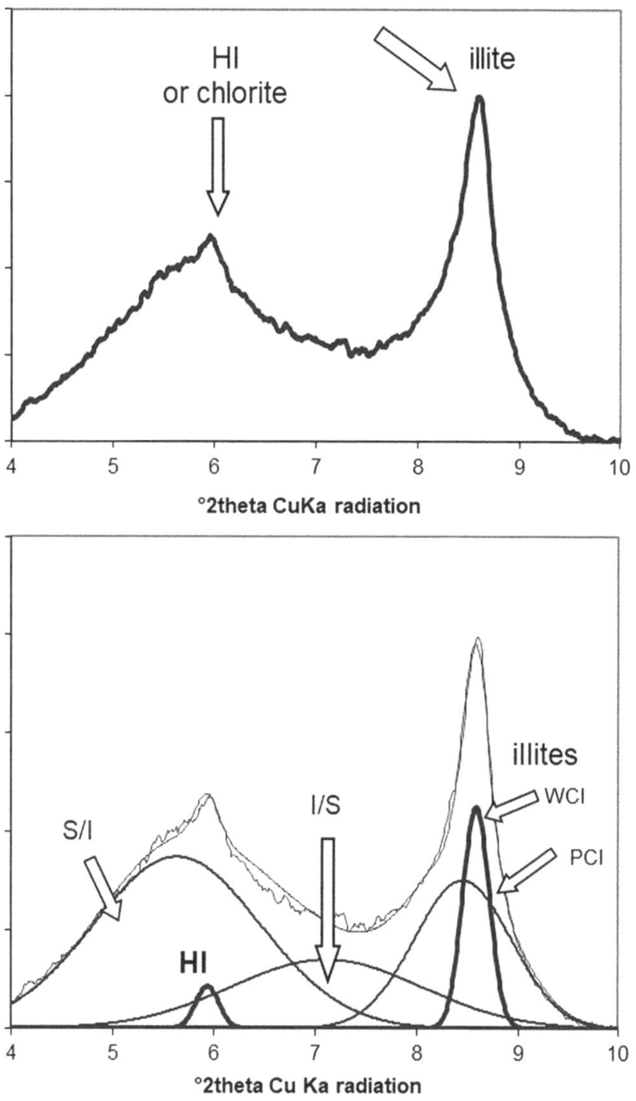

FIGURE A.1 Background-subtracted X-ray diffraction spectrum (top) and its interpretation as a series of overlapping peaks: S/I, smectite-illite; HI, hydroxy-interlayered mineral (Al-chlorite); I/S, illite-smectite; WCI, well-crystallised illite; PCI, poorly crystallised illite

part of the peak can be simulated using a small peak to the low angle, high d value side of the major peak. In the example, the peak is found at 12.5 Å and the asymmetric part of the peak can be simulated with a small band at 14.3 Å. The relative intensity of the small band, simulating the asymmetry of the large peak is 7% of the area of the large peak. Thus, the problem of asymmetry due to physical effect of diffraction is minor. However, asymmetry of peak shape is usually far more important in most clay spectra and thus is due to other effects. *Thus, when there appears to be a small peak widening at the base of a sharp peak, it is most likely due to the presence of another mineral phase.*

Principle 2. It is possible to 'fit' or simulate any complex curve with a suitable number of individual band components. This is the principle of Fourier synthesis analysis. However, there is no use for such results in the identification of specific mineral in a complex assemblage without any basis in mineralogical reality. We are interested in identifying each species present. There is not an infinite number of species present, in fact only a limited number in most cases. We in fact wish to simulate

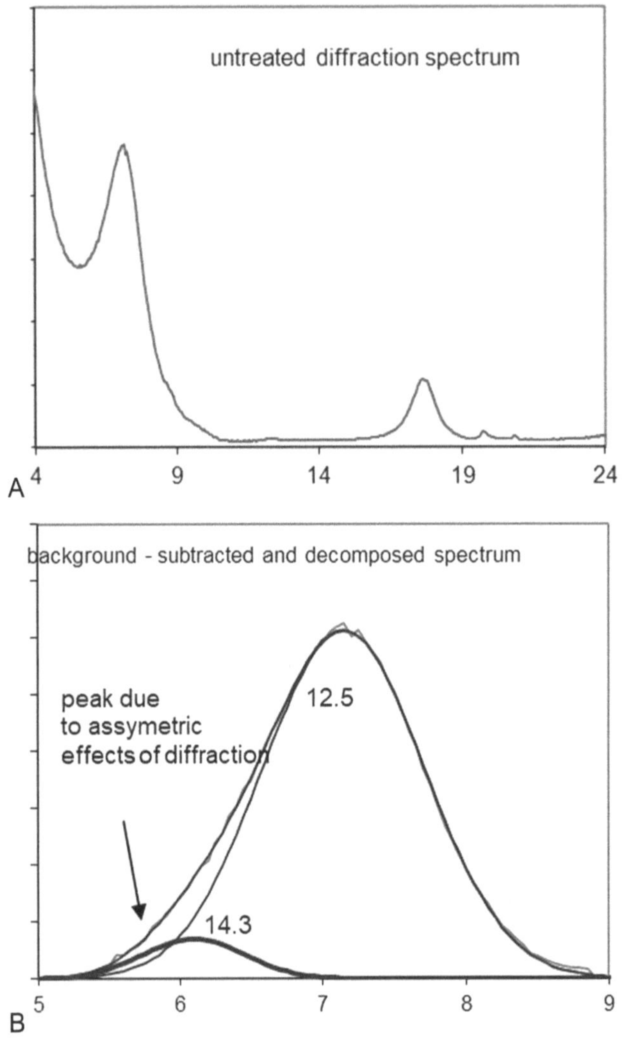

FIGURE A.2 Example of a meta bentonite mono-mineral sample. Background subtraction gives a peak that is almost symmetrical, at near 7° 2 theta Cu radiation, which can be decomposed into one major Gaussian shaped peak and another minor one to the low angle side. The peak is slightly asymmetrical.

a complex spectrum by using the minimum number of components. Hence principle number two is *use the minimum number of bands possible to simulate a complex spectrum.*

Principle 3. Use only bands that could represent a known mineral. Initially this is difficult to follow in that one does not know, *a priori*, what is possible and what is not. Experience is the best teacher, and, above all, use of published, established diffraction components representing different soil clay minerals is recommended.

Principle 4. Peak widths range from 0.2 to 2° 2 theta at half height. Wider peaks are unrealistic. Usually wide peaks are the result of an interstratification of different types of mineral layers in the same crystallite or small crystallite size. If one needs a very large peak to decompose a spectrum, it usually means that there is in fact two peaks present. Peak widening from grain size effects is much less than 2° 2 theta (Figure A.3).

In Figure A.3 the effect of coherent diffracting domain, an approximate estimation of the grain size or order in a crystallite, is shown for illite. This widening effect is especially true of illites in soils. The reason is that the grain size is variable, and some material shows a narrow peak at low

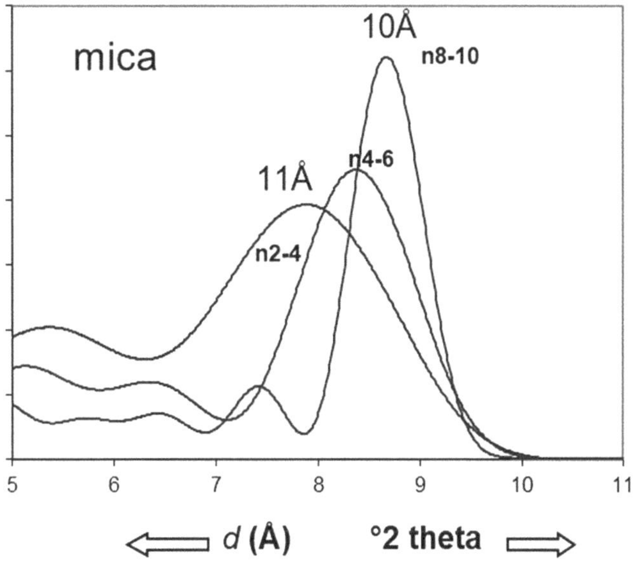

FIGURE A.3 The effect of peak shift due to grain size given as the number of coherent layers diffracting in a crystallite.

d spacings (higher °2 theta) and a larger peak at higher d and lower °2 theta. This effect is usually interpreted (see Meunier and Velde, 2013) as being due to a peak shift and peak broadening due to the dispersion of grain sizes, and perhaps composition of the illite particles. The effect of peak shift due to grain size (number of coherent layers diffracting in a crystallite) is shown in Figure A.3.

The peak shift is not negligible and peak broadening contributes to the well-known apparent asymmetricity of the illite peaks, called in the past illite crystallinity. To our knowledge, peak asymmetry, with a fitted band of more than 15% of the surface area of the major peak, is due to a population of different particle sizes of the same clay species. This is typical of illite and kaolinite but we have not identified it in cases of other minerals (Figure A.4) (see Brindley and Brown, 1980).

In the example shown for a diagenetically formed illite, there is a decided asymmetry of the peak at 10 Å with a width at half height of near 0.7° 2 theta caused by a peak centred near 10.7 Å, which is significantly wider, 1.4° 2 theta. Hence it is common to need two peaks to fit the illite peaks at 10 Å and often the kaolinite peaks at 7.21 Å. In the case of illite the narrower peak is called WCI, or well-crystallised illite, and the wider peak is PCI, or poorly crystallised illite (PCI).

A.3.3 Decomposition Examples

All spectra should be background subtracted before an attempt is made to decompose the complex spectrum into its components. In the following examples we show the untreated spectrum before treatment, and begin the decomposition with the background-subtracted spectrum.

The easiest procedure for decomposing a complex spectrum is to work from the edges to the centre, like eating a sandwich. Once the outermost peaks are defined, one can make an attempt to determine peaks in the centre. This usually means going from the 10 Å illite peak to the 14–15 Å peaks representing the HI or smectite phases. We advise using the divalent ion air-dried state for the initial determination of mineral species present. It is important to use a specific protocol for curve fitting (decomposition) in that the proposition one uses to establish the best fit of a series of peaks to a complex curve depends upon a knowledge of the phases (peaks) likely to be present. In most soils one finds kaolinite present, but the position of kaolinite is such that it is detached (little or no overlap with other peaks) and it can be treated as an isolated example, usually with two peaks as indicated

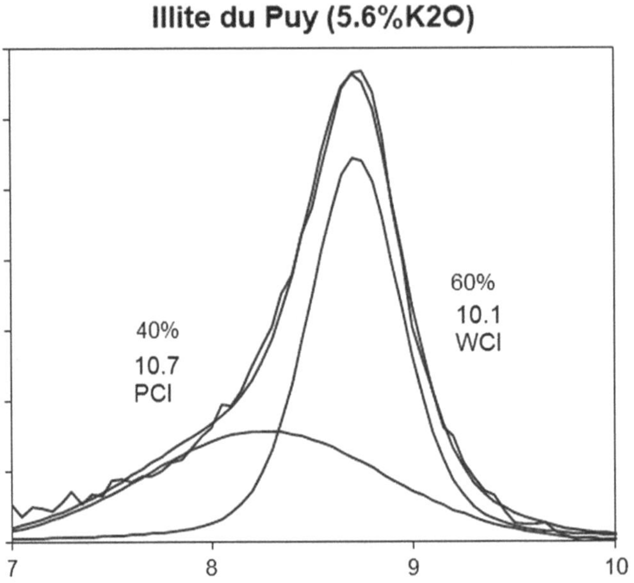

FIGURE A.4 Examples of peak asymmetry for illites. WCI, well-crystallised illite; PCI, poorly crystallised illite.

earlier. We consider the 10–15 Å region (4–12° 2 theta using Cu radiation) as the most complicated and, for plants, the most critical part of an X-ray diffractogram necessary for the identification of the soil clay assemblages. It is here that one finds the 2:1 minerals, which can contain potassium Ca or Mg hydrated ions or Al-OH ions.

In Figure A.5 we show a case of complex mineralogy where two peaks are evident in the spectrum: one at 10 Å and the other at 14.2 Å, indicated by arrows D and B in the figure. In between one sees a distinct but wide hump, arrow C. However, there is one other point to be remarked, indicated by arrow A, which is where the slope to the low angle, high d value side of the 14.2 Å peak shows an apparent asymmetry or spreading at the base. The 14.2 Å peak is sharp and well defined but its base is wider that a normal peak shape would allow. We use a proposition to fit the wide base of peak B

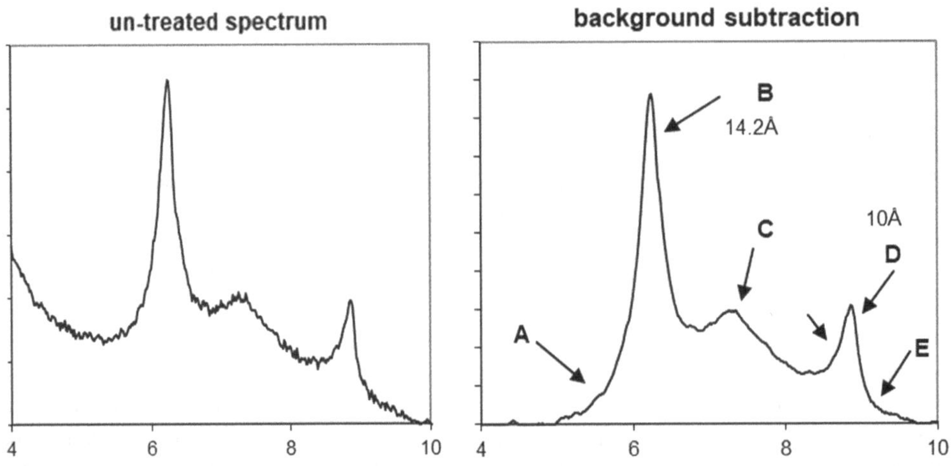

FIGURE A.5 Relative intensity and position in spectrum using Cu K alpha radiation in degrees 2 theta (left) and identification of peaks (right) on the diffraction recording

with curve A. Then one can fit the sharp peak (B) to the 14.2 Å peak. The combined peaks A + B account for much of the diffraction in the region of the first peak.

Now we can look at the high angle, lower d spacing side of the spectrum. Here the peak at 14.2 Å is well expressed and dominant (B), while the peak at 10 Å (illite, peak D) is very small, in fact a shoulder on a larger, broader peak, indicated by arrow C. Even so, the illite peak is evident and can be fitted with a band at 10 Å (peak D). However, there is a slight widening at the high angles side of the peak. Also, there is some asymmetry to the low angles side of the 10 Å. This can be fitted using peak E. As indicated earlier, illite is usually described by two peaks, one at 10 Å (WCI) and the other slightly to lower angles 2 theta and higher d spacings (PCI). Thus, a two-peak proposition is most often necessary to describe the potassic illite component in soils.

Having taken care of the exterior parts of the spectrum one can then look at the obvious peak at near 11.9 Å in the centre of the spectrum (C). This can be simulated by a peak at 11.9 Å. However, there is still some diffraction intensity found between this peak C and the illite peaks (D and E). A peak at 11 Å is needed to complete the simulation. Here we have found two peaks at each end of the spectrum associated with a diffraction maximum: one in the centre associated with another obvious diffraction maximum, and another which is found due to the incomplete description of the initial proposition. The last peak at 11 Å is not to be ignored in that it represents 15% of the total peak diffraction area. The sharp, clearly visible peaks represent only 25% of the surface in the example given. Therefore, it is extremely important to use decomposition in order to have an idea of the total diffraction surface and phases present. The final least squares refinement of the spectrum, based upon the propositions we have proposed, shifts some peaks slightly, for example peak A, which is widened and moved to a higher position in ° 2 theta. This indicates that the whole spectrum must be considered when all elements are fitted to find the best solution to the problem.

In Figure A.6 relative intensity and position in degrees are for 2 theta Cu K alpha radiation. Next we indicate another example of similar mineralogy but with different problems of decomposition.

A.3.4 Peak Shape and Intensity Change: The Comparative Method

The major use of decomposition is to identify the peaks present and to compare their relative areas in order to identify changes in mineral abundance and mineral type. This can be done for similar materials which have been subjected to different chemical changes.

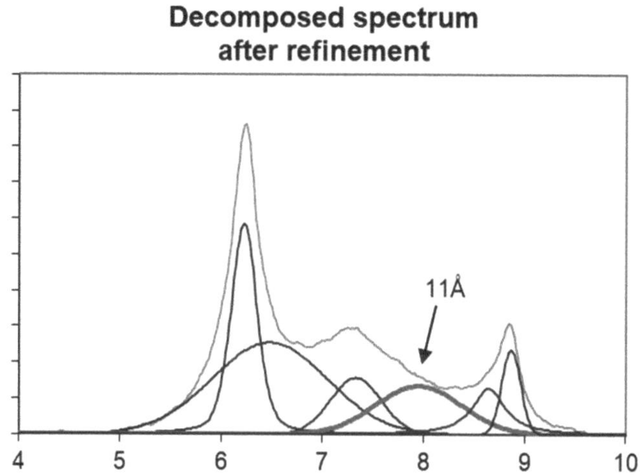

FIGURE A.6 Decomposed spectrum from Figure A.5.

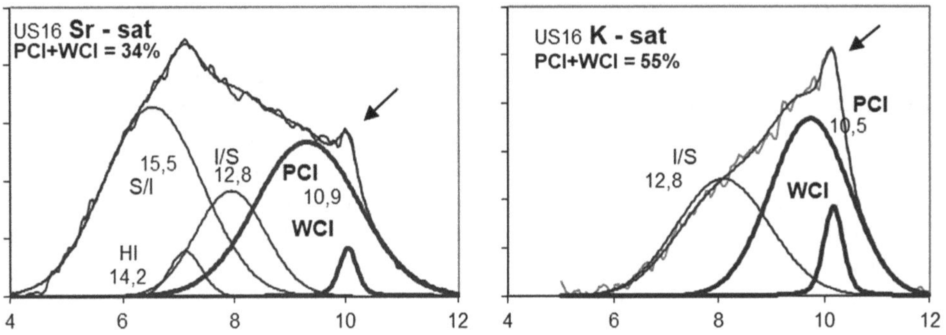

FIGURE A.7 Decomposed XRD trace of clay from an agricultural prairie soil from Illinois in the strontium-saturated (left) and potassium-saturated state (right).

An example of the use of decomposition to follow changes in spectra is given in Figure A.7. A sample of agricultural prairie soil from Illinois is shown in the strontium-saturated state and in the potassium-saturated state. In the Sr-sat diagram, one finds two mixed-layer illite-smectite peaks. The first is the smectite-rich phase (S/I), which is very close to smectite in composition, given its spacing at 15.5 Å. Another illite-smectite mixed-layer mineral is present, whose peak is closer to 10 Å (12.8 Å). The two illite peaks (PCI and WCI) are present, with the PCI of larger peak area than the WCI. The two illite peaks represent 34% of total peak areas.

Potassium saturation shifts the peaks to higher peak positions and lower d values, indicating an increase in potassium content as the peaks approach 10 Å. Potassium in the anhydrous form in the interlayer position produces a 10 Å spacing. In the global experimental spectrum, the irregular line above the component peaks, one sees an increase in the presence of the illite peak, indicated by the arrow. The combined area of the illite peaks now represents 55% of total peak area. Further, the smectite-rich mixed-layer mineral peak has disappeared, with an important I/S phase present as in the initial spectrum. Note that the S/I and I/S peaks define the peak envelope in Sr- and K-saturated states, and that the PCI peaks define the other side of the peak envelope. By comparing the decomposition peaks in the two spectra one can numerically determine changes in the mineralogy (interlayer ion occupancy) of the samples under the two different treatments.

A.3.5 Surface Area

In this example we indicate the relative peak area of the different phases identified. The method of estimating the surface area for Gaussian peaks is simple: one multiplies the width at half height times the peak maximum value, which defines a good approximation of the area. Summing the peaks and dividing the area of each by the total gives an estimation of the relative area for each phase.

Another example of the use of decomposition methods can be seen in a series of three soil samples: (top of A horizon) under a sequoia tree near the trunk, at 2 m from the trunk and in the open prairie (lawn) around the 150-year-old tree (park of the château Saint Fargeau, Vienne, France). The effect of the sequoia, grown on the same initial soil as that of the grass in the park, is to increase the potassium interlayer content of certain clays. One sees in Figure A.8 that the smectite-rich phase (S/I in the figure) maintains its peak position and relative intensity. This peak, defining the left hand side of the peak envelope, is stable. The smectite-rich interlayered mineral (S/I) appears to maintain its peak position also either in the grass or sequoia soils. However, one finds that there is a great increase in the intensity of a peak near 11 Å, an illite-rich interlayered mineral (I/S) and in the intensity of the illite peak. Here it seems that the phases present initially remain essentially of the same composition (ratio of illite and smectite present in the crystallites), while a portion of the initial phases becomes either an illite-rich interlayered mineral or illite.

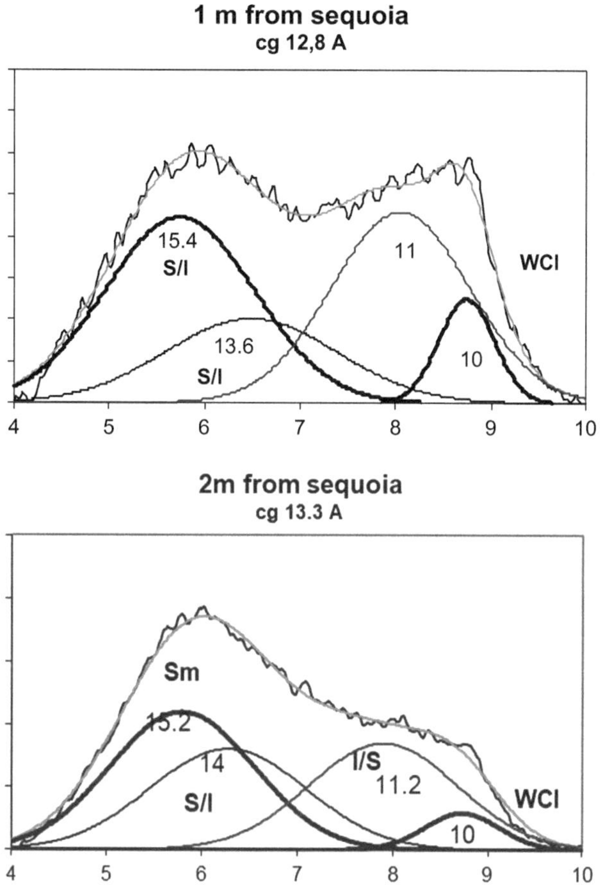

FIGURE A.8 Decomposed XRD traces of clays from soils at different distances from a sequoia tree showing centres of gravity.

A.4 CENTRE OF GRAVITY

If one multiplies the peak area of each phase (width at half height times intensity) and compares it to the total peak areas for all peaks, relative peak area, one can then multiply by the area by the peak position (in Å). This gives the contribution for each peak. By adding the contributions, one makes a weighted average of the peaks in a spectrum. This is the centre of gravity approach (Barré et al., 2008) to numerical comparisons. In the Figure A.8 the centre of gravity measurements (cg) are indicated for each spectrum. One sees that the centre of gravity directly reflects the shift in peak intensities and peak positions described earlier. The cg shifts from a position of 14.2 Å under grass showing the predominance of smectite (two water layer hydrations and a contribution at 15.2 Å), which is decreased under the influence of the sequoia tree where the cg value is 12.8 Å mid-way between smectite and illite positions. This indicates the change in illite content of the mixed-layered minerals and the amount of illite created by potassium saturation. This measure is very useful as a comparative tool for soil assemblages that are similar but have experienced different histories.

Curve decomposition is in fact rather simple when one uses a minimum number of peaks to describe a complex spectrum. The most powerful use is to follow changes from one state to another of a clay assemblage which has been subjected to chemical changes either in the laboratory or those induced by plants in the soils themselves.

In Figure A.9 the average peak position indicates the amount of 10 angstrom layers in the various minerals, either as pure phases or interstratified with other types of mineral.

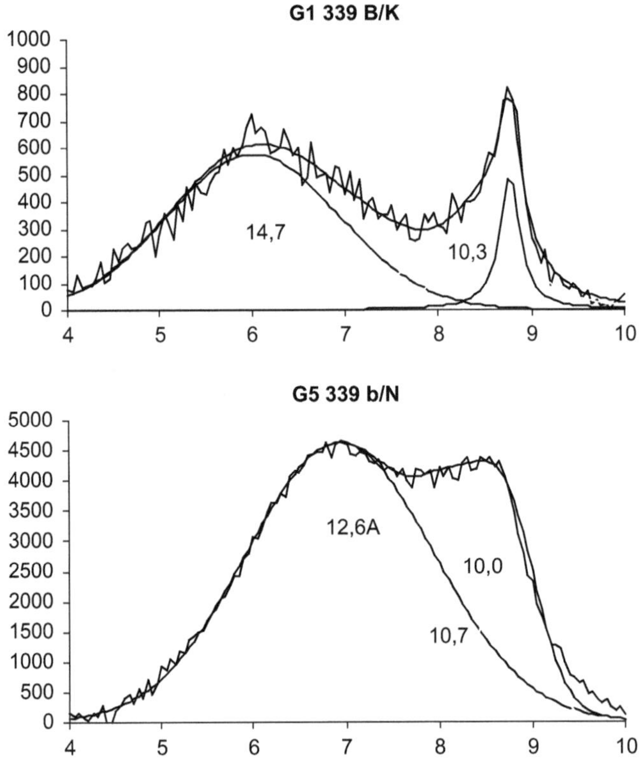

FIGURE A.9 Example of the use of peak position and intensity to determine the mineral content. Ten angstrom peaks (near 9° 2 theta) indicate illite-like clays, which can be either potassic or ammonium-bearing. Potassium content was determined by electron microprobe. (Data from Tyce Herrmann, University of Wisconsin.)

3.5 SAMPLE PREPARATION AND IDENTIFICATION OF PHYLLOSILICATE PHASES

We propose then to use a very simplified method of sample preparation and interpretation concerning the major silicate phases present in soils.

3.5.1 Preparation

Take a small amount of soil sample (a gram or so), and let it dry in the laboratory. Crush it to a powder and put in a small bottle with distilled water to a depth of 10 cm. Let it hydrate overnight. Take the sample in its hydrated form and use an ultrasonic generator to disperse the clays for a minute or two. Shaking for several minutes can do some of this job. Let the sample settle overnight and pour the dispersed clay solution into another bottle. Add several drops of concentrated $SrCl_2$ solution to flocculate the clays making it settle and to ensure that the interlayer ions are all of approximately the same type. Let the flocculating sample settle overnight.

3.5.2 Peaks

Pour the clay-free solution out of the bottle and then pour the clay concentrate on a glass slide to be placed in an X-ray diffractometer. Let the sample dry overnight. You now have an air-dried sample with two water layer ions in the interlayer site (absorbed).

Make an X-ray diffraction spectral record of the sample in the range of 24 to 5 angstroms. This shows the phyllosilicate mineral range of soil clay minerals for the 1:1 and 2:1 minerals. The spectral range of 5 to 2 angstroms shows much of the oxide and hydroxide mineralogy.

The peaks will show the presence of 2:1 and 1:1 phases. The 1:1 minerals (kaolinite especially) are present as relatively sharp peaks in the range of 7.2 angstroms with a band width of near 0.5 angstroms. A peak to the higher spacing side and wider indicates the presence of interstratified smectite/kaolinite. The mineral halloysite has a somewhat wider peak at near 10.2 angstroms, which is wider than the micas illite and muscovite (see later).

A narrow peak at 10 angstroms is typical of the anhydrous potassium interlayer mineral illite or mica. A smaller and somewhat wider band to higher spacings indicates fine-grained illite material or perhaps ammonium minerals.

Narrow peaks at 14.2 angstroms indicate the presence of hydroxyl-interlayered minerals chlorite or hydroxyl vermiculite. Chlorite has a smaller peak here and a more intense one at 14 angstroms. Vermiculite has an intense peak here and almost none at 7 angstroms.

Wide peaks (several degrees 2 theta) between 15.2 and 10 angstroms indicate the presence of smectite (15.4 angstroms) and micas at 10 angstroms in mixed-layered structures of various proportions of each component. The closer the peak is to 15.4 angstroms, the more smectite is present. It is common to see several peaks in the 15.4 to 10 angstrom region.

Normally hydrating the sample overnight in an atmosphere of high humidity will shift the smectite layers to positions near 17 angstroms and an average position for the mixed-layered minerals, while the vermiculite–chlorite minerals remain at 14.2 angstroms.

REFERENCES

Barré, P., B. Velde, C. Fontaine, N. Catel, and L. Abbadie. 2008. Which 2:1 clay minerals are involved in the soil potassium reservoir? Insights from potassium addition or removal experiments on three temperate grassland soil clay assemblages. *Geoderma* 146: 216–223.

Brindley, G.W., and G. Brown (eds.). 1980. *Crystal Structures of Clay Minerals and Their Xray Identification.* Mineralogical Society, London.

Casetou-Gustafson, S., S. Hillier, C. Akselsson, M. Simonsson, J. Stendahl, and B.A. Olsson. 2018. Comparison of measured (XRPD) and modeled (A2M) soil mineralogies: A study of some Swedish forest soils in the context of weathering rate predictions. *Geoderma* 310: 77–88.

Lanson, B. 1997. Decomposition of experimental X-ray diffraction patterns (profile fitting): A convenient way to study clays. *Clays and Clay Minerals* 45: 132–146.

Meunier, A., and B. Velde, 2013. *Illite: Origins, Evolution and Metamorphism.* Springer, Heidelberg.

Index

D

E

F

G

H

I

T

U

V

W